FIELD THEORY

A Path Integral Approach

Third Edition

World Scientific Lecture Notes in Physics

ISSN: 1793-1436

*For the complete list of published titles, please visit
http://www.worldscientific.com/series/wslnp

World Scientific Lecture Notes in Physics – Vol. 83

FIELD THEORY

A Path Integral Approach

Third Edition

Ashok Das

University of Rochester, USA
Institute of Physics, Bhubaneswar, India

W⊖ World Scientific

NEW JERSEY · LONDON · SINGAPORE · BEIJING · SHANGHAI · HONG KONG · TAIPEI · CHENNAI · TOKYO

Published by

World Scientific Publishing Co. Pte. Ltd.

5 Toh Tuck Link, Singapore 596224

USA office: 27 Warren Street, Suite 401-402, Hackensack, NJ 07601

UK office: 57 Shelton Street, Covent Garden, London WC2H 9HE

Library of Congress Cataloging-in-Publication Data
Names: Das, Ashok, 1953–   author.
Title: Field theory: a path integral approach / Ashok Das (University of
    Rochester, USA & Institute of Physics, Bhubaneswar, India).
Other titles: World Scientific lecture notes in physics ; v. 83.
Description: Third edition. | Singapore ; Hackensack, NJ : World Scientific
    Publishing Co. Pte. Ltd., [2019] | Series: World Scientific lecture notes
    in physics ; vol. 83 | Includes bibliographical references and index.
Identifiers: LCCN 2019000260| ISBN 9789811202544 (hardcover ; alk. paper) |
    ISBN 9811202540 (hardcover ; alk. paper)
Subjects: LCSH: Path integrals. | Quantum field theory.
Classification: LCC QC174.52.P37 D37 2019 | DDC 530.14/3--dc23
LC record available at https://lccn.loc.gov/2019000260

British Library Cataloguing-in-Publication Data
A catalogue record for this book is available from the British Library.

For any available supplementary material, please visit
https://www.worldscientific.com/worldscibooks/10.1142/11339#t=suppl

Printed in Singapore

To
Lakshmi
and
Gouri

Preface to the First Edition

Traditionally, field theory had its main thrust of development in high energy physics. Consequently, the conventional field theory courses are taught with a heavy emphasis on high energy physics. Over the years, however, it has become quite clear that the methods and techniques of field theory are widely applicable in many areas of physics. The canonical quantization methods, which is how conventional field theory courses are taught, do not bring out this feature of field theory. A path integral description of field theory is the appropriate setting for this. It is with this goal in mind, namely, to make graduate students aware of the applicability of the field theoretic methods to various areas, that the Department of Physics and Astronomy at the University of Rochester introduced a new one semester course on field theory in Fall 1991.

This course was aimed at second year graduate students who had already taken a one year course on nonrelativistic quantum mechanics but had not necessarily specialized into any area of physics and these lecture notes grew out of this course which I taught. In fact, the lecture notes are identical to what was covered in the class. Even in the published form, I have endeavored to keep as much of the detailed derivations of various results as I could — the idea being that a reader can then concentrate on the logical development of concepts without worrying about the technical details. Most of the concepts were developed within the context of quantum mechanics — which the students were expected to be familiar with — and subsequently these concepts were applied to various branches of physics. In writing these lecture notes, I have added some references at the end of every chapter. They are only intended to be suggestive. There is so much literature that is available in this subject that it would have been impossible to include all of them. The references are not meant to be complete and I apologize to many whose works I have not cited in the references. Since this was developed as a course for general students, the many interesting topics of gauge theories are also not covered in these lectures. It simply would have been impossible to do justice to these topics within a one-semester course.

There are many who were responsible for these lecture notes. I would like to thank our chairman, Paul Slattery, for asking me to teach and design a syllabus for this course. The students deserve the most credit for keeping all the derivations complete and raising . many issues which I, otherwise, would have taken for granted. I am grateful to my students Paulo Bedaque and Wen-Jui Huang as well as to Dr. Zhu Yang for straightening out many little details which were essential in presenting the material in a coherent and consistent way. I would also like to thank Michael Begel for helping out in numerous ways, in particular, in computer-generating all the figures in the book. The support of many colleagues was also vital for the completion of these lecture notes. Judy Mack, as always, has done a superb job as far as the appearance of the book is concerned and I sincerely thank her. Finally, I am grateful to Ammani for being there.

Ashok Das
Rochester.

Preface to the Second Edition

This second edition of the book is an expanded version which contains a chapter on path integral quantization of gauge theories as well as a chapter on anomalies. In addition, Chapter 6 (Supersymmetry) has been expanded to include a section on supersymmetric singular potentials. While these topics were not covered in the original course on path integrals, they are part of my lectures in other courses that I have taught at the University of Rochester and have been incorporated into this new edition at the request of colleagues from all over the world. There are many people who have helped me to complete this edition of the book and I would like to thank, in particular, Judy Mack, Arsen Melikyan, Dave Munson and J. Boersma for all their assistance.

Ashok Das
Rochester.

Preface to the Third Edition

This third edition grew out of lectures I gave when I taught the course last Fall in 2017. As a result of these lectures, almost all chapters have been expanded for more clarity. The chapter on Anomalies and the Schwinger model (Chapter 13) has been completely rewritten for logical clarity. Two new chapters have been added in the new edition which were primarily requested by students and colleagues all over the world. The proper time formalism of Schwinger has been discussed with simple examples at zero as well as at finite temperature. This is followed by the zeta function regularization where, again, simple examples are used to describe the essential features of the method. I would like to thank Sarah Henry and Pushpa Kalauni for going through parts of the book carefully and Dave Munson for sorting out all the computer related glitches.

Ashok Das
Rochester.

Contents

CHAPTER 1

Introduction

1.1 Particles and fields

Classically, there are two kinds of dynamical systems that we encounter. First, there is the motion of a point particle or of finitely many particles (with a finite number of degrees of freedom) which can be described by a finite number of coordinates. And then, there are physical systems where the number of degrees of freedom is non-denumerably (noncountably) infinite. Such systems are described by fields. Familiar examples of classical fields are the electromagnetic fields described by $\mathbf{E}(\mathbf{x}, t)$ and $\mathbf{B}(\mathbf{x}, t)$ or equivalently by the potentials $(\phi(\mathbf{x}, t), \mathbf{A}(\mathbf{x}, t))$. Similarly, the motion of a one-dimensional string is also described by a field $\phi(\mathbf{x}, t)$, namely, the displacement field. Thus, while the coordinates of a particle depend only on time, fields depend continuously on some space variables as well. Correspondingly, a theory describing the dynamics of fields is usually known as a $D + 1$ dimensional field theory where D represents the number of spatial coordinates on which the field variables depend. For example, a theory describing the displacements of the one-dimensional string would constitute a 1+1 dimensional field theory whereas the more familiar Maxwell's equations (in four dimensions) can be regarded as described by a 3+1 dimensional field theory. In this language, then, it is clear that a theory describing the motion of a point particle can be regarded as a special case, namely, we can think of such a theory as a 0+1 dimensional field theory.

1.2 Metric and other notations

In these lectures, we will discuss both non-relativistic as well as relativistic theories. For the relativistic case, we will use the Bjorken-Drell convention. Namely, the contravariant coordinates are assumed to be given by

$$x^\mu = (t, \mathbf{x}), \quad \mu = 0, 1, 2, 3, \tag{1.1}$$

while the covariant coordinates have the form

$$x_\mu = \eta_{\mu\nu} x^\nu = (t, -\mathbf{x}). \tag{1.2}$$

Here we have assumed the speed of light to be unity ($c = 1$). The covariant metric, therefore, follows to have a diagonal form with unit elements and signatures

$$\eta_{\mu\nu} = (+, -, -, -) = \eta_{\nu\mu}. \tag{1.3}$$

The inverse or the contravariant metric clearly also has the same form, namely,

$$\eta^{\mu\nu} = (+, -, -, -) = \eta^{\nu\mu}. \tag{1.4}$$

The invariant length is given by

$$x^2 = x^\mu x_\mu = \eta^{\mu\nu} x_\mu x_\nu = \eta_{\mu\nu} x^\mu x^\nu = t^2 - \mathbf{x}^2. \tag{1.5}$$

The gradients are similarly obtained from Eqs. (1.1) and (1.2) to be

$$\partial_\mu = \frac{\partial}{\partial x^\mu} = \left(\frac{\partial}{\partial t}, \boldsymbol{\nabla} \right), \tag{1.6}$$

$$\partial^\mu = \frac{\partial}{\partial x_\mu} = \left(\frac{\partial}{\partial t}, -\boldsymbol{\nabla} \right), \tag{1.7}$$

so that the D'Alembertian takes the form

$$\Box = \partial^\mu \partial_\mu = \eta^{\mu\nu} \partial_\mu \partial_\nu = \frac{\partial^2}{\partial t^2} - \boldsymbol{\nabla}^2. \tag{1.8}$$

1.3 Functionals

It is evident that in dealing with dynamical systems we are dealing with functions of continuous variables. In fact, most of the time, we are really dealing with functions of functions which are otherwise known as functionals. For example, if we are considering the motion of a particle in a potential in one dimension with potential energy $V(x)$, then the Lagrangian describing the system is given by

$$L(x, \dot{x}) = \frac{1}{2}m\dot{x}^2 - V(x) = T(\dot{x}) - V(x), \tag{1.9}$$

where $x(t)$ and $\dot{x}(t)$ denote the coordinate and the velocity of the particle and the simplest functional we can think of is the action functional defined as

$$S[x] = \int_{t_i}^{t_f} dt\, L(x, \dot{x}). \tag{1.10}$$

Note that unlike a function whose value depends on a particular point in the coordinate space, the value of the action depends on the entire trajectory along which the integration is carried out. For different paths connecting the initial and the final points, the value of the action functional will be different.

A simple functional has the form

$$F[f] = \int dx\, F(f(x)), \tag{1.11}$$

where, for example, we may have

$$F(f(x)) = (f(x))^n. \tag{1.12}$$

Sometimes, one also says, loosely, that $F(f(x))$ is a functional.

The notion of a derivative can be extended to the case of functionals in a natural way through the notion of generalized functions. Thus, one defines the functional derivative or the Gateaux derivative from the linear functional

$$\frac{d}{d\epsilon}F[f + \epsilon v]\Big|_{\epsilon=0} = \int dx \, \frac{\delta F[f]}{\delta f(x)} \, v(x) \,. \tag{1.13}$$

Equivalently, from the working point of view, this simply corresponds to defining (namely, one changes the value of the function infinitesimally only at $x = y$)

$$\frac{\delta F(f(x))}{\delta f(y)} = \lim_{\epsilon \to 0} \frac{F(f(x) + \epsilon \delta(x - y)) - F(f(x))}{\epsilon} \,, \tag{1.14}$$

and identifying

$$\frac{\delta F[f]}{\delta f(y)} = \int dx \, \frac{\delta F(f(x))}{\delta f(y)} \,. \tag{1.15}$$

It now follows from Eq. (1.14) that

$$\frac{\delta f(x)}{\delta f(y)} = \delta(x - y) \,. \tag{1.16}$$

The functional derivative satisfies all the properties of a derivative, namely, it is linear and is associative,

$$\frac{\delta}{\delta f(x)}(F_1[f] + F_2[f]) = \frac{\delta F_1[f]}{\delta f(x)} + \frac{\delta F_2[f]}{\delta f(x)},$$

$$\frac{\delta}{\delta f(x)}(F_1[f]F_2[f]) = \frac{\delta F_1[f]}{\delta f(x)} F_2[f] + F_1[f] \frac{\delta F_2[f]}{\delta f(x)} \,. \tag{1.17}$$

It obeys the chain rule of differentiation. Furthermore, we now see that given a functional $F[f]$, we can Taylor expand it in the form

$$F[f] = \int dx \, F(f(x))$$

$$= \int dx \, P_0(x) + \int dx dx_1 \, P_1(x, x_1)(f(x_1) - \bar{f}(x_1))$$

$$+ \int dx dx_1 dx_2 \, P_2(x, x_1, x_2)(f(x_1) - \bar{f}(x_1))(f(x_2) - \bar{f}(x_2))$$

$$+ \cdots \,, \tag{1.18}$$

where

$$P_0(x) = F(f(x))|_{f(x)=\bar{f}(x)} \, ,$$

$$P_1(x, x_1) = \left. \frac{\delta F(f(x))}{\delta f(x_1)} \right|_{f(x)=\bar{f}(x)} , \qquad (1.19)$$

$$P_2(x, x_1, x_2) = \left. \frac{1}{2!} \frac{\delta^2 F(f(x))}{\delta f(x_1)\delta f(x_2)} \right|_{f(x)=\bar{f}(x)} ,$$

and so on.

As simple examples, let us calculate a few functional derivatives.

(i) Let

$$F[f] = \int dy \, F(f(y)) = \int dy \, (f(y))^n, \qquad (1.20)$$

where n denotes a positive integer. Then, from (1.14) we obtain

$$\frac{\delta F(f(y))}{\delta f(x)} = \lim_{\epsilon \to 0} \frac{F(f(y) + \epsilon\delta(y - x)) - F(f(y))}{\epsilon}$$

$$= \lim_{\epsilon \to 0} \frac{(f(y) + \epsilon\delta(y - x))^n - (f(y))^n}{\epsilon}$$

$$= \lim_{\epsilon \to 0} \frac{(f(y))^n + n\epsilon(f(y))^{n-1}\delta(y - x) + O(\epsilon^2) - (f(y))^n}{\epsilon}$$

$$= n(f(y))^{n-1}\delta(y - x) = F'(f(y))\,\delta(y - x)\,, \qquad (1.21)$$

where we have identified

$$F'(f(y)) = \frac{\partial F(f(y))}{\partial f(y)}\,. \qquad (1.22)$$

Therefore, we obtain

$$\frac{\delta F[f]}{\delta f(x)} = \int dy \; \frac{\delta F(f(y))}{\delta f(x)}$$

$$= \int dy \; n(f(y))^{n-1} \delta(y - x)$$

$$= n(f(x))^{n-1} . \tag{1.23}$$

(ii) Let us next consider the one-dimensional action in Eq. (1.10)

$$S[x] = \int\limits_{t_i}^{t_f} dt' \; L(x(t'), \dot{x}(t')), \tag{1.24}$$

with

$$L(x(t), \dot{x}(t)) = \frac{1}{2} m (\dot{x}(t))^2 - V(x(t))$$

$$= T(\dot{x}(t)) - V(x(t)) . \tag{1.25}$$

In a straightforward manner (see (1.21)), we obtain

$$\frac{\delta V(x(t'))}{\delta x(t)} = \lim_{\epsilon \to 0} \frac{V(x(t') + \epsilon \delta(t' - t)) - V(x(t'))}{\epsilon}$$

$$= V'(x(t')) \delta(t' - t) , \tag{1.26}$$

where we have defined

$$V'(x(t')) = \frac{\partial V(x(t'))}{\partial x(t')} . \tag{1.27}$$

Similarly, we have

$$\frac{\delta T(\dot{x}(t'))}{\delta x(t)} = \lim_{\epsilon \to 0} \frac{T(\dot{x}(t') + \epsilon \frac{d}{dt'} \delta(t' - t)) - T(\dot{x}(t'))}{\epsilon}$$

$$= \lim_{\epsilon \to 0} \frac{\frac{m}{2} (\dot{x}(t') + \epsilon \frac{d}{dt'} \delta(t' - t))^2 - \frac{m}{2} (\dot{x}(t'))^2}{\epsilon}$$

$$= m \dot{x}(t') \frac{d}{dt'} \delta(t' - t) . \tag{1.28}$$

It is clear now that

$$\frac{\delta L(x(t'), \dot{x}(t'))}{\delta x(t)} = \frac{\delta(T(\dot{x}(t')) - V(x(t')))}{\delta x(t)}$$

$$= m\dot{x}(t')\frac{\mathrm{d}}{\mathrm{d}t'}\delta(t' - t) - V'(x(t'))\delta(t' - t) \,. \tag{1.29}$$

Consequently, in this case, we obtain for $t_i \leq t \leq t_f$

$$\frac{\delta S[x]}{\delta x(t)} = \int_{t_i}^{t_f} \mathrm{d}t' \, \frac{\delta L(x(t'), \dot{x}(t'))}{\delta x(t)}$$

$$= \int_{t_i}^{t_f} \mathrm{d}t' \, (m\dot{x}(t')\frac{\mathrm{d}}{\mathrm{d}t'}\delta(t' - t) - V'(x(t'))\delta(t' - t))$$

$$= -m\ddot{x}(t) - V'(x(t))$$

$$= -\frac{\mathrm{d}}{\mathrm{d}t} \frac{\partial L(x(t), \dot{x}(t))}{\partial \dot{x}(t)} + \frac{\partial L(x(t), \dot{x}(t))}{\partial x(t)} \,. \tag{1.30}$$

The right hand side is, of course, reminiscent of the Euler-Lagrange equation. In fact, we note that

$$\frac{\delta S[x]}{\delta x(t)} = -\frac{\mathrm{d}}{\mathrm{d}t} \frac{\partial L}{\partial \dot{x}(t)} + \frac{\partial L}{\partial x(t)} = 0 \,, \tag{1.31}$$

gives the Euler-Lagrange equation as a functional extremum of the action. This is nothing other than the principle of least action expressed in a compact notation in the language of functionals.

1.4 Review of quantum mechanics

In this section, we will describe very briefly the essential features of quantum mechanics assuming that the readers are familiar with the subject. The conventional approach to quantum mechanics starts with the Hamiltonian formulation of classical mechanics and promotes observables to non-commuting operators. The dynamics of the system, in this case, is given by the time-dependent Schrödinger equation

$$i\hbar\frac{\partial|\psi(t)\rangle}{\partial t} = H|\psi(t)\rangle\,, \tag{1.32}$$

where H denotes the Hamiltonian operator of the system. Equivalently, in the one dimensional case, the wave function of a particle satisfies

$$i\hbar\frac{\partial\psi(x,t)}{\partial t} = H(x)\psi(x,t)$$

$$= \left(-\frac{\hbar^2}{2m}\frac{\partial^2}{\partial x^2} + V(x)\right)\psi(x,t)\,, \tag{1.33}$$

where we have identified the wave function as

$$\psi(x,t) = \langle x|\psi(t)\rangle\,, \qquad X|x\rangle = x|x\rangle\,, \tag{1.34}$$

with $|x\rangle$ denoting the coordinate basis states. This, then, defines the time evolution of the system.

The main purpose behind solving the Schrödinger equation lies in determining the time evolution operator which generates time translation of the system. Namely, the time evolution operator transforms the quantum mechanical state at an earlier time t_2 to a later time t_1 as

$$|\psi(t_1)\rangle = U(t_1, t_2)|\psi(t_2)\rangle\,, \tag{1.35}$$

where $U(t_1, t_2)$ denotes the time evolution operator. Clearly, for a time independent Hamiltonian, we see from (the Schrödinger equation) Eq. (1.32) that for $t_1 > t_2$,

$$U(t_1, t_2) = e^{-\frac{i}{\hbar}(t_1-t_2)H}\,, \quad |\psi(t_1)\rangle = e^{-\frac{i}{\hbar}(t_1-t_2)H}|\psi(t_2)\rangle\,. \tag{1.36}$$

More explicitly, we can write

$$U(t_1, t_2) = \theta(t_1 - t_2)e^{-\frac{i}{\hbar}(t_1-t_2)H}\,, \tag{1.37}$$

and it is now obvious that the time evolution operator is nothing other than the Green's function for the time dependent Schrödinger equation and satisfies

$$\left(i\hbar\frac{\partial}{\partial t_1} - H\right)U(t_1, t_2) = i\hbar\delta(t_1 - t_2). \tag{1.38}$$

Determining this operator is equivalent to finding its matrix elements in a given basis. Thus, for example, in the coordinate basis defined by

$$X|x\rangle = x|x\rangle, \tag{1.39}$$

we can write

$$\langle x_1|U(t_1, t_2)|x_2\rangle = U(x_1, t_1; x_2, t_2). \tag{1.40}$$

If we know the function $U(x_1, t_1; x_2, t_2)$ completely, then we know the time evolution operator and the time evolution of the wave function can be written as (see (1.35))

$$\psi(x_1, t_1) = \int dx_2\, U(x_1, t_1; x_2, t_2)\psi(x_2, t_2). \tag{1.41}$$

It is interesting to note that the dependence on the intermediate times drops out in the above equation as can be easily checked.

Our discussion has been within the framework of the Schrödinger picture so far where the quantum states $|\psi(t)\rangle$ carry time dependence while the operators are time independent. On the other hand, in the Heisenberg picture, where the quantum states are time independent, using Eq. (1.36) we can identify

$$|\psi\rangle_H = |\psi(t = 0)\rangle_S = |\psi(t = 0)\rangle$$

$$= e^{\frac{i}{\hbar}tH}|\psi(t)\rangle = e^{\frac{i}{\hbar}tH}|\psi(t)\rangle_S. \tag{1.42}$$

In this picture, the operators carry all the time dependence. For example, the coordinate operator in the Heisenberg picture is related to the coordinate operator in the Schrödinger picture through the relation

$$X_H(t) = e^{\frac{i}{\hbar}tH} X e^{-\frac{i}{\hbar}tH} \,. \tag{1.43}$$

The eigenstates of this operator satisfying

$$X_H(t)|x,t\rangle_H = x|x,t\rangle_H \,, \tag{1.44}$$

are then easily seen to be related to the coordinate basis in the Schrödinger picture through

$$|x,t\rangle_H = e^{\frac{i}{\hbar}tH}|x\rangle \,. \tag{1.45}$$

We comment here parenthetically that the eigenvalues of any Heisenberg operator Ω_H, not depending on time explicitly, is time independent. This follows from the fact that the expectation value of the operator, in its eigenstate, is time independent

$$_S\langle\omega|\Omega_S|\omega\rangle_S = \omega = {}_H\langle\omega,t|\Omega_H|\omega,t\rangle_H. \tag{1.46}$$

This holds true even for the interaction picture which we will comment on shortly (this is how different pictures are defined).

It is clear now that for $t_1 > t_2$ we can write

$$\begin{aligned}
_H\langle x_1,t_1|x_2,t_2\rangle_H &= \langle x_1|e^{-\frac{i}{\hbar}t_1 H}e^{\frac{i}{\hbar}t_2 H}|x_2\rangle \\
&= \langle x_1|e^{-\frac{i}{\hbar}(t_1-t_2)H}|x_2\rangle \\
&= \langle x_1|U(t_1,t_2)|x_2\rangle \\
&= U(x_1,t_1;x_2,t_2) \,. \tag{1.47}
\end{aligned}$$

This shows that the matrix elements of the time evolution operator in the coordinate basis (see (1.40)) are nothing other than the time ordered transition amplitudes between the coordinate basis states in the Heisenberg picture. Therefore, if we know the transition amplitudes of the coordinate basis in the Heisenberg picture completely, we know the time evolution operator.

Finally, there is the interaction picture where both the quantum states as well as the operators carry partial time dependence. Without going into any technical detail, let us simply note here that the interaction picture is quite useful in the study of nontrivially interacting theories. In any case, the goal of the study of quantum mechanics in any of these pictures is to construct the matrix elements of the time evolution operator which (in the coordinate basis) as we have seen can be identified with the time ordered transition amplitudes between the coordinate basis states in the Heisenberg picture.

1.5 References

A. Das, *Lectures on Quantum Mechanics*, Hindustan Book Agency and World Scientific Publishing.

P. A. M. Dirac, *Principles of Quantum Mechanics*, Oxford Univ. Press.

L. I. Schiff, *Quantum Mechanics*, McGraw-Hill Publishing.

Chapter 2

Path integrals and quantum mechanics

2.1 Basis states

Before going into the derivation of the path integral representation for the transition amplitude $U(x_f, t_f; x_i, t_i)$, let us recapitulate some of the basic relations from quantum mechanics. Consider, for simplicity, a one dimensional quantum mechanical system. The eigenstates of the coordinate operator, as we have seen in Eq. (1.39), satisfy

$$X|x\rangle = x|x\rangle. \tag{2.1}$$

These eigenstates define an orthonormal basis. Namely, they satisfy

$$\langle x|x'\rangle = \delta(x - x'),$$

$$\int dx \, |x\rangle\langle x| = 1. \tag{2.2}$$

Similarly, the eigenstates of the momentum operator satisfying

$$P|p\rangle = p|p\rangle, \tag{2.3}$$

also define an orthonormal basis. (In general, the eigenstates of any Hermitian operator define a complete basis.) Namely, the momentum eigenstates satisfy

$$\langle p|p'\rangle = \delta(p - p'),$$

$$\int dp \, |p\rangle\langle p| = 1. \tag{2.4}$$

The inner product of the coordinate and the momentum basis states defines the transition amplitude between the two basis states. Another way of saying this is to note that the inner product between the two basis states defines the coefficients of expansion of one basis in terms of the other. One can readily determine that

$$\langle p|x\rangle = \frac{1}{\sqrt{2\pi\hbar}} e^{-\frac{i}{\hbar}px} = \langle x|p\rangle^* \,. \tag{2.5}$$

In fact, these arise naturally in the defining relations for the Fourier transform of a function. Namely, using the completeness relations of the basis states, the Fourier transform of a function can be defined as

$$f(x) = \langle x|f\rangle = \int \mathrm{d}p \, \langle x|p\rangle\langle p|f\rangle$$

$$= \frac{1}{\sqrt{2\pi\hbar}} \int \mathrm{d}p \, e^{\frac{i}{\hbar}px} \, f(p) = \frac{1}{\sqrt{2\pi}} \int \frac{\mathrm{d}p}{\hbar} \, e^{\frac{i}{\hbar}px} \, \sqrt{\hbar} f(p)$$

$$= \frac{1}{\sqrt{2\pi}} \int \mathrm{d}k \, e^{ikx} \tilde{f}(k) \,, \qquad \tilde{f}(k) = \hbar f(p), \tag{2.6}$$

$$\tilde{f}(k) = \sqrt{\hbar} \, f(p) = \sqrt{\hbar}\langle p|f\rangle = \sqrt{\hbar} \int \mathrm{d}x \, \langle p|x\rangle\langle x|f\rangle$$

$$= \frac{\sqrt{\hbar}}{\sqrt{2\pi\hbar}} \int \mathrm{d}x \, e^{-\frac{i}{\hbar}px} f(x)$$

$$= \frac{1}{\sqrt{2\pi}} \int \mathrm{d}x \, e^{-ikx} f(x) \,. \tag{2.7}$$

The Fourier transforms simply take a function from a given space to its conjugate space or the dual space. Here $k = \frac{p}{\hbar}$ can be thought of as the wave number in the case of a quantum mechanical particle. (Some other authors may define Fourier transform with alternate normalizations. Here, the definition is symmetrical.)

As we have seen in Eq. (1.45), the Heisenberg states are related to the Schrödinger states in a simple way. For the coordinate basis states, for example, we have

$$|x, t\rangle_H = e^{\frac{i}{\hbar} tH} |x\rangle \,. \tag{2.8}$$

It follows now that the coordinate basis states in the Heisenberg picture satisfy

$$\begin{aligned}
{}_H\langle x, t | x', t\rangle_H &= \langle x | e^{-\frac{i}{\hbar} tH} e^{\frac{i}{\hbar} tH} | x'\rangle \\
&= \langle x | x'\rangle = \delta(x - x') \,,
\end{aligned} \tag{2.9}$$

and

$$\begin{aligned}
\int dx \, |x, t\rangle_H \, {}_H\langle x, t| &= \int dx \, e^{\frac{i}{\hbar} tH} |x\rangle \langle x| e^{-\frac{i}{\hbar} tH} \\
&= e^{\frac{i}{\hbar} tH} \int dx \, |x\rangle \langle x| \, e^{-\frac{i}{\hbar} tH} \\
&= e^{\frac{i}{\hbar} tH} \mathbb{1} \, e^{-\frac{i}{\hbar} tH} \\
&= \mathbb{1} \,. \tag{2.10}
\end{aligned}$$

It is worth noting here that the orthonormality as well as the completeness relations hold for the Heisenberg states only at equal times.

2.2 Operator ordering

In the Hamiltonian formalism, the transition from classical mechanics to quantum mechanics is achieved by promoting observables to Hermitian operators (which do not necessarily commute). Consequently, the Hamiltonian of the classical system is supposed to go over to the quantum operator

$$H(x, p) \rightarrow H(x_{\text{op}}, p_{\text{op}}) \,. \tag{2.11}$$

This, however, does not specify what should be done when products of x and p (which are non-commuting as operators) are involved. For example, classically we know that

$$xp = px \,.$$

Therefore, the order of these terms does not matter in the classical Hamiltonian. Quantum mechanically, however, the order of the operators is quite crucial and *a priori* it is not clear what such a term ought to correspond to in the quantum theory. This is the operator ordering problem in quantum mechanics and, unfortunately, there is no well defined principle which specifies the order of operators in the passage from classical to quantum mechanics. There are, however, two prescriptions which one uses conventionally. In normal ordering, one orders the products of x's and p's such that the momenta stand to the left of the coordinates. Thus,

$$xp \xrightarrow{\text{N.O.}} px\,,$$

$$px \xrightarrow{\text{N.O.}} px\,,$$

$$x^2 p \xrightarrow{\text{N.O.}} px^2\,,$$

$$xpx \xrightarrow{\text{N.O.}} px^2\,, \tag{2.12}$$

and so on. (This is commonly used in the operator description of relativistic quantum field theories.) However, the prescription that is much more widely used in quantum mechanics and which is much more satisfactory from various other points of view is the Weyl ordering. Here one symmetrizes the product of operators in all possible combinations with equal weight. Thus,

$$xp \xrightarrow{\text{W.O.}} \frac{1}{2}(xp + px)\,,$$

$$px \xrightarrow{\text{W.O.}} \frac{1}{2}(xp + px)\,,$$

$$x^2 p \xrightarrow{\text{W.O.}} \frac{1}{3}(x^2 p + xpx + px^2)\,,$$

$$xpx \xrightarrow{\text{W.O.}} \frac{1}{3}(x^2 p + xpx + px^2)\,, \tag{2.13}$$

and so on.

For normal ordering, it is easy to see that for any quantum Hamiltonian obtained from the classical Hamiltonian $H(x, p)$

$$\langle x'|H^{\text{N.O.}}|x\rangle = \int dp \, \langle x'|p\rangle\langle p|H^{\text{N.O.}}|x\rangle = \int dp \, \langle x'|p\rangle\langle p|x\rangle \, H(x,p)$$

$$= \int \frac{dp}{2\pi\hbar} \, e^{\frac{i}{\hbar}p(x'-x)} H(x,p) \,. \tag{2.14}$$

Here we have used the completeness relations of the momentum basis states given in Eq. (2.4) as well as the defining relations in Eqs. (2.1), (2.3) and (2.5). (The matrix element of the quantum Hamiltonian is a classical function for which the ordering is irrelevant.) To understand Weyl ordering, on the other hand, let us note that the expansion of

$$(\alpha x_{\text{op}} + \beta p_{\text{op}})^N \,,$$

generates the Weyl ordering of products of the form $x_{\text{op}}^n p_{\text{op}}^m$ naturally if we treat x_{op} and p_{op} as non-commuting operators. In fact, we can easily show that

$$(\alpha x_{\text{op}} + \beta p_{\text{op}})^N = \sum_{n+m=N} \frac{N!}{n!m!} \, \alpha^n \beta^m (x_{\text{op}}^n p_{\text{op}}^m)^{\text{W.O.}} \tag{2.15}$$

The expansion of the exponential operator

$$e^{(\alpha x_{\text{op}} + \beta p_{\text{op}})} \,,$$

would, of course, generate all such powers and by analyzing the matrix elements of this exponential operator, we can learn about the matrix elements of Weyl ordered Hamiltonians.

From the fact that the commutator of x_{op} and p_{op} is a constant, using the Baker-Campbell-Hausdorff formula, we obtain

$$e^{(\frac{\alpha x_{\text{op}}}{2})} e^{\beta p_{\text{op}}} e^{(\frac{\alpha x_{\text{op}}}{2})} = e^{(\frac{\alpha x_{\text{op}}}{2})} e^{(\beta p_{\text{op}} + \frac{\alpha x_{\text{op}}}{2} - \frac{i\hbar\alpha\beta}{4})}$$

$$= e^{(\alpha x_{\text{op}} + \beta p_{\text{op}} + \frac{i\hbar\alpha\beta}{4} - \frac{i\hbar\alpha\beta}{4})}$$

$$= e^{(\alpha x_{\text{op}} + \beta p_{\text{op}})} \,. \tag{2.16}$$

Using this relation, it can now be easily shown that

$$\langle x'|e^{(\alpha x_{\mathrm{op}}+\beta p_{\mathrm{op}})}|x\rangle = \langle x'|e^{(\frac{\alpha x_{\mathrm{op}}}{2})}e^{\beta p_{\mathrm{op}}}e^{(\frac{\alpha x_{\mathrm{op}}}{2})}|x\rangle$$

$$= e^{\frac{\alpha x'}{2}}\langle x'|e^{\beta p_{\mathrm{op}}}|x\rangle\, e^{\frac{\alpha x}{2}}$$

$$= \int \mathrm{d}p\, \langle x'|e^{\beta p_{\mathrm{op}}}|p\rangle\langle p|x\rangle\, e^{\frac{\alpha(x+x')}{2}}$$

$$= \int \frac{\mathrm{d}p}{2\pi\hbar}\, e^{\frac{i}{\hbar}p(x'-x)}e^{(\frac{\alpha(x+x')}{2}+\beta p)}. \qquad (2.17)$$

Once again, we have used here the completeness properties given in Eq. (2.4) as well as the defining relations in Eqs. (2.1), (2.3) and (2.5). It follows from this that for a Weyl ordered quantum Hamiltonian, we have

$$\langle x'|H^{\mathrm{W.O.}}(x_{\mathrm{op}}, p_{\mathrm{op}})|x\rangle = \int \frac{\mathrm{d}p}{2\pi\hbar}\, e^{\frac{i}{\hbar}p(x'-x)}H\left(\frac{x+x'}{2}, p\right). \quad (2.18)$$

As we see, the matrix elements of any Weyl ordered Hamiltonian simply lead to what is known as the mid-point prescription and this is what we will use in all of our discussions.

2.3 Feynman path integral

We are now ready to calculate the transition amplitude between two coordinate basis states in the Heisenberg picture. Let us recall that in the Heisenberg picture, for $t_f > t_i$, we have

$$U(x_f, t_f; x_i, t_i) = {}_H\langle x_f, t_f|x_i, t_i\rangle_H.$$

Let us divide the time interval between the initial and the final time into N equal segments of infinitesimal length ϵ. Namely, let

$$\epsilon = \frac{t_f - t_i}{N}. \qquad (2.19)$$

In other words, for simplicity, we discretize the time interval and in the end, we will take the continuum limit $\epsilon \to 0$ and $N \to \infty$ such that Eq. (2.19) holds true. We can now label the intermediate times as, say,

$$t_n = t_i + n\epsilon, \quad n = 1, 2, \cdots, (N-1). \tag{2.20}$$

Introducing complete sets of coordinate basis states for every intermediate time point (see Eq. (2.10)), we obtain

$$U(x_f, t_f; x_i, t_i) = {}_H\langle x_f, t_f | x_i, t_i \rangle_H$$

$$= \lim_{\substack{\epsilon \to 0 \\ N \to \infty}} \int dx_1 \cdots dx_{N-1}\, {}_H\langle x_f, t_f | x_{N-1}, t_{N-1} \rangle_H$$

$$\times\ {}_H\langle x_{N-1}, t_{N-1} | x_{N-2}, t_{N-2} \rangle_H \cdots {}_H\langle x_1, t_1 | x_i, t_i \rangle_H\,.$$

$$\tag{2.21}$$

In writing this, we have clearly assumed an inherent time ordering from left to right. Let us also note here that while there are N inner products in the above expression, there are only $(N-1)$ intermediate points of integration. Furthermore, we note that any intermediate inner product in Eq. (2.21) has the form

$$_H\langle x_n, t_n | x_{n-1}, t_{n-1} \rangle_H = \langle x_n | e^{-\frac{i}{\hbar} t_n H} e^{\frac{i}{\hbar} t_{n-1} H} | x_{n-1} \rangle$$

$$= \langle x_n | e^{-\frac{i}{\hbar}(t_n - t_{n-1})H} | x_{n-1} \rangle$$

$$= \langle x_n | e^{-\frac{i}{\hbar}\epsilon H} | x_{n-1} \rangle$$

$$= \int \frac{dp_n}{2\pi\hbar}\, e^{\frac{i}{\hbar} p_n (x_n - x_{n-1}) - \frac{i}{\hbar}\epsilon H\left(\frac{x_n + x_{n-1}}{2}, p_n\right)}\,. \tag{2.22}$$

Here we have used the mid-point prescription of Eq. (2.18) corresponding to Weyl ordering.

Substituting this form of the inner product into the transition amplitude, we obtain

$$U(x_f, t_f; x_i, t_i) = \lim_{\substack{\epsilon \to 0 \\ N \to \infty}} \int dx_1 \cdots dx_{N-1} \frac{dp_1}{2\pi\hbar} \cdots \frac{dp_N}{2\pi\hbar}$$

$$\times e^{\frac{i}{\hbar} \sum_{n=1}^{N} \left(p_n(x_n - x_{n-1}) - \epsilon H\left(\frac{x_n + x_{n-1}}{2}, p_n\right) \right)}$$

$$= \lim_{\substack{\epsilon \to 0 \\ N \to \infty}} \int dx_1 \cdots dx_{N-1} \frac{dp_1}{2\pi\hbar} \cdots \frac{dp_N}{2\pi\hbar}$$

$$\times e^{\frac{i\epsilon}{\hbar} \sum_{n=1}^{N} \left(\frac{(p_n(x_n - x_{n-1})}{\epsilon} - H\left(\frac{x_n + x_{n-1}}{2}, p_n\right) \right)} . \qquad (2.23)$$

In writing this, we have identified

$$x_0 = x_i, \qquad x_N = x_f . \qquad (2.24)$$

This is the crudest form of Feynman's path integral and is defined in the phase space of the system. It is worth emphasizing here that the number of intermediate coordinate integrations differs from the number of momentum integrations and has profound consequences in the study of the symmetry properties of the transition amplitudes. Note that in the continuum limit, namely, for $\epsilon \to 0$, we can write the exponent in Eq. (2.23) as

$$\lim_{\substack{\epsilon \to 0 \\ N \to \infty}} \frac{i}{\hbar} \epsilon \sum_{n=1}^{N} \left(p_n \left(\frac{x_n - x_{n-1}}{\epsilon} \right) - H \left(\frac{x_n + x_{n-1}}{2}, p_n \right) \right)$$

$$= \frac{i}{\hbar} \int_{t_i}^{t_f} dt \, (p\dot{x} - H(x, p))$$

$$= \frac{i}{\hbar} \int_{t_i}^{t_f} dt \, L . \qquad (2.25)$$

Namely, it is proportional to the action in the mixed variables $(x, \dot{x}, p)$. (It is worth commenting here that if we take the continuum limit in the exponent itself, the action would appear to be the same, both for the normal ordering and the Weyl ordering since we have neglected order $O(\epsilon)$ terms. However, we should remember that in the calculation of the transition amplitude, the continuum limit is to be taken only at the end. This may lead to different contributions from the two orderings of the Hamiltonian. For example, the integrations may give rise to divergences which may combine with $O(\epsilon)$ or higher terms to give finite contributions.)

To obtain the more familiar form of the path integral involving the Lagrangian in the configuration space, let us specialize to the class of Hamiltonians which are quadratic in the momentum variables. Namely, let us choose

$$H(x, p) = \frac{p^2}{2m} + V(x).$$
(2.26)

In such a case, we have from Eq. (2.23)

$$U(x_f, t_f; x_i, t_i) = \lim_{\substack{\epsilon \to 0 \\ N \to \infty}} \int dx_1 \cdots dx_{N-1} \frac{dp_1}{2\pi\hbar} \cdots \frac{dp_N}{2\pi\hbar}$$

$$\times \, e^{\frac{i\epsilon}{\hbar} \sum_{n=1}^{N} \left(p_n \left(\frac{x_n - x_{n-1}}{\epsilon} \right) - \frac{p_n^2}{2m} - V \left(\frac{x_n + x_{n-1}}{2} \right) \right)}.$$
(2.27)

The momentum integrals are Gaussian and, therefore, can be done readily. We note that

$$\int \frac{dp_n}{2\pi\hbar} \, e^{-\frac{i\epsilon}{\hbar} \left(\frac{p_n^2}{2m} - \frac{p_n (x_n - x_{n-1})}{\epsilon} \right)}$$

$$= \int \frac{dp_n}{2\pi\hbar} \, e^{-\frac{i\epsilon}{2m\hbar} \left(p_n^2 - \frac{2m p_n (x_n - x_{n-1})}{\epsilon} \right)}$$

$$= \int \frac{dp_n}{2\pi\hbar} \, e^{-\frac{i\epsilon}{2m\hbar} \left[\left(p_n - \frac{m(x_n - x_{n-1})}{\epsilon} \right)^2 - \left(\frac{m(x_n - x_{n-1})}{\epsilon} \right)^2 \right]}$$

$$= \frac{1}{2\pi\hbar} \left(\frac{2\pi m\hbar}{i\epsilon} \right)^{\frac{1}{2}} e^{\frac{im\epsilon}{2\hbar} \left(\frac{x_n - x_{n-1}}{\epsilon} \right)^2}$$

$$= \left(\frac{m}{2\pi i\hbar\epsilon} \right)^{\frac{1}{2}} e^{\frac{im\epsilon}{2\hbar} \left(\frac{x_n - x_{n-1}}{\epsilon} \right)^2}.$$
(2.28)

Substituting this back into the transition amplitude in Eq. (2.27), we obtain

$$U(x_f, t_f; x_i, t_i) = \lim_{\substack{\epsilon \to 0 \\ N \to \infty}} \left(\frac{m}{2\pi i \hbar \epsilon}\right)^{\frac{N}{2}}$$

$$\times \int dx_1 \cdots dx_{N-1} e^{\frac{i\epsilon}{\hbar} \sum_{n=1}^{N} \left(\frac{m}{2}\left(\frac{x_n - x_{n-1}}{\epsilon}\right)^2 - V\left(\frac{x_n + x_{n-1}}{2}\right)\right)}$$

$$= A \int \mathcal{D}x \, e^{\frac{i}{\hbar} \int_{t_i}^{t_f} dt \left(\frac{1}{2}m\dot{x}^2 - V(x)\right)}$$

$$= A \int \mathcal{D}x \, e^{\frac{i}{\hbar}S[x]}, \tag{2.29}$$

where A is a constant independent of the dynamics of the system and $S[x]$ is the action for the system given in Eq. (1.10). This is Feynman's path integral for the transition amplitude in quantum mechanics.

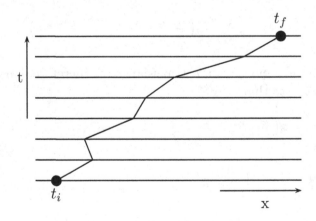

To understand the meaning of (2.29), let us try to understand the meaning of the path integral measure $\mathcal{D}x$. In this integration, the end points are held fixed and only the intermediate coordinates are integrated over the entire space. Any given set of particular values for the intermediate points, of course, defines a trajectory between the initial and the final points. Therefore, integrating over all such values (that is precisely what the integrations over the intermediate points are supposed to do) is equivalent to summing over all the paths connecting the initial and the final points. In other words, Feynman's path integral (2.29) simply says that the transition amplitude between an initial and a final (coordinate) state is the sum over all paths, connecting the two points, of the weight factor $e^{\frac{i}{\hbar}S[x]}$. We know from the study of quantum mechanics that if a process can take place in several distinct ways, then the transition (probability) amplitude is the sum of the individual amplitudes corresponding to every possible way the process can take place. The sum over the paths is, therefore, quite expected. However, it is the weight factor, $e^{\frac{i}{\hbar}S[x]}$, that is quite crucial and unexpected. Classically, we know that it is the classical action (the action for the classical trajectory which satisfies the Euler-Lagrange equation) that determines the classical dynamics. Quantum mechanically, however, what we see is that the values of the action for all the paths contribute to the transition amplitude. It is also worth pointing out here that even though we have derived the path integral representation for the transition amplitude for a special class of Hamiltonians, the expression holds in general. For Hamiltonians which are not quadratic in the momenta, one should simply be more careful in defining the path integral measure $\mathcal{D}x$.

2.4 The classical limit

As we have seen in Eq. (2.29), the transition amplitude can be written as the sum over paths of a phase factor involving the action and in the case of a one dimensional Hamiltonian which is quadratic in the momentum, it has the form

$$
U(x_f, t_f; x_i, t_i) = A \int \mathcal{D}x\, e^{\frac{i}{\hbar}S[x]} = \lim_{\substack{\epsilon \to 0 \\ N \to \infty}} A_N \int dx_1 \cdots dx_{N-1}
$$

$$
\times\, e^{\frac{i\epsilon}{\hbar} \sum_{n=1}^{N} \left(\frac{m}{2}\left(\frac{x_n - x_{n-1}}{\epsilon} \right)^2 - V\left(\frac{x_n + x_{n-1}}{2} \right) \right)},
\tag{2.30}
$$

where

$$
A_N = \left(\frac{m}{2\pi i\hbar\epsilon} \right)^{\frac{N}{2}}.
\tag{2.31}
$$

Even though one can be more quantitative in the discussion of the behavior of the transition amplitude, let us try to be qualitative in the following. We note that for paths where

$$x_n \gg x_{n-1},$$

the first term in the exponential would be quite large, particularly since ϵ is infinitesimally small. Therefore, such paths will lead to a very large phase and consequently, the weight factor can easily be positive or negative. In other words, for every such x_n, there would be a nearby x_n differing only slightly with a phase which would have a cancelling effect. Thus, in the path integral, all such contributions will average out to zero.

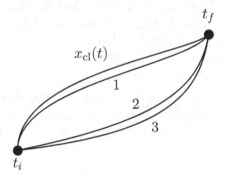

Let us, therefore, concentrate only on paths connecting the initial and the final points that differ from one another only slightly. For simplicity, we only look at continuous paths which are differentiable. (A more careful analysis shows that the paths which contribute nontrivially are the continuous paths which are not necessarily differentiable. But for simplicity of argument, we will ignore this technical point.) The question that we would like to understand is how, from among all the paths which can contribute to the transition amplitude, it is only the classical path that is singled out in the classical limit, namely, when $\hbar \to 0$. We note here that the weight factor in the path integral, namely, $e^{\frac{i}{\hbar}S[x]}$, is a phase which involves the action multiplied by a large quantity when $\hbar \to 0$. Mathematically, therefore, it is clear that the dominant contribution to the path integral would arise from paths near the ones which extremize the phase factor. In other words, only trajectories close to the ones satisfying

$$\frac{\delta S[x]}{\delta x(t)} = 0, \tag{2.32}$$

would contribute significantly to the transition amplitude in the classical limit. But, from the principle of least action, we know that these are precisely the trajectories which a classical particle would follow, namely, the classical trajectories. We can see this more intuitively in the following way. Suppose, we are considering a path, say #3, which is quite far away from the classical trajectory and which contributes a weight factor $e^{\frac{i}{\hbar}S_3}$. For every such path, there will be a nearby path, infinitesimally close, say #2, where the action would differ by a small amount so that it would contribute a weight factor $e^{\frac{i}{\hbar}(S_3+\delta)}$. However, since the action is multiplied by a large constant $(\hbar \to 0)$, the phase difference between the two paths would be large and $e^{\frac{i}{\hbar}\delta}$ can have any sign. As a result, all such paths will average out to zero in the sum. Near the classical trajectory, however, the action is stationary. Consequently, if we choose a path infinitesimally close to the classical path, the action will not change and the weight factor for any two such paths will be the same. Therefore, all such paths will add up coherently and give the dominant contribution as $\hbar \to 0$. It is in this way that the classical trajectory is singled out in the classical limit, not because it contributes the most, but rather because there are paths infinitesimally close to it which add coherently. One can, of course, make various estimates as to how far away a path can be from the classical trajectory before its contribution becomes unimportant. But let us not go into these details here.

2.5 Equivalence with the Schrödinger equation

At this point one may wonder about the Schrödinger equation in the path integral formalism. Namely, it is not clear how we can recover the time dependent Schrödinger equation (see Eq. (1.33)) from the path integral representation of the transition amplitude. Let us recall that the Schrödinger equation is a differential equation. Therefore, it determines changes in the wave function for infinitesimal time intervals. Consequently, to derive the Schrödinger equation, we merely have to examine the infinitesimal form of the transition amplitude or the path integral. From the explicit form of the transition amplitude in Eq. (2.30), we obtain for infinitesimal ϵ

$$U(x_f, t_f = \epsilon; x_i, t_i = 0)$$
$$= \left(\frac{m}{2\pi i \hbar \epsilon}\right)^{\frac{1}{2}} e^{\frac{i\epsilon}{\hbar}\left(\frac{m}{2}\left(\frac{x_f - x_i}{\epsilon}\right)^2 - V\left(\frac{x_f + x_i}{2}\right)\right)}. \tag{2.33}$$

We also know from Eq. (1.41) that the transition amplitude is the propagator which gives the propagation (time evolution) of the wave function in the following way,

$$\psi(x, \epsilon) = \int_{-\infty}^{\infty} dx' \, U(x, \epsilon; x', 0)\psi(x', 0). \tag{2.34}$$

Therefore, substituting the form of the transition amplitude namely, Eq. (2.33) into Eq. (2.34), we obtain

$$\psi(x, \epsilon) = \left(\frac{m}{2\pi i \hbar \epsilon}\right)^{\frac{1}{2}} \int_{-\infty}^{\infty} dx' e^{\frac{im}{2\hbar\epsilon}(x-x')^2 - \frac{i\epsilon}{\hbar}V\left(\frac{x+x'}{2}\right)}\psi(x', 0). \tag{2.35}$$

Let us next change variables to

$$\eta = x' - x, \tag{2.36}$$

so that we can write

$$\psi(x, \epsilon) = \left(\frac{m}{2\pi i \hbar \epsilon}\right)^{\frac{1}{2}} \int_{-\infty}^{\infty} d\eta \, e^{\left(\frac{im}{2\hbar\epsilon}\eta^2 - \frac{i\epsilon}{\hbar}V(x+\frac{\eta}{2})\right)}\psi(x+\eta, 0). \tag{2.37}$$

It is obvious that because ϵ is infinitesimal, if η is large, then the first term in the exponent in (2.37) would lead to rapid oscillations and all such contributions will average out to zero. Therefore, the dominant contribution will come from the region of integration

$$0 \leq |\eta| \leq \left(\frac{2\pi\hbar\epsilon}{m}\right)^{\frac{1}{2}}, \tag{2.38}$$

where the change in the first exponent is of the order of unity. (In other words, $|\eta| \simeq \epsilon^{1/2}$.) Keeping this in mind, we can Taylor expand the integrand and since we are interested in the infinitesimal behavior, we can keep terms consistently up to order ϵ. In this way, we obtain (remember $\epsilon \sim \eta^2$)

$$\psi(x, \epsilon) = \left(\frac{m}{2\pi i\hbar\epsilon}\right)^{\frac{1}{2}} \int_{-\infty}^{\infty} \mathrm{d}\eta \, e^{\frac{im}{2\hbar\epsilon}\eta^2} \left(1 - \frac{i\epsilon}{\hbar}V\left(x + \frac{\eta}{2}\right)\right)\psi(x + \eta, 0)$$

$$= \left(\frac{m}{2\pi i\hbar\epsilon}\right)^{\frac{1}{2}} \int_{-\infty}^{\infty} \mathrm{d}\eta \, e^{\frac{im}{2\hbar\epsilon}\eta^2} \left(1 - \frac{i\epsilon}{\hbar}V(x) + O(\epsilon\eta)\right)$$

$$\times \left(\psi(x,0) + \eta\psi'(x,0) + \frac{\eta^2}{2}\psi''(x,0) + O(\eta^3)\right)$$

$$= \left(\frac{m}{2\pi i\hbar\epsilon}\right)^{\frac{1}{2}} \int_{-\infty}^{\infty} \mathrm{d}\eta \, e^{\frac{im}{2\hbar\epsilon}\eta^2} \left[\psi(x,0) - \frac{i\epsilon}{\hbar}V(x)\psi(x,0)\right.$$

$$\left. + \eta\psi'(x,0) + \frac{\eta^2}{2}\psi''(x,0) + O(\eta^3, \epsilon^2)\right]. \tag{2.39}$$

The individual integrations can be easily done and the results are

$$\int_{-\infty}^{\infty} \mathrm{d}\eta \, e^{\frac{im}{2\hbar\epsilon}\eta^2} = \left(\frac{2\pi i\hbar\epsilon}{m}\right)^{\frac{1}{2}},$$

$$\int_{-\infty}^{\infty} \mathrm{d}\eta \, \eta \, e^{\frac{im}{2\hbar\epsilon}\eta^2} = 0,$$

$$\int_{-\infty}^{\infty} \mathrm{d}\eta \, \eta^2 \, e^{\frac{im}{2\hbar\epsilon}\eta^2} = \frac{i\hbar\epsilon}{m}\left(\frac{2\pi i\hbar\epsilon}{m}\right)^{\frac{1}{2}}. \tag{2.40}$$

Note that these integrals contain oscillatory integrands and the simplest way of evaluating them is through a regularization. For example,

$$
\int_{-\infty}^{\infty} d\eta \, e^{\frac{im}{2\hbar\epsilon}\eta^2} = \lim_{\delta \to 0^+} \int_{-\infty}^{\infty} d\eta \, e^{\left(\frac{im}{2\hbar\epsilon}-\delta\right)\eta^2}
$$

$$
= \lim_{\delta \to 0^+} \left(\frac{\pi}{\delta - \frac{im}{2\hbar\epsilon}}\right)^{\frac{1}{2}} = \left(\frac{2\pi i\hbar\epsilon}{m}\right)^{\frac{1}{2}}, \tag{2.41}
$$

and so on.

Substituting these back into Eq. (2.39), we obtain

$$
\psi(x, \epsilon) = \left(\frac{m}{2\pi i\hbar\epsilon}\right)^{\frac{1}{2}} \left[\left(\frac{2\pi i\hbar\epsilon}{m}\right)^{\frac{1}{2}} \left(\psi(x,0) - \frac{i\epsilon}{\hbar}V(x)\psi(x,0)\right)\right.
$$

$$
\left. + \frac{i\hbar\epsilon}{2m}\left(\frac{2\pi i\hbar\epsilon}{m}\right)^{\frac{1}{2}} \psi''(x,0) + O\left(\epsilon^2\right)\right]
$$

$$
= \psi(x,0) + \frac{i\hbar\epsilon}{2m}\psi''(x,0) - \frac{i\epsilon}{\hbar}V(x)\psi(x,0) + O\left(\epsilon^2\right),
$$

$$
\text{or,}\ \psi(x,\epsilon) - \psi(x,0) = -\frac{i\epsilon}{\hbar}\left(-\frac{\hbar^2}{2m}\frac{\partial^2}{\partial x^2} + V(x)\right)\psi(x,0) + O(\epsilon^2),
$$

$$
\text{or,}\ i\hbar\left(\frac{\psi(x,\epsilon) - \psi(x,0)}{\epsilon}\right) = \left(-\frac{\hbar^2}{2m}\frac{\partial^2}{\partial x^2} + V(x)\right)\psi(x,0) + O(\epsilon^2). \tag{2.42}
$$

In the limit $\epsilon \to 0$, therefore, we obtain the time dependent Schrödinger equation (Eq. (1.33))

$$
i\hbar \frac{\partial \psi(x,t)}{\partial t} = \left(-\frac{\hbar^2}{2m}\frac{\partial^2}{\partial x^2} + V(x)\right)\psi(x,t) .
$$

The path integral representation, therefore, contains the Schrödinger equation and is equivalent to it.

2.6 Free particle

We recognize that the path integral is a functional integral. Namely, the integrand which is the phase factor is a functional of the trajectory between the initial and the final points and we are integrating over all possible paths between these two points. Since we do not have a feeling for such quantities, let us evaluate some of these integrals associated with simple systems. The free particle is probably the simplest of quantum mechanical systems. For a free particle in one dimension, the Lagrangian has the form

$$L = \frac{1}{2}m\dot{x}^2 .\tag{2.43}$$

Therefore, from our definition of the transition amplitude in Eq. (2.29) or (2.30), we obtain

$$U(x_f, t_f; x_i, t_i)$$

$$= \lim_{\substack{\epsilon \to 0 \\ N \to \infty}} \left(\frac{m}{2\pi i\hbar\epsilon}\right)^{\frac{N}{2}} \int dx_1 \cdots dx_{N-1}\, e^{\frac{i\epsilon}{\hbar} \sum_{n=1}^{N} \frac{m}{2}\left(\frac{x_n - x_{n-1}}{\epsilon}\right)^2}$$

$$= \lim_{\substack{\epsilon \to 0 \\ N \to \infty}} \left(\frac{m}{2\pi i\hbar\epsilon}\right)^{\frac{N}{2}} \int dx_1 \cdots dx_{N-1}\, e^{\frac{im}{2\hbar\epsilon} \sum_{n=1}^{N} (x_n - x_{n-1})^2} .$$

$$\tag{2.44}$$

Scaling the variables of integration as

$$y_n = \left(\frac{m}{2\hbar\epsilon}\right)^{\frac{1}{2}} x_n ,\tag{2.45}$$

we have

$$U(x_f, t_f; x_i, t_i) = \lim_{\substack{\epsilon \to 0 \\ N \to \infty}} \left(\frac{m}{2\pi i\hbar\epsilon}\right)^{\frac{N}{2}} \left(\frac{2\hbar\epsilon}{m}\right)^{\frac{N-1}{2}}$$

$$\times \int dy_1 \cdots dy_{N-1}\, e^{i \sum_{n=1}^{N} (y_n - y_{n-1})^2} .\tag{2.46}$$

This is a coupled set of Gaussian integrals which can be evaluated in many different ways. However, the simplest method probably is to work out a few lower order terms and derive a pattern. For a single integration, we note that

$$\int dy_1 \ e^{i((y_1-y_0)^2+(y_2-y_1)^2)} = \int dy_1 \ e^{i(2(y_1-\frac{y_0+y_2}{2})^2+\frac{1}{2}(y_2-y_0)^2)}$$

$$= \left(\frac{i\pi}{2}\right)^{\frac{1}{2}} e^{\frac{i}{2}(y_2-y_0)^2} . \qquad (2.47)$$

If we had two intermediate integrations, then we will have

$$\int dy_1 dy_2 \ e^{i((y_1-y_0)^2+(y_2-y_1)^2+(y_3-y_2)^2)}$$

$$= \left(\frac{i\pi}{2}\right)^{\frac{1}{2}} \int dy_2 \ e^{i(\frac{1}{2}(y_2-y_0)^2+(y_3-y_2)^2)}$$

$$= \left(\frac{i\pi}{2}\right)^{\frac{1}{2}} \int dy_2 \ e^{\left(\frac{3i}{2}(y_2-\frac{y_0+2y_3}{3})^2+\frac{i}{3}(y_3-y_0)^2\right)}$$

$$= \left(\frac{i\pi}{2}\right)^{\frac{1}{2}} \left(\frac{2i\pi}{3}\right)^{\frac{1}{2}} e^{\frac{i}{3}(y_3-y_0)^2}$$

$$= \left(\frac{(i\pi)^2}{3}\right)^{\frac{1}{2}} e^{\frac{i}{3}(y_3-y_0)^2} . \qquad (2.48)$$

A pattern is now obvious and using this we can write

$$U(x_f, t_f; x_i, t_i)$$

$$= \lim_{\substack{\epsilon\to 0 \\ N\to\infty}} \left(\frac{m}{2\pi i\hbar\epsilon}\right)^{\frac{N}{2}} \left(\frac{2\hbar\epsilon}{m}\right)^{\frac{N-1}{2}} \left(\frac{(i\pi)^{N-1}}{N}\right)^{\frac{1}{2}} e^{\frac{i}{N}(y_N-y_0)^2}$$

$$= \lim_{\substack{\epsilon\to 0 \\ N\to\infty}} \left(\frac{m}{2\pi i\hbar\epsilon}\right)^{\frac{N}{2}} \left(\frac{2\pi i\hbar\epsilon}{m}\right)^{\frac{N-1}{2}} \frac{1}{\sqrt{N}} e^{\frac{im}{2\hbar N\epsilon}(x_N-x_0)^2}$$

$$= \lim_{\substack{\epsilon \to 0 \\ N \to \infty}} \left(\frac{m}{2\pi i\hbar N\epsilon} \right)^{\frac{1}{2}} e^{\frac{im}{2\hbar N\epsilon}(x_f - x_i)^2}$$

$$= \left(\frac{m}{2\pi i\hbar\,(t_f - t_i)} \right)^{\frac{1}{2}} e^{\frac{i}{\hbar}\frac{m(x_f - x_i)^2}{2(t_f - t_i)}}. \tag{2.49}$$

Thus, we see that for a free particle, the transition amplitude can be explicitly evaluated. It has the right behavior in the sense that, as $t_f \to t_i$,

$$U(x_f, t_f; x_i, t_i) \to \delta(x_f - x_i), \tag{2.50}$$

which is nothing other than the orthonormality relation for the states in the Heisenberg picture given in Eq. (2.9). Second, all the potentially dangerous singular terms involving ϵ have disappeared. Furthermore, this is exactly the result one would obtain by solving the time dependent Schrödinger equation. It expresses the well known fact that even a well localized wave packet spreads (disperses) with time. Namely, even the simplest of (linear) equations has only dispersive solutions.

Let us note here that since

$$S[x] = \int\limits_{t_i}^{t_f} dt\,\frac{1}{2}m\dot{x}^2, \tag{2.51}$$

the Euler-Lagrange equations for the system give (see Eq. (1.31))

$$\frac{\delta S[x]}{\delta x(t)} = m\ddot{x} = 0. \tag{2.52}$$

This gives as solutions

$$\dot{x}_{\rm cl}(t) = v = {\rm constant}. \tag{2.53}$$

Thus, for the classical trajectory, we have

$$S[x_{\rm cl}] = \int_{t_i}^{t_f} dt \; \frac{1}{2} m \dot{x}_{\rm cl}^2 = \frac{1}{2} m v^2 (t_f - t_i) \,.$$ (2.54)

On the other hand, since v is a constant, we can write

$$x_f - x_i = v(t_f - t_i),$$

or, $\quad v = \dfrac{x_f - x_i}{t_f - t_i} \,.$ (2.55)

Substituting this back into Eq. (2.54), we obtain

$$S[x_{\rm cl}] = \frac{1}{2} m \left(\frac{x_f - x_i}{t_f - t_i} \right)^2 (t_f - t_i) = \frac{m}{2} \frac{(x_f - x_i)^2}{t_f - t_i} \,.$$ (2.56)

We recognize, therefore, that we can also write the quantum transition amplitude, in this case, simply as

$$U(x_f, t_f; x_i, t_i) = \left(\frac{m}{2\pi i \hbar (t_f - t_i)} \right)^{\frac{1}{2}} e^{\frac{i}{\hbar} S[x_{\rm cl}]} \,.$$ (2.57)

This is a particular characteristic of some of the quantum systems which can be solved exactly. Namely, for these systems, the transition amplitude can be written in the form

$$U(x_f, t_f; x_i, t_i) = A \, e^{\frac{i}{\hbar} S[x_{\rm cl}]} \,,$$ (2.58)

where A is a constant.

Finally, let us note from the explicit form of the transition amplitude in Eq. (2.49) that

$$\frac{\partial U}{\partial t_f} = -\frac{U}{2(t_f - t_i)} - \frac{im}{2\hbar}\left(\frac{x_f - x_i}{t_f - t_i}\right)^2 U\,,$$

$$\frac{\partial U}{\partial x_f} = \frac{im}{\hbar}\left(\frac{x_f - x_i}{t_f - t_i}\right) U\,,$$

$$\frac{\partial^2 U}{\partial x_f^2} = \frac{im}{\hbar}\frac{U}{t_f - t_i} + \left(\frac{im}{\hbar}\right)^2\left(\frac{x_f - x_i}{t_f - t_i}\right)^2 U$$

$$= -\frac{2m}{\hbar^2}\left(-i\hbar\frac{U}{2(t_f - t_i)} + \frac{m}{2}\left(\frac{x_f - x_i}{t_f - t_i}\right)^2 U\right)$$

$$= -\frac{2m}{\hbar^2}\left(i\hbar\frac{\partial U}{\partial t_f}\right)\,. \tag{2.59}$$

Therefore, it follows that

$$i\hbar\frac{\partial U}{\partial t_f} = -\frac{\hbar^2}{2m}\frac{\partial^2 U}{\partial x_f^2}\,, \tag{2.60}$$

which is equivalent to saying that the transition amplitude obtained from Feynman's path integral, indeed, solves the Schrödinger equation for a free particle (compare with Eq. (1.38)).

2.7 References

A. Das, *Lectures on Quantum Mechanics*, Hindustan Book Agency and World Scientific Publishing.

R. P. Feynman and A. R. Hibbs, *Quantum Mechanics and Path Integrals*, McGraw-Hill Publishing.

B. Sakita, *Quantum Theory of Many Variable Systems and Fields*, World Scientific Publishing.

L. S. Schulman, *Techniques and Applications of Path Integration*, John Wiley Publishing.

CHAPTER 3

Harmonic oscillator

3.1 Path integral for the harmonic oscillator

As a second example of path integrals, let us consider the one dimensional harmonic oscillator which we know can be solved exactly. In fact, let us consider the oscillator interacting with an external source described by the Lagrangian

$$L = \frac{1}{2}m\dot{x}^2 - \frac{1}{2}m\omega^2 x^2 + Jx\,, \tag{3.1}$$

with the action given by

$$S = \int dt\, L\,. \tag{3.2}$$

Here, for example, we can think of the time dependent external source $J(t)$ as an electric field if the oscillator carries an electric charge. The well known results for the free harmonic oscillator can be obtained from this system in the limit $J(t) \to 0$. Furthermore, we know that if the external source were time independent, then the problem can be solved exactly simply because in this case we can write the Lagrangian of Eq. (3.1) as

$$
\begin{aligned}
L &= \frac{1}{2}m\dot{x}^2 - \frac{1}{2}m\omega^2 x^2 + Jx \\
&= \frac{1}{2}m\dot{x}^2 - \frac{1}{2}m\omega^2 \left(x - \frac{J}{m\omega^2} \right)^2 + \frac{J^2}{2m\omega^2} \\
&= \frac{1}{2}m\dot{\bar{x}}^2 - \frac{1}{2}m\omega^2\bar{x}^2 + \frac{J^2}{2m\omega^2}\,,
\end{aligned} \tag{3.3}
$$

35

where we have defined

$$\overline{x} = x - \frac{J}{m\omega^2} .$$ (3.4)

In other words, in such a case, the classical equilibrium position of the oscillator is shifted by a constant amount, namely, the system behaves like a spring suspended freely under the effect of gravity. The system described by Eq. (3.1) is, therefore, of considerable interest because we can obtain various known special cases in different limits. Besides, as we will see later, a source is a great tool for generating correlation functions in the path integral formalism.

The Euler-Lagrange equation for the action in Eq. (3.2) gives the classical trajectory and takes the form

$$\left. \frac{\delta S[x]}{\delta x(t)} \right|_{x=x_{\text{cl}}} = 0,$$

$$\text{or,} \quad m\ddot{x}_{\text{cl}} + m\omega^2 x_{\text{cl}} - J = 0 ,$$ (3.5)

and the general form of the transition amplitude, as we have seen in Eq. (2.29), is given by

$$U(x_f, t_f; x_i, t_i) = A \int \mathcal{D}x \, e^{\frac{i}{\hbar} S[x]} .$$ (3.6)

To evaluate this functional integral, let us define

$$x(t) = x_{\text{cl}}(t) + \eta(t) ,$$ (3.7)

where the variable $\eta(t)$ represents the quantum fluctuations around the classical path, namely, it measures the deviation of a trajectory from the classical trajectory. Since the end points of the trajectories are fixed, the fluctuations satisfy the boundary conditions

$$\eta(t_i) = \eta(t_f) = 0 .$$ (3.8)

We can Taylor expand the action about the classical path as

$$S[x] = S[x_{\rm cl} + \eta] = S[x_{\rm cl}] + \int dt \, \eta(t) \, \frac{\delta S[x]}{\delta x(t)}\bigg|_{x=x_{\rm cl}}$$

$$+ \frac{1}{2!} \int dt_1 dt_2 \, \eta(t_1)\eta(t_2) \, \frac{\delta^2 S[x]}{\delta x(t_1)\delta x(t_2)}\bigg|_{x=x_{\rm cl}}, \qquad (3.9)$$

and the expansion terminates since the action is at most quadratic in the dynamical variables $x(t)$. We note from Eq. (3.5) that the action is an extremum for the classical trajectory. Therefore, we have

$$\frac{\delta S[x]}{\delta x(t)}\bigg|_{x=x_{\rm cl}} = 0, \qquad (3.10)$$

and consequently, we can also write Eq. (3.9) as

$$S[x] = S[x_{\rm cl}] + \frac{1}{2!} \int dt_1 dt_2 \, \eta(t_1)\eta(t_2) \, \frac{\delta^2 S[x]}{\delta x(t_1)\delta x(t_2)}\bigg|_{x=x_{\rm cl}}. \qquad (3.11)$$

If we evaluate the functional derivative in Eq. (3.11) for the action in Eq. (3.2), we can also rewrite the action as (this can also be obtained by substituting (3.7) directly into the action and using the classical equation after integrating by parts the velocity dependent term linear in η)

$$S[x] = S[x_{\rm cl}] + \frac{1}{2} \int_{t_i}^{t_f} dt \left(m\dot{\eta}^2 - m\omega^2\eta^2 \right). \qquad (3.12)$$

As a technical aside, let us note that given the action (3.2), we have

$$\frac{dS[x + \epsilon v]}{d\epsilon}\bigg|_{\epsilon=0} = \int dt \left(m\dot{x}(t)\dot{v}(t) - m\omega^2 x(t)v(t) + Jv(t) \right)$$

$$= \int dt \left(-m\ddot{x}(t) - m\omega^2 x(t) + J(t) \right) v(t), \qquad (3.13)$$

which determines (see (1.13) and compare with (3.5))

$$S'[x(t)] = \frac{\delta S[x]}{\delta x(t)} = -m\ddot{x}(t) - m\omega^2 x(t) + J(t). \tag{3.14}$$

Taking a second functional derivative, we obtain

$$\left. \frac{dS'[x(t_1) + \bar{\epsilon}\,\bar{v}(t_1)]}{d\bar{\epsilon}} \right|_{\bar{\epsilon}=0} = -m\ddot{\bar{v}}(t_1) - m\omega^2 \bar{v}(t_1)$$

$$= \int dt_2 \left(-m\frac{d^2}{dt_2^2} - m\omega^2 \right) \delta(t_1 - t_2)\,\bar{v}(t_2), \tag{3.15}$$

which determines the second functional derivative of the action to be

$$\frac{\delta^2 S[x]}{\delta x(t_2)\delta x(t_1)} = -m \left(\frac{d^2}{dt_2^2} + \omega^2 \right) \delta(t_1 - t_2), \tag{3.16}$$

and, when substituted into (3.11), this leads to (3.12) after integration by parts.

It is clear that summing over all the paths is equivalent to summing over all possible fluctuations subject to the constraint in (3.8). Consequently, we can rewrite the transition amplitude in this case as

$$U(x_f, t_f; x_i, t_i) = A \int \mathcal{D}\eta \, e^{\left(\frac{i}{\hbar} S[x_{cl}] + \frac{i}{2\hbar} \int\limits_{t_i}^{t_f} dt \, (m\dot{\eta}^2 - m\omega^2 \eta^2) \right)}$$

$$= A \, e^{\frac{i}{\hbar} S[x_{cl}]} \int \mathcal{D}\eta \, e^{\frac{im}{2\hbar} \int\limits_{t_i}^{t_f} dt \, (\dot{\eta}^2 - \omega^2 \eta^2)}. \tag{3.17}$$

This is an integral where the exponent is quadratic in the integration variables (namely, it is a functional Gaussian integral) and such an integral can be done in several ways. Since the harmonic oscillator is a fundamental system in any branch of physics, we will evaluate this integral in three different ways so as to develop a feeling for path integrals.

3.2 Method of Fourier transform

First of all, we note that the integrand in the exponent of the functional integral does not depend on time explicitly. Therefore, we can redefine the variable of integration in the exponent as

$$t \to t - t_i, \quad t_i \to 0, \quad t_f \to t_f - t_i, \tag{3.18}$$

in which case, we can write the transition amplitude as

$$U(x_f, t_f; x_i, t_i) = A \, e^{\frac{i}{\hbar} S[x_{cl}]} \int \mathcal{D}\eta \, e^{\frac{im}{2\hbar} \int_0^T dt(\dot{\eta}^2 - \omega^2 \eta^2)}, \tag{3.19}$$

where we have identified the time interval with

$$T = t_f - t_i. \tag{3.20}$$

The variable $\eta(t)$ satisfies the boundary conditions (see Eq. (3.8))

$$\eta(0) = \eta(T) = 0. \tag{3.21}$$

Consequently, the value of the fluctuation at any point on the trajectory can be represented as a Fourier series of the form

$$\eta(t) = \sum_n a_n \sin\left(\frac{n\pi t}{T}\right), \quad n = 1, 2, \ldots, N - 1. \tag{3.22}$$

We note here that since we have chosen to divide the trajectory into N intervals, namely, since there are $(N-1)$ intermediate time points, there can only be $(N-1)$ independent coefficients a_n in the Fourier expansion in Eq. (3.22). Substituting this back, we find that

$$\int_0^T dt \, \dot{\eta}^2 = \sum_{n,m} \int_0^T dt \, a_n a_m \left(\frac{n\pi}{T}\right)\left(\frac{m\pi}{T}\right) \cos\left(\frac{n\pi t}{T}\right) \cos\left(\frac{m\pi t}{T}\right)$$

$$= \frac{T}{2} \sum_n \left(\frac{n\pi}{T}\right)^2 a_n^2, \tag{3.23}$$

where we have used the orthonormality properties of the cosine functions. Similarly, we also obtain

$$
\int_0^T \mathrm{d}t\, \eta^2(t) = \sum_{n,m} \int_0^T \mathrm{d}t\, a_n a_m \sin\left(\frac{n\pi t}{T}\right) \sin\left(\frac{m\pi t}{T}\right)
$$

$$
= \frac{T}{2} \sum_n a_n^2 . \tag{3.24}
$$

Furthermore, we note that integrating over all possible configurations of $\eta(t)$ or all possible quantum fluctuations is equivalent to integrating over all possible values of the coefficients of expansion a_n. Thus we can write the transition amplitude also as

$$
U(x_f, t_f; x_i, t_i)
$$

$$
= \lim_{\substack{\epsilon \to 0 \\ N \to \infty}} A' e^{\frac{i}{\hbar} S[x_{\mathrm{cl}}]} \int \mathrm{d}a_1 \cdots \mathrm{d}a_{N-1}\, e^{\frac{im}{2\hbar} \sum_{n=1}^{N-1} \left(\frac{T}{2}\left(\frac{n\pi}{T}\right)^2 a_n^2 - \frac{T}{2}\omega^2 a_n^2\right)}
$$

$$
= \lim_{\substack{\epsilon \to 0 \\ N \to \infty}} A' e^{\frac{i}{\hbar} S[x_{\mathrm{cl}}]} \int \mathrm{d}a_1 \cdots \mathrm{d}a_{N-1}\, e^{\frac{imT}{4\hbar} \sum_{n=1}^{N-1} \left(\left(\frac{n\pi}{T}\right)^2 - \omega^2\right) a_n^2} . \tag{3.25}
$$

Here we note that any possible factor arising from the Jacobian in the change of variables from η to the coefficients a_n has been lumped into A' whose form we will determine shortly.

We note here that the transition amplitude, in this case, is a product of a set of decoupled integrals each of which has the form of a Gaussian integral that can be easily evaluated. In fact, the individual integrals have the values (see Eq. (2.41))

$$
\int \mathrm{d}a_n\, e^{\frac{imT}{4\hbar}\left(\left(\frac{n\pi}{T}\right)^2 - \omega^2\right) a_n^2}
$$

$$
= \left(\frac{4\pi i\hbar}{mT}\right)^{\frac{1}{2}} \left(\left(\frac{n\pi}{T}\right)^2 - \omega^2\right)^{-\frac{1}{2}}
$$

$$
= \left(\frac{4\pi i\hbar}{mT}\right)^{\frac{1}{2}} \left(\frac{n\pi}{T}\right)^{-1} \left(1 - \left(\frac{\omega T}{n\pi}\right)^2\right)^{-\frac{1}{2}} . \tag{3.26}
$$

Substituting this form of the individual integrals into the expression for the transition amplitude in Eq. (3.25), we obtain (A'' contains the additional constant factors)

$$U(x_f, t_f; x_i, t_i) = \lim_{\substack{\epsilon \to 0 \\ N \to \infty}} A'' e^{\frac{i}{\hbar} S[x_{\text{cl}}]} \prod_{n=1}^{N-1} \left(1 - \left(\frac{\omega T}{n\pi} \right)^2 \right)^{-\frac{1}{2}}. \quad (3.27)$$

If we now use the identity,

$$\lim_{N \to \infty} \prod_{n=1}^{N-1} \left(1 - \left(\frac{\omega T}{n\pi} \right)^2 \right) = \frac{\sin \omega T}{\omega T}, \quad (3.28)$$

we obtain

$$U(x_f, t_f; x_i, t_i) = \lim_{\substack{\epsilon \to 0 \\ N \to \infty}} A'' e^{\frac{i}{\hbar} S[x_{\text{cl}}]} \left(\frac{\sin \omega T}{\omega T} \right)^{-\frac{1}{2}}. \quad (3.29)$$

We can determine the constant A'' by simply noting that when $\omega = 0$, the harmonic oscillator reduces to a free particle for which we have already evaluated the transition amplitude. In fact, recalling from Eq. (2.57) that

$$U_{\text{F.P.}}(x_f, t_f; x_i, t_i) = \left(\frac{m}{2\pi i \hbar (t_f - t_i)} \right)^{\frac{1}{2}} e^{\frac{i}{\hbar} S[x_{\text{cl}}]}, \quad (3.30)$$

and comparing with Eq. (3.29), we obtain

$$\lim_{\substack{\epsilon \to 0 \\ N \to \infty}} A'' = \left(\frac{m}{2\pi i \hbar T} \right)^{\frac{1}{2}}. \quad (3.31)$$

Therefore, we determine the complete form of the transition amplitude for the harmonic oscillator to be

$$U(x_f, t_f; x_i, t_i) = \left(\frac{m}{2\pi i \hbar T} \right)^{\frac{1}{2}} \left(\frac{\sin \omega T}{\omega T} \right)^{-\frac{1}{2}} e^{\frac{i}{\hbar} S[x_{\text{cl}}]}$$

$$= \left(\frac{m\omega}{2\pi i \hbar \sin \omega T} \right)^{\frac{1}{2}} e^{\frac{i}{\hbar} S[x_{\text{cl}}]}. \quad (3.32)$$

It is quite straightforward to see that this expression indeed reduces to the transition amplitude for the free particle in the limit of $\omega \to 0$.

3.3 Matrix method

If the evaluation of the path integral by the method of Fourier transforms appears less satisfactory in any way, then let us evaluate the path integral in the conventional manner by discretizing the time interval. Let us parameterize time on the trajectory as

$$t_n = t_i + n\epsilon, \quad n = 0, 1, \ldots, N, \quad t_0 = t_i, \quad t_N = t_f.$$

Correspondingly, let us define the values of the fluctuations at these points as

$$\eta(t_n) = \eta_n. \tag{3.33}$$

Then, we can write the transition amplitude in Eq. (3.19) in the explicit form

$$U(x_f, t_f; x_i, t_i) = A \, e^{\frac{i}{\hbar} S[x_{cl}]} \int \mathcal{D}\eta \, e^{\frac{im}{2\hbar} \int_{t_i}^{t_f} dt \, \left(\dot{\eta}^2 - \omega^2 \eta^2\right)}$$

$$= \lim_{\substack{\epsilon \to 0 \\ N \to \infty}} \left(\frac{m}{2\pi i \hbar \epsilon}\right)^{\frac{N}{2}} e^{\frac{i}{\hbar} S[x_{cl}]} \int d\eta_1 \cdots d\eta_{N-1}$$

$$\times \, e^{\frac{im\epsilon}{2\hbar} \sum_{n=1}^{N} \left(\left(\frac{\eta_n - \eta_{n-1}}{\epsilon}\right)^2 - \omega^2 \left(\frac{\eta_n + \eta_{n-1}}{2}\right)^2\right)}. \tag{3.34}$$

In this expression, we are supposed to identify

$$\eta_0 = \eta_N = 0, \tag{3.35}$$

corresponding to the boundary conditions in Eq. (3.8), namely,

$$\eta(t_i) = \eta(t_f) = 0.$$

To simplify the integral, let us rescale the variables as

$$\eta_n \to \left(\frac{m}{2\hbar\epsilon}\right)^{-\frac{1}{2}} \eta_n . \tag{3.36}$$

The transition amplitude, in this case, will take the form

$$U(x_f, t_f; x_i, t_i)$$

$$= \lim_{\substack{\epsilon \to 0 \\ N \to \infty}} \left(\frac{m}{2\pi i \hbar\epsilon}\right)^{\frac{N}{2}} \left(\frac{2\hbar\epsilon}{m}\right)^{\frac{N-1}{2}} e^{\frac{i}{\hbar}S[x_{\text{cl}}]} \int d\eta_1 \cdots d\eta_{N-1}$$

$$\times e^{i \sum_{n=1}^{N} \left((\eta_n - \eta_{n-1})^2 - \frac{\epsilon^2 \omega^2}{4}(\eta_n + \eta_{n-1})^2\right)} . \tag{3.37}$$

If we think of the η_n's (there are $(N-1)$ of them) as forming a column matrix, namely,

$$\eta = \begin{pmatrix} \eta_1 \\ \eta_2 \\ \vdots \\ \eta_{N-1} \end{pmatrix}, \tag{3.38}$$

then, we can also write the transition amplitude in terms of matrices as

$$U(x_f, t_f; x_i, t_i) = \lim_{\substack{\epsilon \to 0 \\ N \to \infty}} \left(\frac{m}{2\pi i \hbar\epsilon}\right)^{\frac{N}{2}} \left(\frac{2\hbar\epsilon}{m}\right)^{\frac{N-1}{2}} e^{\frac{i}{\hbar}S[x_{\text{cl}}]}$$

$$\times \int d\eta \, e^{i\eta^T B \eta} . \tag{3.39}$$

Here η^T represents the transpose of the column matrix in Eq. (3.38) and the $(N-1) \times (N-1)$ matrix B has the form

$$B = \begin{pmatrix} 2 & -1 & 0 & 0 & \cdots \\ -1 & 2 & -1 & 0 & \cdots \\ 0 & -1 & 2 & -1 & \cdots \\ \vdots & \vdots & \vdots & \vdots & \end{pmatrix} - \frac{\epsilon^2 \omega^2}{4} \begin{pmatrix} 2 & 1 & 0 & 0 & \cdots \\ 1 & 2 & 1 & 0 & \cdots \\ 0 & 1 & 2 & 1 & \cdots \\ \vdots & \vdots & \vdots & \vdots & \end{pmatrix} . \tag{3.40}$$

This is a symmetric matrix and, therefore, we can write it as

$$
B = \begin{pmatrix} x & y & 0 & 0 & \cdots \\ y & x & y & 0 & \cdots \\ 0 & y & x & y & \cdots \\ \vdots & \vdots & \vdots & \vdots & \end{pmatrix},
\tag{3.41}
$$

where we have defined

$$
x = 2\left(1 - \frac{\epsilon^2\omega^2}{4}\right),
$$

$$
y = -\left(1 + \frac{\epsilon^2\omega^2}{4}\right).
\tag{3.42}
$$

The matrix B is clearly Hermitian (both x and y are real) and, therefore, can be diagonalized by a unitary matrix (more precisely by an orthogonal matrix) which we denote by $\mathcal{U}$. In other words,

$$
B_D = \begin{pmatrix} b_1 & 0 & 0 & \cdots \\ 0 & b_2 & 0 & \cdots \\ 0 & 0 & b_3 & \cdots \\ \vdots & \vdots & \vdots & \end{pmatrix} = \mathcal{U}B\mathcal{U}^\dagger.
\tag{3.43}
$$

Therefore, defining a new vector

$$
\zeta = \mathcal{U}\eta,
\tag{3.44}
$$

we obtain

$$\int d\eta \, e^{i\eta^T B \eta} = \int d\zeta \, e^{i\zeta^T B_D \zeta}$$

$$= \int d\zeta_1 \cdots d\zeta_{N-1} \, e^{i \sum\limits_{n=1}^{N-1} b_n \zeta_n^2}$$

$$= \prod_{n=1}^{N-1} \left(\frac{i\pi}{b_n} \right)^{\frac{1}{2}}$$

$$= (i\pi)^{\frac{N-1}{2}} (\det B)^{-\frac{1}{2}} . \qquad (3.45)$$

Here we have used the familiar fact that the Jacobian for a change of variables by a unitary matrix is unity. Using this result in Eq. (3.39), therefore, we determine the form of the transition amplitude for the harmonic oscillator to be

$$U(x_f, t_f; x_i, t_i)$$

$$= \lim_{\substack{\epsilon \to 0 \\ N \to \infty}} \left(\frac{m}{2\pi i \hbar \epsilon} \right)^{\frac{N}{2}} \left(\frac{2\hbar \epsilon}{m} \right)^{\frac{N-1}{2}} (i\pi)^{\frac{N-1}{2}} (\det B)^{-\frac{1}{2}} e^{\frac{i}{\hbar} S[x_{cl}]}$$

$$= \lim_{\substack{\epsilon \to 0 \\ N \to \infty}} \left(\frac{m}{2\pi i \hbar \epsilon} \right)^{\frac{1}{2}} (\det B)^{-\frac{1}{2}} e^{\frac{i}{\hbar} S[x_{cl}]}$$

$$= \lim_{\substack{\epsilon \to 0 \\ N \to \infty}} \left(\frac{m}{2\pi i \hbar \epsilon \det B} \right)^{\frac{1}{2}} e^{\frac{i}{\hbar} S[x_{cl}]} . \qquad (3.46)$$

It is clear from this analysis that the transition amplitude can be defined only if the matrix B does not have any vanishing eigenvalue.

We note here that the main quantity to calculate in order to determine the transition amplitude is

$$\lim_{\substack{\epsilon \to 0 \\ N \to \infty}} \epsilon \det B .$$

Let us note from the special structure of B in Eqs. (3.41) and (3.42), that if we denote the determinant of the $n \times n$ sub-matrices of B as I_n, then it is easy to check that they satisfy the recursion relation

$$I_{n+1} = xI_n - y^2 I_{n-1}, \quad n = 0, 1, 2, \ldots, \tag{3.47}$$

where we restrict

$$I_{-1} = 0, \quad I_0 = 1. \tag{3.48}$$

This recursion relation can be checked trivially for low orders of the matrix determinants. Substituting the form of x and y, we obtain

$$I_{n+1} = 2 \left(1 - \frac{\epsilon^2 \omega^2}{4} \right) I_n - \left(1 + \frac{\epsilon^2 \omega^2}{4} \right)^2 I_{n-1},$$

$$\text{or,} \quad I_{n+1} - 2I_n + I_{n-1} = -\frac{\epsilon^2 \omega^2}{2} \left(I_n + I_{n-1} + \frac{\epsilon^2 \omega^2}{8} I_{n-1} \right),$$

$$\text{or,} \quad \frac{I_{n+1} - 2I_n + I_{n-1}}{\epsilon^2} = -\frac{\omega^2}{2} \left(I_n + I_{n-1} + \frac{\epsilon^2 \omega^2}{8} I_{n-1} \right).$$

$$\tag{3.49}$$

We are, of course, interested in the continuum limit. In order to do so, let us define a function

$$\phi(t_n - t_i) = \phi(n\epsilon) = \epsilon I_n. \tag{3.50}$$

In the continuum limit, we can think of this as a continuous function $\phi(t)$ of t. In other words, we can identify $t = n\epsilon$ as a continuous variable as $\epsilon \to 0$. We note then, that

$$\lim_{\substack{\epsilon \to 0 \\ N \to \infty}} \epsilon \det B = \lim_{\substack{\epsilon \to 0 \\ N \to \infty}} \epsilon I_{N-1} = \phi(t_f - t_i) = \phi(T). \tag{3.51}$$

We also note from Eqs. (3.48) and (3.50) that, in the continuum limit,

$$\phi(0) = \lim_{\epsilon \to 0} \epsilon I_0 = \lim_{\epsilon \to 0} \epsilon = 0, \tag{3.52}$$

and similarly,

$$\dot{\phi}(0) = \lim_{\epsilon \to 0} \epsilon \left(\frac{I_1 - I_0}{\epsilon} \right) = \lim_{\epsilon \to 0} (x - 1)$$

$$= \lim_{\epsilon \to 0} \left(2 - \frac{\epsilon^2 \omega^2}{2} - 1 \right) = 1 . \tag{3.53}$$

Furthermore, from the recursion relation for the I_n's in Eq. (3.49), we conclude that in the limit $\epsilon \to 0$, the function $\phi(t)$ satisfies the second order differential equation

$$\frac{d^2 \phi(t)}{dt^2} = -\omega^2 \phi(t) . \tag{3.54}$$

We recognize this to be the harmonic oscillator equation and the solution subject to the initial conditions (Eqs. (3.52) and (3.53)) is clearly

$$\phi(t) = \frac{\sin \omega t}{\omega} . \tag{3.55}$$

It now follows from this that

$$\lim_{\substack{\epsilon \to 0 \\ N \to \infty}} \epsilon \det B = \lim_{\substack{\epsilon \to 0 \\ N \to \infty}} \epsilon I_{N-1} = \phi(T) = \frac{\sin \omega T}{\omega} . \tag{3.56}$$

Consequently, for the harmonic oscillator, we obtain the transition amplitude in Eq. (3.46) to be

$$U(x_f, t_f; x_i, t_i) = \lim_{\substack{\epsilon \to 0 \\ N \to \infty}} \left(\frac{m}{2\pi i \hbar \epsilon \det B} \right)^{\frac{1}{2}} e^{\frac{i}{\hbar} S[x_{\mathrm{cl}}]}$$

$$= \left(\frac{m\omega}{2\pi i \hbar \sin \omega T} \right)^{\frac{1}{2}} e^{\frac{i}{\hbar} S[x_{\mathrm{cl}}]} . \tag{3.57}$$

This is, of course, what we had already derived in Eq. (3.32) using the method of Fourier transforms.

Let us next describe an alternate way of determining $(\epsilon \det B)$ which is quite useful in studying some specific problems. Let us recall from Eq. (3.47) that the determinant of the $n \times n$ matrices, I_n, satisfy the recursion relations

$$I_{n+1} = x I_n - y^2 I_{n-1} \,.$$

We note here that we can write these recursion relations also in the simple matrix form as

$$\begin{pmatrix} I_{n+1} \\ I_n \end{pmatrix} = \begin{pmatrix} x & -y^2 \\ 1 & 0 \end{pmatrix} \begin{pmatrix} I_n \\ I_{n-1} \end{pmatrix}. \tag{3.58}$$

Iterating this $(N-2)$ times, for $n = N - 2$, we obtain

$$\begin{pmatrix} I_{N-1} \\ I_{N-2} \end{pmatrix} = \underbrace{\begin{pmatrix} x & -y^2 \\ 1 & 0 \end{pmatrix} \cdots}_{(N-2)\,\text{factors}} \begin{pmatrix} I_1 \\ I_0 \end{pmatrix}$$

$$= \begin{pmatrix} x & -y^2 \\ 1 & 0 \end{pmatrix}^{(N-2)} \begin{pmatrix} x \\ 1 \end{pmatrix}. \tag{3.59}$$

We can determine the eigenvalues of the fundamental 2×2 matrix in Eq. (3.59) in a straightforward manner. From

$$\det \begin{vmatrix} x - \lambda & -y^2 \\ 1 & -\lambda \end{vmatrix} = 0 \,,$$

we obtain

$$\lambda^2 - \lambda x + y^2 = 0,$$

$$\text{or,} \quad \lambda_\pm = \frac{x \pm \sqrt{x^2 - 4y^2}}{2} \,. \tag{3.60}$$

Furthermore, the 2×2 matrix can be trivially diagonalized by a similarity transformation. In fact, if we define

$$S = \begin{pmatrix} c\lambda_+ & d\lambda_- \\ c & d \end{pmatrix}, \tag{3.61}$$

with

$$S^{-1} = \frac{1}{\lambda_+ - \lambda_-} \begin{pmatrix} \frac{1}{c} & -\frac{\lambda_-}{c} \\ -\frac{1}{d} & \frac{\lambda_+}{d} \end{pmatrix}, \tag{3.62}$$

where c and d are arbitrary parameters, then it is easy to check that

$$\begin{pmatrix} x & -y^2 \\ 1 & 0 \end{pmatrix} = S \begin{pmatrix} \lambda_+ & 0 \\ 0 & \lambda_- \end{pmatrix} S^{-1}. \tag{3.63}$$

Using this in Eq. (3.59), we obtain

$$\begin{pmatrix} I_{N-1} \\ I_{N-2} \end{pmatrix} = S \begin{pmatrix} \lambda_+^{N-2} & 0 \\ 0 & \lambda_-^{N-2} \end{pmatrix} S^{-1} \begin{pmatrix} x \\ 1 \end{pmatrix}. \tag{3.64}$$

We recall from Eq. (3.60) that

$$x = \lambda_+ + \lambda_- . \tag{3.65}$$

Using this as well as the forms for S and S^{-1} in Eqs. (3.61) and (3.62), we obtain from Eq. (3.64)

$$I_{N-1} = \frac{1}{\lambda_+ - \lambda_-} \left(\lambda_+^N - \lambda_-^N \right) . \tag{3.66}$$

We can easily check now that

$$I_{-1} = 0 ,$$

$$I_0 = 1 ,$$

$$I_1 = \frac{1}{\lambda_+ - \lambda_-} \left(\lambda_+^2 - \lambda_-^2 \right)$$

$$= \lambda_+ + \lambda_- = x , \tag{3.67}$$

which is consistent with our earlier observations in Eqs. (3.47) and (3.48).

Let us next note from Eqs. (3.60) and (3.42) that

$$\lambda_+ - \lambda_- = \left(x^2 - 4y^2\right)^{\frac{1}{2}}$$

$$= \left(4\left(1 - \frac{\epsilon^2\omega^2}{4}\right)^2 - 4\left(1 + \frac{\epsilon^2\omega^2}{4}\right)^2\right)^{\frac{1}{2}}$$

$$= \left(-4\epsilon^2\omega^2\right)^{\frac{1}{2}} = 2i\epsilon\omega\,,$$

$$\lambda_+^N = \left(\frac{x + \sqrt{x^2 - 4y^2}}{2}\right)^N$$

$$= \left(\frac{2\left(1 - \frac{\epsilon^2\omega^2}{4}\right) + 2i\epsilon\omega}{2}\right)^N$$

$$= \left(1 + i\epsilon\omega + O\left(\epsilon^2\right)\right)^N \simeq \left(1 + i\epsilon\omega\right)^N\,,$$

$$\lambda_-^N = \left(\frac{x - \sqrt{x^2 - 4y^2}}{2}\right)^N$$

$$= \left(\frac{2(1 - \frac{\epsilon^2\omega^2}{4}) - 2i\epsilon\omega}{2}\right)^N$$

$$= \left(1 - i\epsilon\omega + O\left(\epsilon^2\right)\right)^N \simeq \left(1 - i\epsilon\omega\right)^N\,. \tag{3.68}$$

Consequently, substituting these relations into Eq. (3.66), we obtain

$$\lim_{\substack{\epsilon \to 0 \\ N \to \infty}} \epsilon I_{N-1} = \lim_{\substack{\epsilon \to 0 \\ N \to \infty}} \epsilon \frac{1}{\lambda_+ - \lambda_-} \left(\lambda_+^N - \lambda_-^N\right)$$

$$= \lim_{\substack{\epsilon \to 0 \\ N \to \infty}} \epsilon \frac{1}{2i\epsilon\omega} \left((1 + i\epsilon\omega)^N - (1 - i\epsilon\omega)^N\right)$$

$$= \lim_{\substack{\epsilon \to 0 \\ N \to \infty}} \frac{1}{2i\omega} \left(\left(1 + \frac{i\omega T}{N}\right)^N - \left(1 - \frac{i\omega T}{N}\right)^N \right)$$

$$= \frac{1}{2i\omega} \left(e^{i\omega T} - e^{-i\omega T} \right) = \frac{\sin \omega T}{\omega}. \tag{3.69}$$

Here $T = t_f - t_i$ is the time interval between the initial and the final times and this is, of course, what we had obtained earlier in Eq. (3.56), namely, that

$$\lim_{\substack{\epsilon \to 0 \\ N \to \infty}} \epsilon \det B = \lim_{\substack{\epsilon \to 0 \\ N \to \infty}} \epsilon I_{N-1} = \frac{\sin \omega T}{\omega}. \tag{3.70}$$

We note here that a different sign for the square-root in $\lambda_+ - \lambda_-$ in (3.68) would have still led to the same value for the determinant. In any case, we obtain the transition amplitude for the harmonic oscillator to be

$$U(x_f, t_f; x_i, t_i) = \left(\frac{m\omega}{2\pi i \hbar \sin \omega T} \right)^{\frac{1}{2}} e^{\frac{i}{\hbar} S[x_{\text{cl}}]}. \tag{3.71}$$

3.4 The classical action

Once again we see from Eq. (3.32) or (3.71) that the transition amplitude has a generic form similar to the one found for the case of the free particle. Namely, it is proportional to $e^{\frac{i}{\hbar} S[x_{\text{cl}}]}$. A complete determination of the transition amplitude in this case, therefore, would require us to evaluate the classical action for the system. This can be done in a simple manner as follows. We recall that the Euler-Lagrange equation for the present system is given by (see Eq. (3.5))

$$m\ddot{x}_{\text{cl}} + m\omega^2 x_{\text{cl}} - J = 0.$$

In other words, the classical trajectory is a solution of the inhomogeneous differential equation

$$\left(\frac{d^2}{dt^2} + \omega^2 \right) x_{\text{cl}}(t) = \frac{J(t)}{m}. \tag{3.72}$$

The general solution, obviously, would consist of a homogeneous and an inhomogeneous part and can be written as

$$x_{\text{cl}}(t) = x_{\text{H}}(t) + x_{\text{I}}(t),\tag{3.73}$$

where the homogeneous solution is of the form

$$x_{\text{H}}(t) = Ae^{i\omega t} + Be^{-i\omega t},\tag{3.74}$$

with A and B arbitrary constants to be determined from the boundary conditions. To determine the inhomogeneous solution, we use the method of Green's function. Here the Green's function for Eq. (3.72) is defined by the equation

$$\left(\frac{\text{d}^2}{\text{d}t^2} + \omega^2\right) G(t - t') = -\delta(t - t').\tag{3.75}$$

It is clear that if we know the Green's function $G(t - t')$, then the inhomogeneous solution can be written as

$$x_{\text{I}}(t) = -\int_{t_i}^{t_f} \text{d}t'\, G(t - t')\frac{J(t')}{m}.\tag{3.76}$$

The Green's function can be easily determined by transforming to the Fourier space. Defining

$$G(t - t') = \int \frac{\text{d}k}{2\pi}\, e^{-ik(t-t')}G(k),$$

$$\delta(t - t') = \int \frac{\text{d}k}{2\pi}\, e^{-ik(t-t')},\tag{3.77}$$

where $G(k)$ is the Fourier transform of $G(t - t')$ (we are using here the conventional notation that the Fourier transform of the Green's function is not defined in the symmetric manner that we have introduced in (2.6) and (2.7)), and substituting these into Eq. (3.75) we obtain (note also from Eq. (2.6) that the Fourier transform in time is defined with an opposite phase compared to that for the space coordinates)

$$\left(\frac{d^2}{dt^2} + \omega^2\right) G(t - t') = -\delta(t - t'),$$

or, $\quad \frac{1}{2\pi}\left(-k^2 + \omega^2\right) G(k) = -\frac{1}{2\pi},$

or, $\quad G(k) = \frac{1}{k^2 - \omega^2}.$ $\qquad\qquad$ (3.78)

Consequently, the Green's function takes the form

$$G(t - t') = \int \frac{dk}{2\pi} \, e^{-ik(t-t')} G(k)$$

$$= \int \frac{dk}{2\pi} \frac{e^{-ik(t-t')}}{k^2 - \omega^2}. \qquad\qquad (3.79)$$

The integrand in Eq. (3.79) clearly has poles on the real axis at $k = \pm\omega$. Therefore, we must specify a contour in the complex k-plane in order to evaluate the integral. Normally, in classical mechanics, the Green's functions that are of fundamental interest are the retarded and the advanced Green's functions. But the Green's function that is of fundamental significance in quantum theories is the Feynman Green's function and corresponds to choosing a contour as shown in Fig. 3.1.

Equivalently, it corresponds to defining (see Eq. (3.78))

$$G_F(k) = \lim_{\epsilon \to 0^+} \frac{1}{k^2 - \omega^2 + i\epsilon}$$

$$= \lim_{\delta \to 0^+} \frac{1}{(k + \omega - i\delta)(k - \omega + i\delta)}, \qquad (3.80)$$

where we have defined

$$\delta = \frac{\epsilon}{2\omega}. \qquad\qquad (3.81)$$

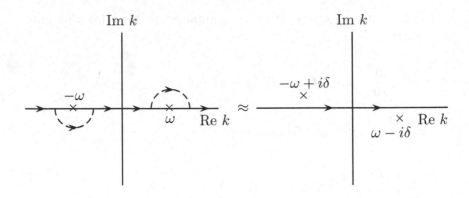

Figure 3.1: Contour in the complex k-plane for the Feynman Green's function.

In other words, we can think of the Feynman Green's function in Eq. (3.80) as the Fourier transform of the function which satisfies the differential equation (this would correspond to letting $S \to S + \frac{i\epsilon m}{2} \int dt\, x^2$ in the exponent, as we will see later, which would lead to a damping in the weight factor)

$$\lim_{\epsilon \to 0^+} \left(\frac{d^2}{dt^2} + \omega^2 - i\epsilon \right) G_F(t - t') = -\delta(t - t'). \qquad (3.82)$$

We note here, for completeness, that the retarded and the advanced Green's functions, in this language, correspond respectively to choosing the Fourier transforms as

$$G^{R,A}(k) = \lim_{\epsilon \to 0^+} \frac{1}{(k \pm i\epsilon)^2 - \omega^2},$$

with the respective contours shown in Fig. 3.2.

With the choice of contour for the Feynman Green's function in Fig. 3.1, enclosing the contour in the lower half plane for $t - t' > 0$, we obtain

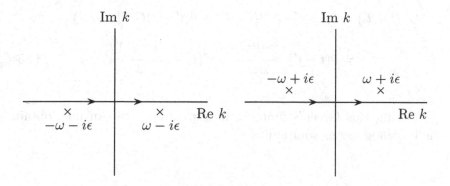

Figure 3.2: The left contour in the complex k-plane is associated with the retarded Green's function while the right contour is that for the advanced Green's function.

$$G^{(+)}(t-t') = \lim_{\delta\to 0^+} \frac{1}{2\pi}\int dk\, \frac{e^{-ik(t-t')}}{(k+\omega-i\delta)(k-\omega+i\delta)}$$

$$= \frac{1}{2\pi}(-2\pi i)\frac{e^{-i\omega(t-t')}}{2\omega}$$

$$= \frac{1}{2i\omega}\,e^{-i\omega(t-t')}\,. \tag{3.83}$$

On the other hand, for $t-t' < 0$, enclosing the contour in the upper half plane, we find

$$G^{(-)}(t-t') = \lim_{\delta\to 0^+} \frac{1}{2\pi}\int dk\, \frac{e^{-ik(t-t')}}{(k+\omega-i\delta)(k-\omega+i\delta)}$$

$$= \frac{1}{2\pi}(2\pi i)\frac{e^{i\omega(t-t')}}{(-2\omega)}$$

$$= \frac{1}{2i\omega}\,e^{i\omega(t-t')}\,. \tag{3.84}$$

Thus, the Feynman Green's function has the form

$$G_F(t - t') = \theta(t - t')G^{(+)}(t - t') + \theta(t' - t)G^{(-)}(t - t')$$

$$= \theta(t - t')\frac{e^{-i\omega(t-t')}}{2i\omega} + \theta(t' - t)\frac{e^{i\omega(t-t')}}{2i\omega}. \tag{3.85}$$

Using this Green's function in Eq. (3.76), we can now obtain the inhomogeneous solution as

$$x_I(t) = -\int_{t_i}^{t_f} dt' \, G_F(t - t')\frac{J(t')}{m}$$

$$= -\frac{1}{m}\left(\int_{t_i}^{t} dt' \, \frac{e^{-i\omega(t-t')}}{2i\omega}J(t') + \int_{t}^{t_f} dt' \, \frac{e^{i\omega(t-t')}}{2i\omega}J(t')\right)$$

$$= -\frac{1}{2im\omega}\left(\int_{t_i}^{t} dt' e^{-i\omega(t-t')}J(t') + \int_{t}^{t_f} dt' e^{i\omega(t-t')}J(t')\right). \tag{3.86}$$

Thus, substituting Eqs. (3.74) and (3.86) into Eq. (3.73) we can write the classical trajectory as

$$x_{cl}(t) = x_H(t) + x_I(t)$$

$$= Ae^{i\omega t} + Be^{-i\omega t}$$

$$- \frac{1}{2im\omega}\left(\int_{t_i}^{t} dt' e^{-i\omega(t-t')}J(t') + \int_{t}^{t_f} dt' e^{i\omega(t-t')}J(t')\right). \tag{3.87}$$

Imposing the boundary conditions

$$x_{\mathrm{cl}}(t_i) = x_i$$

$$= Ae^{i\omega t_i} + Be^{-i\omega t_i} - \frac{1}{2im\omega} \int\limits_{t_i}^{t_f} dt' e^{i\omega(t_i - t')} J(t'),$$

$$x_{\mathrm{cl}}(t_f) = x_f$$

$$= Ae^{i\omega t_f} + Be^{-i\omega t_f} - \frac{1}{2im\omega} \int\limits_{t_i}^{t_f} dt' e^{-i\omega(t_f - t')} J(t'),$$

$$(3.88)$$

we can solve for A and B in terms of the initial and the final coordinates of the trajectory as

$$A = \frac{1}{2i\sin\omega T} \Bigg[\left(x_f e^{-i\omega t_i} - x_i e^{-i\omega t_f} \right)$$

$$+ \frac{e^{-i\omega t_f}}{m\omega} \int\limits_{t_i}^{t_f} dt' \sin\omega(t' - t_i) J(t') \Bigg],$$

$$B = \frac{1}{2i\sin\omega T} \Bigg[\left(x_i e^{i\omega t_f} - x_f e^{i\omega t_i} \right)$$

$$+ \frac{e^{i\omega t_i}}{m\omega} \int\limits_{t_i}^{t_f} dt' \sin\omega(t_f - t') J(t') \Bigg]. \qquad (3.89)$$

Substituting these relations into Eq. (3.87), we determine the classical trajectory to be

$$x_{\mathrm{cl}}(t) = \frac{1}{\sin\omega T} \Bigg[x_f \sin\omega(t - t_i) + x_i \sin\omega(t_f - t)$$

$$+ \frac{1}{2m\omega} \int\limits_{t_i}^{t_f} dt' J(t') (e^{-i\omega T} \cos\omega(t - t') - \cos\omega(t_f + t_i - t - t')) \Bigg]$$

$$
-\frac{1}{2im\omega}\left(\int\limits_{t_i}^{t}dt'J(t')e^{-i\omega(t-t')}+\int\limits_{t}^{t_f}dt'J(t')e^{i\omega(t-t')}\right). \qquad (3.90)
$$

We can now derive the classical action from Eqs. (3.2) and (3.1) in a straightforward manner to be

$$
S[x_{\mathrm{cl}}]=\frac{m\omega}{2\sin\omega T}\left[(x_i^2+x_f^2)\cos\omega T-2x_ix_f\right]
$$

$$
+\frac{x_i}{\sin\omega T}\int\limits_{t_i}^{t_f}dt\,J(t)\sin\omega(t_f-t)+\frac{x_f}{\sin\omega T}\int\limits_{t_i}^{t_f}dt\,J(t)\sin\omega(t-t_i)
$$

$$
-\frac{1}{m\omega\sin\omega T}\int\limits_{t_i}^{t_f}dt\int\limits_{t_i}^{t}dt'J(t)\sin\omega(t_f-t)\sin\omega(t'-t_i)J(t').
$$

$$
(3.91)
$$

This, therefore, completes the derivation of the transition amplitude for the harmonic oscillator interacting with a time dependent external source.

3.5 References

R. P. Feynman and A. R. Hibbs, *Quantum Mechanics and Path Integrals*, McGraw-Hill Publishing.

H. Kleinert, *Path Integrals*, World Scientific Publishing.

L. S. Schulman, *Techniques and Applications of Path Integration*, John Wiley Publishing.

CHAPTER 4

Generating functional

4.1 Euclidean rotation

We have seen in Eq. (2.41) that the standard Gaussian integral
(where the exponent is quadratic) is given by

$$\int_{-\infty}^{\infty} dx \, e^{i\alpha x^2} = \left(\frac{i\pi}{\alpha}\right)^{\frac{1}{2}},$$

(4.1)

and it generalizes to the case of a $n \times n$ matrix as (see Eq. (3.45))

$$\int_{-\infty}^{\infty} d\eta \, e^{i\eta^T A\eta} = \left(\frac{(i\pi)^n}{\det A}\right)^{\frac{1}{2}},$$

(4.2)

provided A is a Hermitian matrix. In fact, we will now see explicitly
that this result holds true even when we replace the Hermitian matrix
A by a Hermitian operator. In other words, we will see that we can
write

$$\int \mathcal{D}\eta \, e^{i\int_{t_i}^{t_f} dt \, \eta(t)O(t)\eta(t)} = N \left[\det O(t)\right]^{-\frac{1}{2}},$$

(4.3)

where $O(t)$ is a Hermitian operator and N a normalization constant
whose explicit form is not important.

To establish this identification, let us go back to the harmonic
oscillator and note that the quantity of fundamental importance in
this case is given by the integral (see Eq. (3.17))

59

$$\int \mathcal{D}\eta \; e^{\frac{i}{\hbar} \int\limits_{t_i}^{t_f} dt \; (\frac{1}{2} m\dot{\eta}^2 - \frac{1}{2} m\omega^2 \eta^2)}$$

$$= \int \mathcal{D}\eta \; e^{\frac{im}{2\hbar} \int\limits_{t_i}^{t_f} dt \; (\dot{\eta}^2 - \omega^2 \eta^2)}$$

$$= \int \mathcal{D}\eta \; e^{-\frac{im}{2\hbar} \int\limits_{t_i}^{t_f} dt \; \eta(t)(\frac{d^2}{dt^2} + \omega^2)\eta(t)} \;, \tag{4.4}$$

with the boundary conditions

$$\eta(t_i) = \eta(t_f) = 0 \,. \tag{4.5}$$

The value of this integral was determined earlier (see Eq. (3.29)) to be

$$A'' \left(\frac{\sin \omega T}{\omega T} \right)^{-\frac{1}{2}} \,, \tag{4.6}$$

where $T = t_f - t_i$ represents the total time interval. We evaluated this integral earlier by carefully discretizing the time interval and calculating the determinant of a matrix (see Eqs. (3.34) and (3.39)) whose matrix elements were nothing other than the discrete form of the matrix elements of the operator in the exponent in Eq. (4.4). This would already justify our claim. But let us, in fact, calculate the determinant of the operator in the exponent of Eq. (4.4) explicitly and compare with the result obtained earlier.

The first problem that we face in evaluating the functional integral is that the exponential in the integrand is oscillatory and, therefore, we have to define the integral in some manner. One can, of course, use the same trick as we employed earlier in defining ordinary oscillatory Gaussian integrals (see Eq. (2.41)). Namely, let us define (we had commented on this earlier before Eq. (3.82))

$$\int \mathcal{D}\eta \; e^{\frac{im}{2\hbar} \int_{t_i}^{t_f} dt \, (\dot{\eta}^2 - \omega^2 \eta^2)}$$

$$= \lim_{\epsilon \to 0^+} \int \mathcal{D}\eta \; e^{-\frac{im}{2\hbar} \int_{t_i}^{t_f} dt \, \eta(t)(\frac{d^2}{dt^2} + \omega^2 - i\epsilon)\eta(t)} . \tag{4.7}$$

This provides proper damping to the integrand and, as we will see, leads to the Feynman Green's functions for the theory. In fact, as we have already seen in Eq. (3.82), the inverse of the operator in the exponent in Eq. (4.7) gives the Feynman Green's function which plays the role of the causal propagator in the quantum theory. It is in this sense that one says that the path integral naturally incorporates causal boundary conditions.

There is an alternate but equivalent way of defining the path integral which is quite pleasing and which gives some sense of rigor to all the manipulations involving the path integral. Very simply, it corresponds to analytically continuing all the integrals to imaginary times in the complex t-plane as shown in Fig. 4.1. More explicitly, we let

$$t \to -i\tau, \quad \tau \text{ real.} \tag{4.8}$$

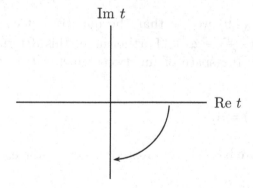

Figure 4.1: Euclidean rotation in the complex t-plane.

With this analytic continuation, the integral in Eq. (4.4) becomes

$$\int \mathcal{D}\eta \; e^{-\frac{im}{2\hbar} \int_{t_i}^{t_f} dt \; \eta(t)(\frac{d^2}{dt^2}+\omega^2)\eta(t)}$$

$$\to \int \mathcal{D}\eta \; e^{-\frac{m}{2\hbar} \int_{\tau_i}^{\tau_f} d\tau \; \eta(\tau)(-\frac{d^2}{d\tau^2}+\omega^2)\eta(\tau)}$$

$$= N' \int \mathcal{D}\eta \; e^{-\frac{1}{2} \int_{\tau_i}^{\tau_f} d\tau \; \eta(\tau)(-\frac{d^2}{d\tau^2}+\omega^2)\eta(\tau)} \,. \tag{4.9}$$

Here we have scaled the variables in the last step and N' represents the Jacobian for the change of variables. Furthermore, we have to evaluate the integral in Eq. (4.9) subject to the boundary conditions

$$\eta(\tau_i) = \eta(\tau_f) = 0 \,. \tag{4.10}$$

The right hand side of Eq. (4.9) is now a well defined quantity since the integrand is exponentially damped. (The analytically continued operator has a positive definite spectrum, as we will see shortly.) We can now evaluate this integral and at the end of our calculations, we are supposed to analytically continue back to real time by letting

$$\tau \to it, \quad t \quad \text{real} \,. \tag{4.11}$$

From Eq. (4.9) we see that the quantity which we are interested in is $\det(-\frac{d^2}{d\tau^2} + \omega^2)$. Furthermore, this determinant has to be evaluated in the space of functions which satisfy the boundary conditions

$$\eta(\tau_i) = \eta(\tau_f) = 0 \,. \tag{4.12}$$

Therefore, we are basically interested in solving the eigenvalue equation

$$\left(-\frac{d^2}{d\tau^2} + \omega^2\right) \psi_n = \lambda_n \psi_n \,, \tag{4.13}$$

subject to the boundary conditions in Eq. (4.10), namely,

$$\psi_n(\tau_i) = \psi_n(\tau_f) = 0. \tag{4.14}$$

The normalized eigenfunctions of Eq. (4.13) are easily obtained to be

$$\psi_n(\tau) = \sqrt{\frac{2}{(\tau_f - \tau_i)}} \, \sin \frac{n\pi(\tau - \tau_i)}{(\tau_f - \tau_i)}, \tag{4.15}$$

with n a positive integer. The corresponding eigenvalues are given by

$$\lambda_n = \left(\frac{n\pi}{\tau_f - \tau_i}\right)^2 + \omega^2. \tag{4.16}$$

Thus, we see that the determinant of the operator in Eq. (4.9) or (4.13) has the form

$$\begin{aligned}
\det\left(-\frac{d^2}{d\tau^2} + \omega^2\right) &= \prod_{n=1}^{\infty} \lambda_n \\
&= \prod_{n=1}^{\infty} \left(\left(\frac{n\pi}{\tau_f - \tau_i}\right)^2 + \omega^2\right) \\
&= \prod_{n=1}^{\infty} \left(\frac{n\pi}{\tau_f - \tau_i}\right)^2 \prod_{n=1}^{\infty} \left(1 + \left(\frac{\omega(\tau_f - \tau_i)}{n\pi}\right)^2\right) \\
&= B \, \frac{\sinh \omega(\tau_f - \tau_i)}{\omega(\tau_f - \tau_i)}, \tag{4.17}
\end{aligned}$$

where we have used a relation similar to the one given in Eq. (3.28) and B is a constant representing the first product whose value can be absorbed into the normalization of the path integral. Analytically continuing this back to real time, we obtain

$$\det\left(-\frac{d^2}{d\tau^2} + \omega^2\right) \rightarrow \det\left(\frac{d^2}{dt^2} + \omega^2\right)$$

$$= A\,\frac{\sin\omega(t_f - t_i)}{\omega(t_f - t_i)} = A\,\frac{\sin\omega T}{\omega T}. \tag{4.18}$$

Here, as before, we have identified $T = t_f - t_i$ with the total time interval and A denotes the constant B analytically continued to real time. This is, of course, related to the value of the path integral which we had obtained earlier in Eq. (3.32) through a careful evaluation. Therefore, we conclude that

$$\int \mathcal{D}\eta\, e^{\frac{im}{2\hbar}\int_{t_i}^{t_f} dt\,(\dot{\eta}^2 - \omega^2\eta^2)} = A''\left(\frac{\sin\omega T}{\omega T}\right)^{-\frac{1}{2}}$$

$$= N\left(\det\left(\frac{d^2}{dt^2} + \omega^2\right)\right)^{-\frac{1}{2}}. \tag{4.19}$$

Later, we will generalize this result to field theories or systems with an infinite number of degrees of freedom.

Let us next discuss in some detail the analytic continuation to the imaginary time. Consider a Minkowski space with coordinates

$$x^\mu = (t, \mathbf{x}),$$

where we leave the dimensionality of space arbitrary. Then, under the analytic continuation,

$$x^\mu = (t, \mathbf{x}) \rightarrow (-i\tau, \mathbf{x}),$$
$$x^2 = (t^2 - \mathbf{x}^2) \rightarrow -(\tau^2 + \mathbf{x}^2) - x_E^2. \tag{4.20}$$

Therefore, we note that τ is nothing other than the Euclidean time. The analytic continuation, consequently, corresponds to a rotation from Minkowski space to Euclidean space. (Although, in one dimension, it does not make sense to talk about a Euclidean space, in

higher dimensional field theories it is quite meaningful.) The sense
of the rotation is completely fixed by the singularity structure of the
theory. Let us note that an analytic continuation is meaningful only
if no singularity is crossed in the process. We know from our study
of the Feynman Green's function in the last chapter (see Eq. (3.80))
that in the complex energy plane, the singularities occur at points
shown in Fig. 4.2.

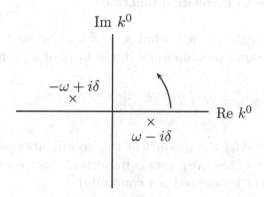

Figure 4.2: Euclidean rotation in the complex energy plane. The
crosses denote the location of poles of the Feynman Green's function.

Therefore, it is clear that an analytic continuation from $\operatorname{Re} k^0$
to $\operatorname{Im} k^0$ is meaningful only if the rotation is anti-clockwise as shown
in Fig. 4.2, namely, only if we let

$$k^0 \to i\kappa, \quad \kappa \text{ real}. \tag{4.21}$$

Since we can represent

$$k^0 \to i\hbar \frac{\partial}{\partial t}, \tag{4.22}$$

it follows now that in the complex t-plane, the consistent rotation
will be

$$t \to -i\tau, \quad \tau \text{ real}. \tag{4.23}$$

Namely, in this case, we will have

$$k^0 \to i\hbar\frac{\partial}{\partial t} \to i\hbar\frac{\partial}{\partial(-i\tau)} = i\left(i\hbar\frac{\partial}{\partial\tau}\right) \to i\kappa.$$

4.2 Time ordered correlation functions

Let us recapitulate quickly what we have done so far. We have obtained the transition amplitude in the form of a path integral as

$$_H\langle x_f, t_f | x_i, t_i \rangle_H = N \int \mathcal{D}x \, e^{\frac{i}{\hbar}S[x]}. \tag{4.24}$$

Let us next consider the product of two coordinate operators of the form (the order of the operators is important since coordinate operators at different times need not commute)

$$X_H(t_1)X_H(t_2), \quad t_1 > t_2, \tag{4.25}$$

and evaluate the matrix element

$$_H\langle x_f, t_f | X_H(t_1)X_H(t_2) | x_i, t_i \rangle_H, \quad t_f \geq t_1 > t_2 \geq t_i. \tag{4.26}$$

Since $t_1 > t_2$, then we can insert complete sets of coordinate basis states and write

$$_H\langle x_f, t_f | X_H(t_1)X_H(t_2) | x_i, t_i \rangle_H$$

$$= \int dx_1 dx_2 \, _H\langle x_f, t_f | X_H(t_1) | x_1, t_1 \rangle_H$$

$$\times \, _H\langle x_1, t_1 | X_H(t_2) | x_2, t_2 \rangle_H \, _H\langle x_2, t_2 | x_i, t_i \rangle_H$$

$$= \int dx_1 dx_2 \, x_1 x_2 \, _H\langle x_f, t_f | x_1, t_1 \rangle_H$$

$$\times \, _H\langle x_1, t_1 | x_2, t_2 \rangle_H \, _H\langle x_2, t_2 | x_i, t_i \rangle_H. \tag{4.27}$$

Here we have used the relation in Eq. (1.44). We note that each inner product in the integrand represents a transition amplitude and, therefore, can be written as a path integral. Combining the products, we can write (for $t_1 > t_2$)

$$
{}_H\langle x_f, t_f | X_H(t_1) X_H(t_2) | x_i, t_i \rangle_H = N \int \mathcal{D}x \; x(t_1) x(t_2) \; e^{\frac{i}{\hbar} S[x]} \,.
$$
(4.28)

Here we have used the identification

$$
x_1 = x(t_1), \quad x_2 = x(t_2),
$$
(4.29)

and we note that the path integral (4.28) should be viewed in the sense of (4.27). Namely, the system evolves in time from (x_i, t_i) to (x_2, t_2) where an insertion of a factor x_2 is made. Then the system evolves from (x_2, t_2) to (x_1, t_1) where a second insertion of a factor x_1 is made. Finally, the system evolves from (x_1, t_1) to (x_f, t_f) and we integrate over x_1, x_2, namely, over all possible locations for the insertions. Similarly, we note that for $t_2 > t_1$, we can write

$$
{}_H\langle x_f, t_f | X_H(t_2) X_H(t_1) | x_i, t_i \rangle_H
$$

$$
= \int dx_1 dx_2 \; {}_H\langle x_f, t_f | X_H(t_2) | x_2, t_2 \rangle_H
$$

$$
\times \; {}_H\langle x_2, t_2 | X_H(t_1) | x_1, t_1 \rangle_H \; {}_H\langle x_1, t_1 | x_i, t_i \rangle_H
$$

$$
= N \int \mathcal{D}x \; x(t_1) x(t_2) \; e^{\frac{i}{\hbar} S[x]} \,.
$$
(4.30)

In the last step, we have used the fact that factors in the integrand such as $x(t_1)$ and $x(t_2)$ are classical quantities and, therefore, their product is commutative.

Thus, we see from Eqs. (4.28) and (4.30) that the path integral naturally gives the time ordered correlation functions as the moments

$$
{}_H\langle x_f, t_f | T(X_H(t_1) X_H(t_2)) | x_i, t_i \rangle_H = N \int \mathcal{D}x \; x(t_1) x(t_2) \; e^{\frac{i}{\hbar} S[x]} \,,
$$
(4.31)

where the time ordering is defined as

$$T(X_H(t_1)X_H(t_2))$$
$$= \theta(t_1 - t_2)X_H(t_1)X_H(t_2) + \theta(t_2 - t_1)X_H(t_2)X_H(t_1) \,. \quad (4.32)$$

In fact, it is obvious now that the time ordered product of any set of operators leads to correlation functions in the path integral formalism as

$$_H\langle x_f, t_f | T(O_1(X_H(t_1)) \cdots O_n(X_H(t_n))) | x_i, t_i \rangle_H$$
$$= N \int \mathcal{D}x \, O_1(x(t_1)) \cdots O_n(x(t_n)) e^{\frac{i}{\hbar}S[x]} \,. \quad (4.33)$$

Furthermore, the beauty of path integrals lies in the fact that all the factors on the right hand side are c-numbers (classical quantities). There are no operators any more.

4.3 Correlation functions in definite states

So far, we have calculated the transition amplitude between two co-ordinate states. In physical applications, however, we are often interested in transitions between physical states. Namely, we would like to know the probability amplitude for a system in a state $|\psi_i\rangle_H$ at time t_i to make a transition to a state $|\psi_f\rangle_H$ at time t_f. This is what the S-matrix elements are supposed to give. Let us note that by definition, this transition amplitude is given by

$$_H\langle \psi_f | \psi_i \rangle_H = \int \mathrm{d}x_f \mathrm{d}x_i \, {}_H\langle \psi_f | x_f, t_f \rangle_H \, {}_H\langle x_f, t_f | x_i, t_i \rangle_H \, {}_H\langle x_i, t_i | \psi_i \rangle_H$$
$$= N \int \mathrm{d}x_f \mathrm{d}x_i \, \psi_f^*(x_f, t_f) \psi_i(x_i, t_i) \int \mathcal{D}x \, e^{\frac{i}{\hbar}S[x]} \,. \quad (4.34)$$

Here we have used the standard definition of the wavefunction. Namely,

$$_H\langle x, t | \psi \rangle_H = \psi(x, t) \,. \quad (4.35)$$

We comment here, parenthetically, that a Heisenberg state is independent of time, it will not change with time unless there is a perturbation. However, if a system in the state $|\psi_i\rangle_H$ is subjected to an external perturbation like the interaction with an external source (for a finite amount of time), it can make a transition to another state $|\psi_f\rangle_H$. Of course, it will also have a finite transition amplitude for remaining in the same state. It is in that sense that we are talking about a transition.

Following our earlier discussion (see Eq. (4.33)), we see that the time ordered correlation functions between such physical states can also be written as ($t_i \leq t_1, t_2, \cdots, t_n \leq t_f$)

$$_H\langle\psi_f|T(O_1(X_H(t_1))\cdots O_n(X_H(t_n)))|\psi_i\rangle_H$$

$$= N \int dx_f dx_i \, \psi_f^*(x_f, t_f)\psi_i(x_i, t_i)$$

$$\times \int \mathcal{D}x \, O_1(x(t_1))\cdots O_n(x(t_n))e^{\frac{i}{\hbar}S[x]}. \tag{4.36}$$

In dealing with physical systems, we are often interested in calculating expectation values of operators in a given state (a better way to say is the normalized correlations in the transition from $|\psi_i\rangle_H$ to $|\psi_i\rangle_H$, but we will use the sloppy language here for convenience). This is simply obtained by noting that ($t_i \leq t_1, t_2, \cdots, t_n \leq t_f$)

$$_H\langle\psi_i|T(O_1(X_H(t_1))\cdots O_n(X_H(t_n)))|\psi_i\rangle_H$$

$$= N \int dx_f dx_i \, \psi_i^*(x_f, t_f)\psi_i(x_i, t_i)$$

$$\times \int \mathcal{D}x \, O_1(x(t_1))\cdots O_n(x(t_n))e^{\frac{i}{\hbar}S[x]}. \tag{4.37}$$

Since the states need not necessarily be normalized, we obtain the expectation value to be

$$\langle T(O_1(X_H(t_1))\cdots O_n(X_H(t_n)))\rangle$$

$$= \frac{_H\langle\psi_i|T(O_1(X_H(t_1))\cdots O_n(X_H(t_n)))|\psi_i\rangle_H}{_H\langle\psi_i|\psi_i\rangle_H} \qquad (4.38)$$

$$= \frac{\int \mathrm{d}x_f \mathrm{d}x_i \psi_i^*(x_f,t_f)\psi_i(x_i,t_i)\int \mathcal{D}x\, O_1(x(t_1))\cdots O_n(x(t_n))e^{\frac{i}{\hbar}S[x]}}{\int \mathrm{d}x_f \mathrm{d}x_i\, \psi_i^*(x_f,t_f)\psi_i(x_i,t_i)\int \mathcal{D}x\, e^{\frac{i}{\hbar}S[x]}}.$$

Note that the normalization constant N cancels out in the ratio and it is for this reason that we do not often worry about the explicit form of the normalization constant. In most field theory questions one is primarily interested in calculating the expectation values of time ordered products of operators in the ground state and, consequently, one tends to be sloppy about this factor in such cases. (We comment here that the true expectation value of any operator can not be calculated within Feynman's path integral formalism where time flows forward. Rather, one needs Schwinger's formalism of closed time path formalism which allows time to run forwards as well as backwards. We will not discuss this in this book. It is better discussed in books on finite temperature field theory.)

From now on let us suppress, for convenience, the subscript H signifying the description in the Heisenberg picture. Let us next note that we can generate various correlation functions in a simple way in the path integral formalism by adding appropriate external sources. Thus, for example, if we want matrix elements of a product of the coordinate operators, we can define a modified action of the form (source coupled linearly to the dynamical variable whose correlations we are interested in)

$$S[x,J] = S[x] + \int_{t_i}^{t_f} \mathrm{d}t\, x(t)J(t)\,, \qquad (4.39)$$

where $J(t)$ denotes an external source and this action satisfies

$$S[x,0] = S[x]\,, \qquad (4.40)$$

where $S[x]$ defines the dynamics of the system. Let us further define

$$\langle\psi_i|\psi_i\rangle_J = N\int \mathrm{d}x_f \mathrm{d}x_i\, \psi_i^*(x_f,t_f)\psi_i(x_i,t_i)\int \mathcal{D}x\, e^{\frac{i}{\hbar}S[x,J]}\,. \qquad (4.41)$$

Clearly, then,

$$\langle\psi_i|\psi_i\rangle_{J=0} = N \int dx_f dx_i \, \psi_i^*(x_f, t_f)\psi_i(x_i, t_i) \int \mathcal{D}x \, e^{\frac{i}{\hbar}S[x]}$$

$$= \langle\psi_i|\psi_i\rangle. \tag{4.42}$$

It is clear now from Eqs. (4.39) and (4.41) that $(t_f \geq t_1 \geq t_i)$

$$\frac{\delta\langle\psi_i|\psi_i\rangle_J}{\delta J(t_1)}$$

$$= N \int dx_f dx_i \, \psi_i^*(x_f, t_f)\psi_i(x_i, t_i) \int \mathcal{D}x \, \frac{i}{\hbar}\frac{\delta S[x, J]}{\delta J(t_1)} \, e^{\frac{i}{\hbar}S[x, J]}$$

$$= N \int dx_f dx_i \, \psi_i^*(x_f, t_f)\psi_i(x_i, t_i) \int \mathcal{D}x \, \frac{i}{\hbar}x(t_1) \, e^{\frac{i}{\hbar}S[x, J]}. \tag{4.43}$$

It follows, therefore, that

$$\frac{\delta\langle\psi_i|\psi_i\rangle_J}{\delta J(t_1)}\bigg|_{J=0}$$

$$= N \int dx_f dx_i \, \psi_i^*(x_f, t_f)\psi_i(x_i, t_i) \int \mathcal{D}x \, \frac{i}{\hbar}x(t_1)e^{\frac{i}{\hbar}S[x]}$$

$$= \frac{i}{\hbar}\langle\psi_i|X(t_1)|\psi_i\rangle, \tag{4.44}$$

where we have used the relation in Eq. (4.37). Similarly, for $t_f \geq t_1, t_2 \geq t_i$, we have

$$\frac{\delta^2\langle\psi_i|\psi_i\rangle_J}{\delta J(t_1)\delta J(t_2)}\bigg|_{J=0}$$

$$= N \int dx_f dx_i \, \psi_i^*(x_f, t_f)\psi_i(x_i, t_i)$$

$$\times \int \mathcal{D}x \left(\frac{i}{\hbar}\right)^2 \frac{\delta S[x, J]}{\delta J(t_1)}\frac{\delta S[x, J]}{\delta J(t_2)}e^{\frac{i}{\hbar}S[x, J]}\bigg|_{J=0}$$

$$= N \int \mathrm{d}x_f \mathrm{d}x_i \, \psi_i^*(x_f, t_f)\psi_i(x_i, t_i) \int \mathcal{D}x \left(\frac{i}{\hbar}\right)^2 x(t_1)x(t_2)e^{\frac{i}{\hbar}S[x]}$$

$$= \left(\frac{i}{\hbar}\right)^2 \langle \psi_i | T(X(t_1)X(t_2)) | \psi_i \rangle . \tag{4.45}$$

In general, it is quite straightforward to show that

$$\frac{\delta^n \langle \psi_i | \psi_i \rangle_J}{\delta J(t_1) \cdots \delta J(t_n)}\bigg|_{J=0} = \left(\frac{i}{\hbar}\right)^n \langle \psi_i | T(X(t_1) \cdots X(t_n)) | \psi_i \rangle . \tag{4.46}$$

Consequently, we can write the normalized time ordered correlations in the state $|\psi_i\rangle$ as

$$\langle T(X(t_1) \cdots X(t_n)) \rangle = \frac{\langle \psi_i | T(X(t_1) \cdots X(t_n)) | \psi_i \rangle}{\langle \psi_i | \psi_i \rangle}$$

$$= \frac{(-i\hbar)^n}{\langle \psi_i | \psi_i \rangle_J} \frac{\delta^n \langle \psi_i | \psi_i \rangle_J}{\delta J(t_1) \cdots \delta J(t_n)}\bigg|_{J=0} . \tag{4.47}$$

It is for this reason that $\langle \psi_i | \psi_i \rangle_J$ is also known as the generating functional for the time ordered correlation functions in a given state.

4.4 Vacuum functional

An object of great interest in quantum field theories is the vacuum to vacuum transition amplitude in the presence of an external source. The simplest way to obtain this is to go back to the definition of the transition amplitude in the coordinate space

$$\langle x_f, t_f | x_i, t_i \rangle_J = N \int \mathcal{D}x \, e^{\frac{i}{\hbar}S[x,J]}$$

$$= N \int \mathcal{D}x \, e^{\frac{i}{\hbar}S[x] + \frac{i}{\hbar} \int\limits_{t_i}^{t_f} \mathrm{d}t \, J(t)x(t)} . \tag{4.48}$$

It is clear from our earlier discussion in Eq. (4.33) that we can also think of this quantity as the matrix element of the operator

$$\langle x_f, t_f | x_i, t_i \rangle_J = N \int \mathcal{D}x \, e^{\frac{i}{\hbar} S[x,J]}$$

$$= \langle x_f, t_f | T \left(e^{\frac{i}{\hbar} \int\limits_{t_i}^{t_f} dt \, J(t) X(t)} \right) | x_i, t_i \rangle. \tag{4.49}$$

Let us next take the limits

$$t_i \to -\infty, \quad t_f \to \infty. \tag{4.50}$$

That is, let us calculate the amplitude for the system to make a transition from the coordinate state in the infinite past labeled by the coordinate x_i to the coordinate state in the infinite future labeled by the coordinate x_f in the presence of an external source which switches on adiabatically. We will enforce the adiabatic condition by assuming that the external source is nonzero within a large but finite interval of time. That is, let us assume that

$$J(t) = 0, \quad \text{for } |t| > \tau, \tag{4.51}$$

where τ is assumed to be large and we will take the limit $\tau \to \infty$ at the end. In such a case, we can write

$$\lim_{\substack{t_i \to -\infty \\ t_f \to \infty}} \langle x_f, t_f | x_i, t_i \rangle_J = N \int \mathcal{D}x \, e^{\frac{i}{\hbar} \int\limits_{-\infty}^{\infty} dt \, (L(x,\dot{x}) + Jx)}. \tag{4.52}$$

Alternatively, we can write from Eq. (4.49)

$$\lim_{\substack{t_i \to -\infty \\ t_f \to \infty}} \langle x_f, t_f | x_i, t_i \rangle_J = \lim_{\tau \to \infty} \lim_{\substack{t_i \to -\infty \\ t_f \to \infty}} \langle x_f, t_f | T \left(e^{\frac{i}{\hbar} \int\limits_{-\tau}^{\tau} dt \, JX} \right) | x_i, t_i \rangle. \tag{4.53}$$

Let us further assume that the ground state energy of our Hamiltonian is normalized to zero so that

$$H|0\rangle = 0,$$

$$H|n\rangle = E_n|n\rangle, \quad E_n > 0 . \tag{4.54}$$

(We wish to point out here that in a relativistic quantum field theory, Poincaré invariance requires that

$$P_\mu|0\rangle = 0, \tag{4.55}$$

which leads to a vanishing ground state energy. In quantum mechanics, however, the ground state energy does not vanish in general and in such a case, the asymptotic limits are not well defined and the present derivation becomes involved. Therefore, for simplicity we choose a derivation parallel to that of a relativistic quantum field theory and assume that the ground state energy is zero (we basically have to add a constant to the quantum theory to make the ground state energy vanish and this can be done since only differences of energy levels are measurable).) Although for simplicity of discussion we have assumed the energy eigenstates to be discrete, it is not essential for our arguments. Introducing complete sets of energy eigenstates into the transition amplitude, we obtain

$$\lim_{\substack{t_i \to -\infty \\ t_f \to \infty}} \langle x_f, t_f | x_i, t_i \rangle_J$$

$$= \lim_{\tau \to \infty} \lim_{\substack{t_i \to -\infty \\ t_f \to \infty}} \sum_{n,m} \langle x_f, t_f | n \rangle \langle n | T \left(e^{\frac{i}{\hbar} \int\limits_{-\tau}^{\tau} dt\, JX} \right) |m\rangle \langle m | x_i, t_i \rangle$$

$$= \lim_{\tau \to \infty} \lim_{\substack{t_i \to -\infty \\ t_f \to \infty}} \sum_{n,m} \langle x_f | e^{-\frac{i}{\hbar} H t_f} | n \rangle \langle n | T \left(e^{\frac{i}{\hbar} \int\limits_{-\tau}^{\tau} dt\, JX} \right) |m\rangle \langle m | e^{\frac{i}{\hbar} H t_i} | x_i \rangle$$

$$= \lim_{\tau \to \infty} \lim_{\substack{t_i \to -\infty \\ t_f \to \infty}} \sum_{n,m} e^{-\frac{i}{\hbar} E_n t_f + \frac{i}{\hbar} E_m t_i} \langle x_f | n \rangle \langle n | T \left(e^{\frac{i}{\hbar} \int\limits_{-\tau}^{\tau} dt\, JX} \right) |m\rangle \langle m | x_i \rangle, \tag{4.56}$$

where we have used Eqs. (1.45) and (4.54). In the limit $t_i \to -\infty$ and $t_f \to \infty$, the exponentials oscillate out to zero except for the ground state where the energy is zero. One can alternatively see this also by looking at the exponentials in a regularized manner or by analytically continuing to the imaginary time axis (Euclidean space in the case of field theories). Thus, in this asymptotic limit, we obtain

$$
\lim_{\substack{t_i \to -\infty \\ t_f \to \infty}} \langle x_f, t_f | x_i, t_i \rangle_J = \lim_{\tau \to \infty} \langle x_f | 0 \rangle \langle 0 | T \left(e^{\frac{i}{\hbar} \int_{-\tau}^{\tau} dt \, JX} \right) | 0 \rangle \langle 0 | x_i \rangle
$$

$$
= \langle x_f | 0 \rangle \langle 0 | x_i \rangle \langle 0 | T \left(e^{\frac{i}{\hbar} \int_{-\infty}^{\infty} dt \, JX} \right) | 0 \rangle . \tag{4.57}
$$

Consequently, we can write

$$
\langle 0 | T \left(e^{\frac{i}{\hbar} \int_{-\infty}^{\infty} dt \, JX} \right) | 0 \rangle = \lim_{\substack{t_i \to -\infty \\ t_f \to \infty}} \frac{\langle x_f, t_f | x_i, t_i \rangle_J}{\langle x_f | 0 \rangle \langle 0 | x_i \rangle} . \tag{4.58}
$$

The left hand side is independent of the end points and, therefore, the right hand side must also be independent of the end points. Furthermore, the right hand side has the structure of a path integral and we can write Eq. (4.58) also as

$$
\langle 0 | T \left(e^{\frac{i}{\hbar} \int_{-\infty}^{\infty} dt \, JX} \right) | 0 \rangle = \langle 0 | 0 \rangle_J = N \int \mathcal{D}x \, e^{\frac{i}{\hbar} S[x,J]} , \tag{4.59}
$$

without the end point constraints and with

$$
S[x, J] = \int_{-\infty}^{\infty} dt \, (L(x, \dot{x}) + Jx) . \tag{4.60}
$$

Let us note that if we define

$$
Z[J] = \langle 0 | 0 \rangle_J = N \int \mathcal{D}x \, e^{\frac{i}{\hbar} S[x,J]} , \tag{4.61}
$$

then, it follows from Eq. (4.47) that, in the vacuum state,

$$\langle T(X(t_1) \cdots X(t_n))) \rangle = \frac{(-i\hbar)^n}{Z[J]} \left. \frac{\delta^n Z[J]}{\delta J(t_1) \cdots \delta J(t_n)} \right|_{J=0}. \qquad (4.62)$$

Namely, $Z[J]$ generates time ordered correlation functions or the Green's functions in the vacuum. If one knows all the vacuum Green's functions, one can construct all the elements of the Scattering matrix of the theory and, therefore, can solve the theory. In quantum field theory, therefore, these correlation functions or the vacuum Green's functions play a central role. $Z[J]$ is correspondingly known as the vacuum functional or the generating functional for vacuum Green's functions.

In quantum mechanics, we are often interested in various statistical deviations from the mean values. These can also be obtained in the path integral formalism in the following way. Let us define

$$Z[J] = e^{\frac{i}{\hbar} W[J]},$$

$$\text{or,} \quad W[J] = -i\hbar \ln Z[J]. \qquad (4.63)$$

We have already seen in the case of the free particle as well as the harmonic oscillator that the path integral for the transition amplitude is proportional to the exponential of the classical action (see Eqs. (2.57) and (3.32)). It is for this reason that $W[J]$ is also called an effective action. Let us note that by definition

$$\left. \frac{\delta W[J]}{\delta J(t_1)} \right|_{J=0} = (-i\hbar) \frac{1}{Z[J]} \left. \frac{\delta Z[J]}{\delta J(t_1)} \right|_{J=0} = \langle X(t_1) \rangle, \qquad (4.64)$$

where $\langle \cdots \rangle$ stands for the vacuum expectation value (vacuum correlation function) from now on. Next, we note that

$$(-i\hbar) \left. \frac{\delta^2 W[J]}{\delta J(t_1) \delta J(t_2)} \right|_{J=0}$$

$$= (-i\hbar)^2 \left(\frac{1}{Z[J]} \frac{\delta^2 Z[J]}{\delta J(t_1) \delta J(t_2)} - \frac{1}{Z^2[J]} \frac{\delta Z[J]}{\delta J(t_1)} \frac{\delta Z[J]}{\delta J(t_2)} \right) \bigg|_{J=0}$$

$$= \langle T(X(t_1)X(t_2))\rangle - \langle X(t_1)\rangle\langle X(t_2)\rangle$$

$$= \langle T\left((X(t_1) - \langle X(t_1)\rangle)(X(t_2) - \langle X(t_2)\rangle)\right)\rangle. \tag{4.65}$$

We recognize this to be the second order deviation from the mean and we note that we can similarly, obtain

$$(-i\hbar)^2 \left.\frac{\delta^3 W[J]}{\delta J(t_1)\delta J(t_2)\delta J(t_3)}\right|_{J=0}$$

$$= (-i\hbar)^3 \left(\frac{1}{Z[J]} \frac{\delta^3 Z[J]}{\delta J(t_1)\delta J(t_2)\delta J(t_3)} - \frac{1}{Z^2[J]} \frac{\delta^2 Z[J]}{\delta J(t_1)\delta J(t_2)} \frac{\delta Z[J]}{\delta J(t_3)} \right.$$

$$- \frac{1}{Z^2[J]} \frac{\delta^2 Z[J]}{\delta J(t_3)\delta J(t_1)} \frac{\delta Z[J]}{\delta J(t_2)} - \frac{1}{Z^2[J]} \frac{\delta^2 Z[J]}{\delta J(t_2)\delta J(t_3)} \frac{\delta Z[J]}{\delta J(t_1)}$$

$$\left. + \frac{2}{Z^3} \frac{\delta Z[J]}{\delta J(t_1)} \frac{\delta Z[J]}{\delta J(t_2)} \frac{\delta Z[J]}{\delta J(t_3)} \right)\Bigg|_{J=0}$$

$$= \langle T\left(X(t_1)X(t_2)X(t_3)\right)\rangle - \langle T\left(X(t_1)X(t_2)\right)\rangle\langle X(t_3)\rangle$$

$$- \langle T\left(X(t_3)X(t_1)\right)\rangle\langle X(t_2)\rangle - \langle T\left(X(t_2)X(t_3)\right)\rangle\langle X(t_1)\rangle$$

$$+ 2\langle X(t_1)\rangle\langle X(t_2)\rangle\langle X(t_3)\rangle \tag{4.66}$$

$$= \langle T\left((X(t_1) - \langle X(t_1)\rangle)(X(t_2) - \langle X(t_2)\rangle)(X(t_3) - \langle X(t_3)\rangle)\right)\rangle.$$

We can go on and the expressions start to take a more complicated form starting with the fourth functional derivative of the effective action $W[J]$. However, $W[J]$ can still be shown to generate various statistical deviations and their moments. In quantum field theory, $W[J]$ is known as the generating functional for the connected vacuum Green's functions.

Let us next go back to the example of the harmonic oscillator which we have studied in some detail. In this case, the generating functional is given by

$$Z[J] = N \int \mathcal{D}x \, e^{\frac{i}{\hbar}S[x,J]}, \tag{4.67}$$

where

$$S[x, J] = \int\limits_{-\infty}^{\infty} dt \left(\frac{1}{2}m\dot{x}^2 - \frac{1}{2}m\omega^2 x^2 + Jx \right). \tag{4.68}$$

Obviously, in this case, we have

$$\langle X(t_1) \rangle = (-i\hbar) \frac{1}{Z[J]} \frac{\delta Z[J]}{\delta J(t_1)} \bigg|_{J=0}$$

$$= \frac{N \int \mathcal{D}x \, x(t_1) \, e^{\frac{i}{\hbar}S[x]}}{N \int \mathcal{D}x \, e^{\frac{i}{\hbar}S[x]}} = 0. \tag{4.69}$$

This vanishes because the integrand in the numerator is odd. Therefore, for the harmonic oscillator, we obtain from Eqs. (4.62) and (4.65) that

$$\langle T(X(t_1)X(t_2)) \rangle = (-i\hbar)^2 \frac{1}{Z[J]} \frac{\delta^2 Z[J]}{\delta J(t_1)\delta J(t_2)} \bigg|_{J=0}$$

$$= (-i\hbar) \frac{\delta^2 W[J]}{\delta J(t_1)\delta J(t_2)} \bigg|_{J=0}. \tag{4.70}$$

Let us also note that because the action for the harmonic oscillator is quadratic in the dynamical variables, we can write

$$Z[J] = N \int \mathcal{D}x \, e^{\frac{i}{\hbar} \int\limits_{-\infty}^{\infty} dt(\frac{1}{2}m\dot{x}^2 - \frac{1}{2}m\omega^2 x^2 + Jx)}$$

$$= \lim_{\epsilon \to 0^+} N \int \mathcal{D}x \, e^{-\frac{im}{2\hbar} \int\limits_{-\infty}^{\infty} dt \, (x(t)(\frac{d^2}{dt^2} + \omega^2 - i\epsilon)x(t) - \frac{2}{m}J(t)x(t))}. \tag{4.71}$$

This a Gaussian integral (involving an operator) and let us recall from (see Eq. (3.82)) that

$$\lim_{\epsilon \to 0^+} \left(\frac{d^2}{dt^2} + \omega^2 - i\epsilon \right) G_F(t - t') = -\delta(t - t'). \tag{4.72}$$

Using this, we can define (this is the generalization of completing the square to the case of a Gaussian involving operators)

$$\tilde{x}(t) = x(t) + \frac{1}{m} \int\limits_{-\infty}^{\infty} dt' \, G_F(t - t') J(t') \,, \tag{4.73}$$

and the generating functional then takes the form

$$Z[J] = \lim_{\epsilon \to 0^+} N \int \mathcal{D}\tilde{x} \, e^{-\frac{im}{2\hbar} \int\limits_{-\infty}^{\infty} dt \, \tilde{x}(t)(\frac{d^2}{dt^2} + \omega^2 - i\epsilon)\tilde{x}(t)}$$

$$\times \, e^{-\frac{i}{2\hbar m} \iint\limits_{-\infty}^{\infty} dt \, dt' \, J(t) G_F(t-t') J(t')}$$

$$= \lim_{\epsilon \to 0^+} N \left[\det(\frac{d^2}{dt^2} + \omega^2 - i\epsilon) \right]^{-\frac{1}{2}}$$

$$\times \, e^{-\frac{i}{2\hbar m} \iint\limits_{-\infty}^{\infty} dt \, dt' \, J(t) G_F(t-t') J(t')}$$

$$= Z[0] \, e^{-\frac{i}{2\hbar m} \iint\limits_{-\infty}^{\infty} dt \, dt' \, J(t) G_F(t-t') J(t')} \,, \tag{4.74}$$

where we have identified

$$Z[0] = N \det \left[\det(\frac{d^2}{dt^2} + \omega^2 - i\epsilon) \right]^{-\frac{1}{2}} \,.$$

We now obtain in a straightforward manner

$$\frac{\delta^2 Z[J]}{\delta J(t_1) \delta J(t_2)} \bigg|_{J=0} = -\frac{i}{\hbar m} G_F(t_1 - t_2) Z[0] \,. \tag{4.75}$$

Consequently, for the harmonic oscillator, we have (see Eq. (4.70))

$$\langle T(X(t_1)X(t_2))\rangle = (-i\hbar)^2 \frac{1}{Z[J]} \left.\frac{\delta^2 Z[J]}{\delta J(t_1)\delta J(t_2)}\right|_{J=0}$$

$$= (-i\hbar)^2 \left(-\frac{i}{\hbar m}\right) G_F(t_1 - t_2) = \frac{i\hbar}{m} G_F(t_1 - t_2). \qquad (4.76)$$

In other words, the two-point time ordered vacuum correlation function, in the present case, gives the Feynman Green's function. This is a general feature of all quantum mechanical theories, namely, that the two-point connected vacuum Green's function is nothing other than the Feynman propagator of the theory.

4.5 Anharmonic oscillator

Just as there are a handful of quantum mechanical systems which can be solved analytically, similarly, there are only a few path integrals that can be evaluated exactly. (There is a one to one correspondence between the quantum mechanical problems that can be analytically solved and the path integrals that can be exactly evaluated.) The Gaussian (recall the free particle and the harmonic oscillator) is the simplest of the path integrals which can be exactly evaluated. However, we also know that if we perturb the harmonic oscillator even slightly by an additional interaction, say a quartic interaction, the problem cannot be solved analytically. In other words, the quantum mechanical system corresponding to the Lagrangian

$$L = \frac{1}{2}m\dot{x}^2 - \frac{1}{2}m\omega^2 x^2 - \frac{\lambda}{4}x^4, \quad \lambda > 0, \qquad (4.77)$$

is impossible to solve exactly even when $\lambda \ll 1$. (This model is known as the anharmonic oscillator and the restriction on the coupling constant, $\lambda > 0$, ensures that the potential is bounded from below so that a ground state of the theory exists.) In such a case, we use perturbation theory and calculate corrections to the unperturbed system in a power series in the coupling constant λ. In the Feynman path integral approach, the manifestation of this lies in the fact that the path integral corresponding to the Lagrangian for this anharmonic oscillator cannot be evaluated exactly and has to be calculated perturbatively in the following way.

Let us introduce an external source and write the vacuum functional for this theory as

$$Z[J] = N \int \mathcal{D}x \, e^{\frac{i}{\hbar}S[x,J]}$$

$$= N \int \mathcal{D}x \, e^{\frac{i}{\hbar} \int\limits_{-\infty}^{\infty} dt \, (\frac{1}{2}m\dot{x}^2 - \frac{1}{2}m\omega^2 x^2 - \frac{\lambda}{4}x^4 + Jx)}. \tag{4.78}$$

If we separate the action into a free (quadratic) part and an interaction part, we can write

$$S[x, J] = S_0[x, J] - \frac{\lambda}{4} \int\limits_{-\infty}^{\infty} dt \, x^4, \tag{4.79}$$

where

$$S_0[x, J] = \int\limits_{-\infty}^{\infty} dt \left(\frac{1}{2}m\dot{x}^2 - \frac{1}{2}m\omega^2 x^2 + Jx \right), \tag{4.80}$$

is the action for the harmonic oscillator in the presence of a source. It follows now that

$$\frac{\delta S_0[x, J]}{\delta J(t)} = x(t), \tag{4.81}$$

leading to

$$(-i\hbar)\frac{\delta}{\delta J(t)} e^{\frac{i}{\hbar} S_0[x,J]} = \frac{\delta S_0[x, J]}{\delta J(t)} e^{\frac{i}{\hbar} S_0[x,J]} = x(t) \, e^{\frac{i}{\hbar} S_0[x,J]}. \tag{4.82}$$

In other words, operationally we can identify

$$(-i\hbar)\frac{\delta}{\delta J(t)} \to x(t), \tag{4.83}$$

when the functional derivative acts on $e^{\frac{i}{\hbar}S_0[x,J]}$. Now, we can use this identification to write

$$Z[J] = N \int \mathcal{D}x \left(e^{-\frac{i\lambda}{4\hbar} \int\limits_{-\infty}^{\infty} dt\, x^4} \right) e^{\frac{i}{\hbar}S_0[x,J]}$$

$$= N \int \mathcal{D}x \left(e^{-\frac{i\lambda}{4\hbar} \int\limits_{-\infty}^{\infty} dt\, (-i\hbar\frac{\delta}{\delta J(t)})^4} \right) e^{\frac{i}{\hbar}S_0[x,J]}$$

$$= \left(e^{-\frac{i\lambda}{4\hbar} \int\limits_{-\infty}^{\infty} dt\, (-i\hbar\frac{\delta}{\delta J(t)})^4} \right) N \int \mathcal{D}x\, e^{\frac{i}{\hbar}S_0[x,J]}$$

$$= \left(e^{-\frac{i\lambda}{4\hbar} \int\limits_{-\infty}^{\infty} dt\, (-i\hbar\frac{\delta}{\delta J(t)})^4} \right) Z_0[J], \tag{4.84}$$

where $Z_0[J]$ is the vacuum functional for the harmonic oscillator interacting with an external source. We have already seen in Eq. (4.74) that it has the form

$$Z_0[J] = Z_0[0]\, e^{-\frac{i}{2\hbar m} \iint\limits_{-\infty}^{\infty} dt'\, dt''\, J(t')G_F(t'-t'')J(t'')}. \tag{4.85}$$

Substituting this back then, we obtain

$$Z[J] = Z_0[0] \left(e^{-\frac{i\lambda}{4\hbar} \int\limits_{-\infty}^{\infty} dt\, (-i\hbar\frac{\delta}{\delta J(t)})^4} \right)$$

$$\times\ e^{-\frac{i}{2\hbar m} \iint\limits_{-\infty}^{\infty} dt'\, dt''\, J(t')G_F(t'-t'')J(t'')}. \tag{4.86}$$

It is clear that if λ is small, i.e., for weak coupling, we can Taylor expand the first exponential and we will be able to obtain the vacuum functional as a power series in λ. Consequently, all the vacuum Green's functions can be calculated perturbatively. This is basically perturbation theory within the framework of Feynman's path integral and we will discuss more about it later.

4.6 References

S. Coleman, *Secret Symmetry*, Erice Lectures (1973).

A. Das, *Finite Temperature Field Theory*, World Scientific Publishing.

K. Huang, *Quarks, Leptons and Gauge Fields*, World Scientific Publishing.

J. Schwinger, Phys. Rev. 82, 914 (1951); *ibid* 91, 713 (1953).

CHAPTER 5

Path integrals for fermions

5.1 Fermionic oscillator

As we know, there are two kinds of particles in nature, namely, bosons and fermions. They are described by quantum mechanical operators with very different properties. The operators describing bosons, for example, obey commutation relations whereas the fermionic operators (i.e., operators describing fermions) satisfy anti-commutation relations. As a preparation for such systems, let us study a prototype example, namely, the fermionic oscillator.

There are many ways to introduce the fermionic oscillator. Let us discuss the one that is the most intuitive. Let us recall that the bosonic harmonic oscillator in one dimension with a natural frequency ω has the Hamiltonian which, written in terms of creation and annihilation operators, takes the form

$$H_B = \frac{\omega}{2} \left(a_B^\dagger a_B + a_B a_B^\dagger \right). \tag{5.1}$$

Here, for simplicity, we are assuming that $\hbar = 1$. The creation and the annihilation operators are supposed to satisfy the commutation relations

$$\left[a_B, a_B^\dagger \right] = a_B a_B^\dagger - a_B^\dagger a_B = 1, \tag{5.2}$$

with all others vanishing. The symmetric structure of the Hamiltonian, in this case, is a reflection of the fact that we are dealing with Bose particles and, consequently, the states (or wave functions) must have a symmetric form.

Fermionic systems, on the other hand, have an inherent anti-symmetry. Therefore, let us try a Hamiltonian for a fermionic oscillator with frequency ω of the form

$$H_F = \frac{\omega}{2}\left(a_F^\dagger a_F - a_F a_F^\dagger\right). \tag{5.3}$$

If a_F and $a_F^\dagger$ were to satisfy commutation relations like the bosonic oscillator in Eq. (5.2), namely, if we had

$$\left[a_F, a_F^\dagger\right] = a_F a_F^\dagger - a_F^\dagger a_F = 1, \tag{5.4}$$

with all others vanishing, then using this, we can rewrite the fermionic Hamiltonian in Eq. (5.3) to be

$$H_F = \frac{\omega}{2}\left(a_F^\dagger a_F - \left(a_F^\dagger a_F + 1\right)\right) = -\frac{\omega}{2}. \tag{5.5}$$

In other words, in such a case, there would be no dynamics associated with the Hamiltonian. Let us assume, therefore, that the fermionic operators a_F and $a_F^\dagger$ instead satisfy anti-commutation relations. Namely, let

$$[a_F, a_F]_+ \equiv a_F^2 + a_F^2 = 2a_F^2 = 0,$$

$$\left[a_F^\dagger, a_F^\dagger\right]_+ \equiv \left(a_F^\dagger\right)^2 + \left(a_F^\dagger\right)^2 = 2\left(a_F^\dagger\right)^2 = 0,$$

$$\left[a_F, a_F^\dagger\right]_+ \equiv a_F a_F^\dagger + a_F^\dagger a_F = 1 = \left[a_F^\dagger, a_F\right]_+. \tag{5.6}$$

In contrast to the commutators, the anti-commutators are by definition symmetric.

An immediate consequence of the anti-commutation relations in Eq. (5.6) is that in such a system, the particles (quanta) must obey Pauli exclusion principle (Fermi-Dirac statistics). To see this, let us note that if we identify the operators a_F and $a_F^\dagger$ with the annihilation and creation operators for such a system, then we can define a number operator as usual as

$$N_F = a_F^\dagger a_F. \tag{5.7}$$

From the anti-commutation relations in Eq. (5.6) we note that

$$N_F^2 = a_F^\dagger a_F a_F^\dagger a_F$$

$$= a_F^\dagger \left(1 - a_F^\dagger a_F\right) a_F$$

$$= a_F^\dagger a_F = N_F,$$

or, $\quad N_F \left(N_F - 1\right) = 0$. $\hfill (5.8)$

Therefore, the eigenvalues of the number operator can only be zero or one. This is the reflection of the Pauli principle or the Fermi-Dirac statistics, namely, we can at the most have one fermion in a given quantum mechanical state. This shows that the anti-commutation relations are the natural choice for a fermionic system.

Given this, let us rewrite the Hamiltonian for the fermionic oscillator in Eq. (5.3) as

$$H_F = \frac{\omega}{2} \left(a_F^\dagger a_F - \left(1 - a_F^\dagger a_F\right)\right) = \omega \left(a_F^\dagger a_F - \frac{1}{2}\right)$$

$$= \omega \left(N_F - \frac{1}{2}\right). \hfill (5.9)$$

We note that this has a form similar to the Hamiltonian of the bosonic oscillator except for the sign of the constant term (the zero point energy). Furthermore, the commutation relations between $a_F, a_F^\dagger$ and N_F can now be calculated in a straightforward manner

$$[a_F, N_F] = \left[a_F, a_F^\dagger a_F\right] = \left[a_F, a_F^\dagger\right]_+ a_F = a_F,$$

$$\left[a_F^\dagger, N_F\right] = \left[a_F^\dagger, a_F^\dagger a_F\right] = -a_F^\dagger \left[a_F^\dagger, a_F\right]_+ = -a_F^\dagger. \hfill (5.10)$$

(As a technical aside, we note here that a commutator involving products of operators can, of course, be written in terms of commutators, but can also be expressed in terms of anti-commutators. For example,

$$[A, BC] = ABC - BCA = (AB + BA) C - B (AC + CA)$$

$$= [A, B]_+ C - B [A, C]_+, \hfill (5.11)$$

and this is what we have used in Eq. (5.10).) As a result, we can think of $a_F, a_F^\dagger$ as lowering and raising operators respectively.

We note that if an eigenstate of N_F is denoted by $|n_F\rangle$, then it satisfies

$$N_F|n_F\rangle = n_F|n_F\rangle \,, \tag{5.12}$$

with $n_F = 0, 1$. The ground state with no quantum is denoted by $|0\rangle$ and satisfies

$$N_F|0\rangle = 0 \,,$$
$$H_F|0\rangle = \omega\left(N_F - \frac{1}{2}\right)|0\rangle = -\frac{\omega}{2}|0\rangle \,. \tag{5.13}$$

Similarly, the state with one quantum is denoted by $|1\rangle$ and satisfies

$$N_F|1\rangle = |1\rangle \,,$$
$$H_F|1\rangle = \omega\left(N_F - \frac{1}{2}\right)|1\rangle = \frac{\omega}{2}|1\rangle \,. \tag{5.14}$$

The ground state is annihilated by a_F and we have

$$a_F|0\rangle = 0 \,,$$
$$a_F^\dagger|0\rangle = |1\rangle \,. \tag{5.15}$$

It is clear from the anti-commutation relations in Eq. (5.6) that

$$a_F^\dagger|1\rangle = a_F^\dagger a_F^\dagger|0\rangle = 0 \,. \tag{5.16}$$

Therefore, the Hilbert space of the theory, in this case, is two dimensional and we note here that the ground state energy has opposite sign from the ground state energy (zero point energy) of a bosonic oscillator.

5.2 Grassmann variables

Since fermions have no classical analog, we cannot directly write down a Lagrangian for the fermionic oscillator in terms of the usual coordinates and momenta. Obviously, we need the notion of anti-commuting classical variables. (This is essential to develop the path integral for such systems since path integrals involve only classical variables.) Such variables have been well studied in mathematics and go under the name of Grassmann variables. As one can readily imagine, they have very uncommon properties and let us recall here only some of these properties which we will need for our discussions. For example, if θ_i, $i = 1, 2, \ldots, n$, define a set of Grassmann variables (classical), then they satisfy

$$\theta_i \theta_j + \theta_j \theta_i = 0, \quad i, j = 1, 2, \ldots, n. \tag{5.17}$$

This, in particular, implies that for any given i,

$$\theta_i^2 = 0, \quad i \text{ not summed}. \tag{5.18}$$

In other words, the Grassmann variables are nilpotent. This has the immediate consequence that if $f(\theta)$ is a function of only one Grassmann variable, then it has the simple Taylor expansion

$$f(\theta) = a + b\,\theta. \tag{5.19}$$

Since θ_i's are anti-commuting, the derivatives have to be defined carefully in the sense that the direction in which the derivatives operate needs to be specified. Thus, for example, a right derivative for Grassmann variables would give

$$\frac{\partial}{\partial \theta_i} (\theta_j \theta_k) = \theta_j \left(\frac{\partial \theta_k}{\partial \theta_i} \right) - \left(\frac{\partial \theta_j}{\partial \theta_i} \right) \theta_k = \delta_{ik} \theta_j - \delta_{ij} \theta_k, \tag{5.20}$$

whereas a left derivative would lead to

$$\frac{\partial}{\partial \theta_i} (\theta_j \theta_k) = \left(\frac{\partial \theta_j}{\partial \theta_i} \right) \theta_k - \theta_j \left(\frac{\partial \theta_k}{\partial \theta_i} \right) = \delta_{ij} \theta_k - \delta_{ik} \theta_j. \tag{5.21}$$

Thus, the sense of the derivative is crucial and in all our discussions involving Grassmann variables, we will use left derivatives.

Let us note that like the Grassmann variables, the derivatives with respect to these variables also anti-commute. Namely,

$$\frac{\partial}{\partial \theta_i} \frac{\partial}{\partial \theta_j} + \frac{\partial}{\partial \theta_j} \frac{\partial}{\partial \theta_i} = 0 \,. \tag{5.22}$$

These derivatives, in fact, behave quite like the exterior derivatives in differential geometry. We note in particular that for a fixed i

$$\left(\frac{\partial}{\partial \theta_i} \right)^2 = 0 \,. \tag{5.23}$$

In other words, the derivatives, in this case, are nilpotent just like the variables themselves. Furthermore, the conventional commutation relation between derivatives and coordinates now takes the form

$$\left[\frac{\partial}{\partial \theta_i}, \theta_j \right]_+ = \frac{\partial}{\partial \theta_i} \theta_j + \theta_j \frac{\partial}{\partial \theta_i}$$

$$= \frac{\partial \theta_j}{\partial \theta_i} - \theta_j \frac{\partial}{\partial \theta_i} + \theta_j \frac{\partial}{\partial \theta_i} = \delta_{ij} \,. \tag{5.24}$$

The notion of integration can also be generalized to Grassmann variables. Denoting by D the operation of differentiation with respect to one Grassmann variable and by I the operation of integration, we note that these must satisfy the relations

$$ID = 0 \,,$$

$$DI = 0 \,. \tag{5.25}$$

Namely, the integral of a total derivative must vanish if we ignore surface terms and furthermore, an integral, being independent of the variable, must give zero upon differentiation. Note that since differentiation with respect to a Grassmannn variable is nilpotent (see Eq. (5.23)), it satisfies the above properties and hence for Grassmann variables integration can be naturally identified with differentiation. Namely, in this case, we have

$$I = D.$$ (5.26)

Thus, for a function of a single Grassmann variable, we have

$$\int d\theta \, f(\theta) = \frac{\partial f(\theta)}{\partial \theta}.$$ (5.27)

This immediately leads to the fundamental result that for Grassmann variables

$$\int d\theta = 0,$$

$$\int d\theta \, \theta = 1.$$ (5.28)

This is an essential difference between ordinary variables and Grassmann variables and has far reaching consequences.

An immediate consequence of the definition of the integral in Eq. (5.27) is that if we redefine the variable of integration as

$$\theta' = a\theta, \quad a \neq 0,$$ (5.29)

then, from Eq. (5.27) we obtain

$$\int d\theta \, f(\theta) = \frac{\partial f(\theta)}{\partial \theta} = \frac{\partial \theta'}{\partial \theta} \frac{\partial f\left(\frac{\theta'}{a}\right)}{\partial \theta'} = a \int d\theta' \, f\left(\frac{\theta'}{a}\right).$$ (5.30)

But this is precisely the inverse of what happens for ordinary variables. Namely, we note that the Jacobian in the case of redefinition of Grassmann variables is the inverse of what one would naively expect for ordinary (commuting) variables. This result can be generalized to integrations involving many Grassmann variables and it can be shown that if

$$\theta_i' = a_{ij}\theta_j \,, \tag{5.31}$$

with $\det a_{ij} \neq 0$ and repeated indices being summed, then

$$\int \prod_{i=1}^{n} d\theta_i \, f(\theta_i) = (\det a_{ij}) \int \prod_{i=1}^{n} d\theta_i' f\left(a_{ij}^{-1}\theta_j'\right) . \tag{5.32}$$

We can also define a delta function in the space of Grassmann variables as

$$\delta(\theta) = \theta \,. \tag{5.33}$$

That this satisfies all the properties of the delta function can be easily seen by noting that

$$\int d\theta \, \delta(\theta) = \int d\theta \, \theta = 1, \tag{5.34}$$

which follows from Eq. (5.28). In addition, if

$$f(\theta) = a + b \, \theta \,, \tag{5.35}$$

then,

$$\int d\theta \, \delta(\theta) \, f(\theta) = \int d\theta \, \theta \, f(\theta) = \int d\theta \, \theta(a + b\theta)$$

$$= \int d\theta \, \theta a = \frac{\partial(\theta a)}{\partial \theta} = a = f(0) \,. \tag{5.36}$$

Here we have used the nilpotency of the Grassmann variables. Furthermore, consistent with the rule for change of variables for the Grassmann variables, if

$$g(\theta) = a\theta \,, \tag{5.37}$$

then, we obtain

$$\delta(g(\theta)) = \delta(a\theta) = a\theta = a\delta(\theta) = \frac{\partial g(\theta)}{\partial \theta}\delta(\theta) , \tag{5.38}$$

where $g(\theta)$ is assumed to be Grassmann odd. An integral representation for the delta function can be obtained simply by noting that if ζ is also a Grassmann variable, then

$$\int d\zeta \, e^{i\zeta\theta} = \int d\zeta \, (1 + i\zeta\theta)$$

$$= \frac{\partial}{\partial\zeta}(1 + i\zeta\theta) = i\theta = i\delta(\theta) . \tag{5.39}$$

Let us next evaluate the basic Gaussian integral for Grassmann variables. Let us consider two sets of independent Grassmann variables, namely, $(\theta_1, \theta_2, \ldots, \theta_n)$ and $(\theta_1^*, \theta_2^*, \ldots, \theta_n^*)$ and analyze the integral

$$I = \int \prod_{i,j} d\theta_i^* \, d\theta_j \, e^{-(\theta_i^* M_{ij}\theta_j + c_i^*\theta_i + \theta_i^* c_i)} , \tag{5.40}$$

where we are assuming that c_i and c_i^* are independent Grassmann variables (sources). Furthermore, the convention for summation over repeated indices is always assumed. Note that if we make the change of variables (we are assuming that M^{-1} exists)

$$\theta_i' = M_{ij}\theta_j + c_i,$$

$$\text{or,} \quad \theta_i = M_{ij}^{-1}(\theta_j' - c_j) , \tag{5.41}$$

and

$$\theta_i^{*'} = \theta_i^* + c_j^* M_{ji}^{-1} , \tag{5.42}$$

then, using Eqs. (5.32), (5.41) and (5.42) we obtain

$$I = \int \prod_{i,j} d\theta_i^* \, d\theta_j \, e^{-\left(\theta_i^*(M_{ij}\theta_j + c_i) + c_i^*\theta_i\right)}$$

$$= \det M_{ij} \int \prod_{ij} d\theta_i^* \, d\theta_j' \, e^{-\left(\theta_i^*\theta_i' + c_i^* M_{ij}^{-1}(\theta_j' - c_j)\right)}$$

$$= \det M_{ij} \int \prod_{ij} d\theta_i^* \, d\theta_j' \, e^{-\left((\theta_i^* + c_j^* M_{ji}^{-1})\theta_i' - c_i^* M_{ij}^{-1}c_j\right)}$$

$$= \det M_{ij} \int \prod_{ij} d\theta_i^{*\prime} \, d\theta_j' \, e^{-\theta_i^{*\prime}\theta_i' + c_i^* M_{ij}^{-1}c_j}$$

$$= N \, \det M_{ij} \, e^{c_i^* M_{ij}^{-1}c_j}. \tag{5.43}$$

Here N is a constant (basically, it corresponds to a sign factor arising from the orders of the integrations of Grassmann variables) and we note that the Gaussian integral in the case of Grassmann variables has the same form as the integral for ordinary (commuting) variables except for the positive power of the determinant. This leads to an essential difference between quantum mechanical bosonic and fermionic theories.

5.3 Generating functional for fermions

With all this background on Grassmann variables, we can now ask whether it is possible to write a Lagrangian for the fermionic oscillator. Indeed, let us consider the classical Lagrangian

$$L = \frac{i}{2}\left(\bar{\psi}\dot{\psi} - \dot{\bar{\psi}}\psi\right) - \frac{\omega}{2}\left[\bar{\psi}, \psi\right], \tag{5.44}$$

where ψ and $\bar{\psi}$ are two independent Grassmann variables which are functions of time. Quite often one eliminates a total derivative to write an equivalent Lagrangian of the form

$$L = i\bar{\psi}\dot{\psi} - \frac{\omega}{2}[\bar{\psi}, \psi] = i\bar{\psi}\dot{\psi} - \omega\bar{\psi}\psi. \tag{5.45}$$

However, to begin with we will continue with the first form of the Lagrangian, namely, Eq. (5.44). This Lagrangian ((5.44) or (5.45)) is first order in the time derivative and one can define canonical conjugate momenta associated with the Grassmann variables ψ and $\bar{\psi}$ as usual (recall that we are using left derivatives)

$$\Pi_\psi = \frac{\partial L}{\partial \dot{\psi}} = -\frac{i}{2}\bar{\psi},$$

$$\Pi_{\bar{\psi}} = \frac{\partial L}{\partial \dot{\bar{\psi}}} = -\frac{i}{2}\psi. \tag{5.46}$$

(Such a system is known as a constrained system and one should use the formalism of Dirac quantization to quantize such a system. We will, however, not go into these technical details which are not quite relevant to our present discussion.) With the convention of left derivatives, the proper definition of the Hamiltonian (which is only a function of coordinates and momenta and which also leads to the correct dynamical equations of motion) is

$$H = \dot{\psi}\Pi_\psi + \dot{\bar{\psi}}\Pi_{\bar{\psi}} - L$$

$$= -\frac{i}{2}\dot{\psi}\bar{\psi} - \frac{i}{2}\dot{\bar{\psi}}\psi - \frac{i}{2}\left(\bar{\psi}\dot{\psi} - \dot{\bar{\psi}}\psi\right) + \frac{\omega}{2}[\bar{\psi},\psi]$$

$$= \frac{i}{2}\left(\bar{\psi}\dot{\psi} - \dot{\bar{\psi}}\psi\right) - \frac{i}{2}\left(\bar{\psi}\dot{\psi} - \dot{\bar{\psi}}\psi\right) + \frac{\omega}{2}[\bar{\psi},\psi]$$

$$= \frac{\omega}{2}[\bar{\psi},\psi]. \tag{5.47}$$

It is clear, therefore, that this simple Lagrangian will yield the quantum mechanical Hamiltonian of Eq. (5.3) for the fermionic oscillator if we identify (as quantum operators)

$$\psi \to a_F,$$

$$\bar{\psi} \to a_F^\dagger. \tag{5.48}$$

With the identification of ψ and $\bar{\psi}$ with the annihilation and the creation operators respectively, the hermiticity properties for these variables now follow. Namely, we note that (we should use complex conjugation for classical quantities, but complex conjugation for products of Grassmann variables also transposes the order which is why we use "†" for simplicity)

$$\psi^\dagger = \bar{\psi} \,,$$

$$\bar{\psi}^\dagger = \psi \,. \tag{5.49}$$

We note that with this convention, the number operator defined as

$$N_F = a_F^\dagger a_F \to \bar{\psi}\psi \,, \tag{5.50}$$

is Hermitian and, therefore, $\bar{\psi}\psi$ should be real which requires

$$N_F^\dagger = N_F \to \left(\bar{\psi}\psi\right)^\dagger = \psi^\dagger \bar{\psi}^\dagger = \bar{\psi}\psi \,. \tag{5.51}$$

Namely, since Grassmann variables have no classical analog, even when we are dealing with the reality question of products of ordinary Grassmann variables (not operators), we follow the above prescription in defining complex conjugation. Namely, for any pair of Grassmann variables η and χ, we define

$$(\eta\chi)^* = \chi^*\eta^*. \tag{5.52}$$

In other words, even classically, we continue to treat Grassmann variables like operators. This is the only way a consistent transition from a classical to a quantum Lagrangian involving fermions is possible.

   With this prescription, let us note that the Lagrangian for the fermionic oscillator given in Eq. (5.44) is Hermitian (real)

$$\begin{aligned} L^\dagger &= \left(\frac{i}{2} \left(\bar{\psi}\dot{\psi} - \dot{\bar{\psi}}\psi \right) - \frac{\omega}{2} [\bar{\psi}, \psi] \right)^\dagger \\ &= -\frac{i}{2} \left(\dot{\psi}^\dagger \bar{\psi}^\dagger - \psi^\dagger \dot{\bar{\psi}}^\dagger \right) - \frac{\omega}{2} [\psi^\dagger, \bar{\psi}^\dagger] \\ &= -\frac{i}{2} \left(\dot{\bar{\psi}}\psi - \bar{\psi}\dot{\psi} \right) - \frac{\omega}{2} [\bar{\psi}, \psi] \\ &= \frac{i}{2} \left(\bar{\psi}\dot{\psi} - \dot{\bar{\psi}}\psi \right) - \frac{\omega}{2} [\bar{\psi}, \psi] \\ &= L. \end{aligned} \tag{5.53}$$

We can now write the vacuum functional for the fermionic oscillator in the presence of linear external sources as ($\hbar = 1$)

$$Z[\eta, \bar{\eta}] = \langle 0|0 \rangle_{\eta, \bar{\eta}} = N \int \mathcal{D}\bar{\psi} \mathcal{D}\psi \, e^{iS[\psi, \bar{\psi}, \eta, \bar{\eta}]} , \tag{5.54}$$

where we have denoted the two Grassmann sources for ψ and $\bar{\psi}$ by $\bar{\eta}$ and η respectively. The complete action for the oscillator, in this case, has the form

$$S[\psi, \bar{\psi}, \eta, \bar{\eta}] = S[\psi, \bar{\psi}] + \int\limits_{-\infty}^{\infty} dt \, (\bar{\eta}\psi + \bar{\psi}\eta) , \tag{5.55}$$

with

$$S[\psi, \bar{\psi}] = \int\limits_{-\infty}^{\infty} dt \, L = \int\limits_{-\infty}^{\infty} dt \left(\frac{i}{2} \left(\bar{\psi}\dot{\psi} - \dot{\bar{\psi}}\psi \right) - \frac{\omega}{2} [\bar{\psi}, \psi] \right) . \tag{5.56}$$

Once again, we will assume the hermiticity conditions for the sources similar to the ones given in Eq. (5.49), namely,

$$\eta^\dagger = \bar{\eta} ,$$
$$\bar{\eta}^\dagger = \eta , \tag{5.57}$$

in order that the complete action in Eq. (5.55) is Hermitian.

Just as an ordinary derivative with respect to a Grassmann variable is directional, similarly, there are right and left functional derivatives with respect to fermionic variables. The definition of the functional derivative is still the same as given in Eq. (1.14), namely,

$$\frac{\delta F(\psi(t))}{\delta \psi(t')} = \lim_{\epsilon \to 0} \frac{F(\psi(t) + \epsilon\delta(t - t')) - F(\psi(t))}{\epsilon} . \tag{5.58}$$

However, since ϵ is now a Grassmann variable, the position of ϵ^{-1} in the expression defines the direction of the derivative. (Incidentally, since $\epsilon^2 = 0$, ϵ^{-1} does not really exist and one should think of ϵ^{-1} simply as $\frac{\partial}{\partial \epsilon}$. Secondly, we note here that for polynomial functionals, the limit $\epsilon \to 0$ is redundant since $\epsilon^2 = 0$ and the highest power of ϵ in the expansion of the functional is linear.) Thus, a left functional derivative corresponds to defining

$$\frac{\delta F\left(\psi(t)\right)}{\delta \psi(t')} = \lim_{\epsilon \to 0} \epsilon^{-1} \left[F\left(\psi(t) + \epsilon \delta(t - t')\right) - F\left(\psi(t)\right) \right], \qquad (5.59)$$

whereas a right functional derivative would be defined as

$$\frac{\delta F\left(\psi(t)\right)}{\delta \psi(t')} = \lim_{\epsilon \to 0} \left[F\left(\psi(t) + \epsilon \delta(t - t')\right) - F\left(\psi(t)\right) \right] \epsilon^{-1}. \qquad (5.60)$$

As we have mentioned earlier, we will always work with left derivatives even when we are dealing with functionals involving fermionic variables.

5.4 Feynman propagator

Let us next go back to the Lagrangian in Eq. (5.44) for the fermionic oscillator and note that when fermions are involved, we essentially have a matrix structure. This is another reflection of the fact that the Grassmann variables inherently behave like operators. Let us define the following two component matrices.

$$\Psi = \begin{pmatrix} \psi \\ \bar{\psi} \end{pmatrix},$$

$$\bar{\Psi} = \Psi^\dagger \sigma_3 = \overbrace{\bar{\psi} \quad \psi}\, \sigma_3 = \overbrace{\bar{\psi} \quad -\psi},$$

$$\Theta = \begin{pmatrix} \eta \\ \bar{\eta} \end{pmatrix},$$

$$\bar{\Theta} = \Theta^\dagger \sigma_3 = \overbrace{\bar{\eta} \quad \eta}\, \sigma_3 = \overbrace{\bar{\eta} \quad -\eta}, \qquad (5.61)$$

where σ_3 denotes the diagonal Pauli matrix, namely,

$$\sigma_3 = \begin{pmatrix} 1 & 0 \\ 0 & -1 \end{pmatrix}.$$ (5.62)

We note, then, that

$$\bar{\Psi}\sigma_3 i\frac{d}{dt}\Psi = \overbrace{\begin{pmatrix} \bar{\psi} & -\psi \end{pmatrix}} \begin{pmatrix} i\frac{d}{dt} & 0 \\ 0 & -i\frac{d}{dt} \end{pmatrix} \begin{pmatrix} \psi \\ \bar{\psi} \end{pmatrix}$$

$$= \overbrace{\begin{pmatrix} \bar{\psi} & -\psi \end{pmatrix}} \begin{pmatrix} i\dot{\psi} \\ -i\dot{\bar{\psi}} \end{pmatrix} = i\left(\bar{\psi}\dot{\psi} + \psi\dot{\bar{\psi}}\right)$$

$$= i\left(\bar{\psi}\dot{\psi} - \dot{\bar{\psi}}\psi\right),$$

$$\bar{\Psi}\Psi = \overbrace{\begin{pmatrix} \bar{\psi} & -\psi \end{pmatrix}} \begin{pmatrix} \psi \\ \bar{\psi} \end{pmatrix}$$

$$= \left(\bar{\psi}\psi - \psi\bar{\psi}\right) = [\bar{\psi}, \psi],$$

$$\bar{\Psi}\Theta = \overbrace{\begin{pmatrix} \bar{\psi} & -\psi \end{pmatrix}} \begin{pmatrix} \eta \\ \bar{\eta} \end{pmatrix} = \left(\bar{\psi}\eta - \psi\bar{\eta}\right)$$

$$= \left(\bar{\psi}\eta + \bar{\eta}\psi\right) = \bar{\Theta}\Psi,$$ (5.63)

where we have used the anti-commuting properties of the Grassmann variables.

It is now straightforward to show that the action for the fermionic oscillator (including sources) in Eq. (5.55) can be written as

$$S = \int_{-\infty}^{\infty} dt \left(\frac{i}{2}\bar{\Psi}\sigma_3\frac{d}{dt}\Psi - \frac{\omega}{2}\bar{\Psi}\Psi + \bar{\Theta}\Psi\right)$$

$$= \int_{-\infty}^{\infty} dt \left(\frac{1}{2}\bar{\Psi}(i\sigma_3\frac{d}{dt} - \omega)\Psi + \bar{\Theta}\Psi\right).$$ (5.64)

We note here that, using Eq. (5.63), we could have written the source term also as

$$\int\limits_{-\infty}^{\infty} dt \ \bar{\Psi}\Theta \,, \tag{5.65}$$

so that Θ and $\bar{\Theta}$ are not really independent and we need to be careful in taking functional derivatives. We can write the vacuum functional as

$$Z[\Theta] = N \int \mathcal{D}\Psi \, e^{iS}$$

$$= N \int \mathcal{D}\Psi \, e^{\,i \int\limits_{-\infty}^{\infty} dt \left(\frac{1}{2}\bar{\Psi}(i\sigma_3 \frac{d}{dt} - \omega)\Psi + \bar{\Theta}\Psi \right)} . \tag{5.66}$$

This is a Gaussian integral and we know that we can evaluate the generating functional if we know the Green's function associated with the operator in the quadratic term in the exponent (see Eq. (5.43)). Let us, therefore, study the equation

$$\left(i\sigma_3 \frac{d}{dt} - \omega \right) G(t - t') = \delta(t - t') \,. \tag{5.67}$$

This is a matrix equation and it can be solved easily in the momentum space. Thus, we define

$$G(t - t') = \int \frac{dk}{2\pi} \, G(k) \, e^{-ik(t-t')} \,,$$

$$\delta(t - t') = \int \frac{dk}{2\pi} \, e^{-ik(t-t')} \,, \tag{5.68}$$

where $G(k)$ is a matrix in the Fourier transformed space. Substituting these expressions back into Eq. (5.67), we obtain

$$\frac{1}{2\pi}(\sigma_3 k - \omega)G(k) = \frac{1}{2\pi} \,,$$

$$\text{or,} \quad G(k) = \frac{1}{\sigma_3 k - \omega} = \frac{\sigma_3 k + \omega}{k^2 - \omega^2} \,. \tag{5.69}$$

Consequently, the Green's function in Eq. (5.67) has the form

$$G(t - t') = \int \frac{dk}{2\pi} G(k) e^{-ik(t-t')}$$

$$= \int \frac{dk}{2\pi} \frac{\sigma_3 k + \omega}{k^2 - \omega^2} e^{-ik(t-t')}. \tag{5.70}$$

The singularity structure of the integrand in Eq. (5.70) is obvious and the Feynman prescription, in this case, will lead to the propagator (see Eq. (3.80))

$$G_F(t - t') = \lim_{\epsilon \to 0^+} \int \frac{dk}{2\pi} \frac{\sigma_3 k + \omega}{k^2 - \omega^2 + i\epsilon} e^{-ik(t-t')}$$

$$= \lim_{\epsilon \to 0^+} \int \frac{dk}{2\pi} \frac{\sigma_3 k + \omega}{k^2 - (\omega - \frac{i\epsilon}{2\omega})^2} e^{-ik(t-t')}. \tag{5.71}$$

This can also be written in the alternate form

$$G_F(t - t') = \lim_{\epsilon \to 0^+} \int \frac{dk}{2\pi} \frac{1}{\sigma_3 k - \omega + i\epsilon} e^{-ik(t-t')}, \tag{5.72}$$

and satisfies the equation (see Eq. (3.82))

$$\lim_{\epsilon \to 0^+} \left(i\sigma_3 \frac{d}{dt} - \omega + i\epsilon \right) G_F(t - t') = \delta(t - t'). \tag{5.73}$$

This defines the Feynman propagator in the present case.

Going back to the vacuum functional, we note that we can write

$$Z[\Theta] = \lim_{\epsilon \to 0^+} N \int \mathcal{D}\Psi \, e^{i \int dt (\frac{1}{2} \bar{\Psi}(i\sigma_3 \frac{d}{dt} - \omega + i\epsilon)\Psi + \bar{\Theta}\Psi)}. \tag{5.74}$$

Therefore, the one point function can be obtained to be

$$\frac{\delta Z[\Theta]}{\delta \bar{\Theta}(t_1)}\bigg|_{\Theta = \bar{\Theta} = 0} = \lim_{\epsilon \to 0^+} N \int \mathcal{D}\Psi \, (i\Psi(t_1)) \, e^{\frac{i}{2} \int dt \, \bar{\Psi}(i\sigma_3 \frac{d}{dt} - \omega + i\epsilon)\Psi}$$

$$= 0. \tag{5.75}$$

This follows from the fact that the integrand in Eq. (5.75) is odd under (this is also related to the rules for integrations for Grassmann variables)

$$\Psi \to -\Psi, \quad \bar{\Psi} \to -\bar{\Psi}. \tag{5.76}$$

Similarly, we can obtain from Eq. (5.74)

$$\frac{\delta Z[\Theta]}{\delta \Theta(t_1)}\bigg|_{\Theta=\bar{\Theta}=0} = 0. \tag{5.77}$$

Thus, we see that if we write

$$Z[\Theta] = e^{iW[\Theta]}, \tag{5.78}$$

then, in this case,

$$(-i)\frac{\delta^2 W[\Theta]}{\delta\bar{\Theta}(t_1)\delta\Theta(t_2)}\bigg|_{\Theta=\bar{\Theta}=0} = (-i)^2 \frac{1}{Z[\Theta]}\frac{\delta^2 Z[\Theta]}{\delta\bar{\Theta}(t_1)\delta\Theta(t_2)}\bigg|_{\Theta=\bar{\Theta}=0}$$
$$= -\langle T\left(\Psi(t_1)\bar{\Psi}(t_2)\right)\rangle. \tag{5.79}$$

(Compare this with Eq. (4.70) with $\hbar = 1$.) Incidentally, in the case of fermionic variables, time ordering is defined as

$$T\left(\Psi(t_1)\bar{\Psi}(t_2)\right) = \theta(t_1 - t_2)\Psi(t_1)\bar{\Psi}(t_2) - \theta(t_2 - t_1)\bar{\Psi}(t_2)\Psi(t_1). \tag{5.80}$$

The relative negative sign between the two terms in Eq. (5.80) arises from the change in the order of the fermionic variables, which anti-commute, in the second term. Going back to the vacuum functional, we note that since the exponent is quadratic in the variables (namely, it is a Gaussian integral), it can be explicitly evaluated using Eq. (5.43) to be

$$Z[\Theta] = \lim_{\epsilon \to 0^+} N \int \mathcal{D}\Psi e^{i \int \mathrm{d}t(\frac{1}{2}\bar{\Psi}(i\sigma_3\frac{\mathrm{d}}{\mathrm{d}t}-\omega+i\epsilon)\Psi+\bar{\Theta}\Psi)}$$

$$= \overline{N} e^{-\frac{i}{2} \iint \mathrm{d}t_1 \mathrm{d}t_2\, \bar{\Theta}(t_1) G_F(t_1-t_2)\Theta(t_2)}$$

$$= Z[0]\, e^{-\frac{i}{2} \iint \mathrm{d}t_1 \mathrm{d}t_2\, \bar{\Theta}(t_1) G_F(t_1-t_2)\Theta(t_2)}\,, \tag{5.81}$$

where the determinant arising from the integration of the Grassmann variables has been absorbed into the definition of the normalization constant $\overline{N}$ and $Z[0]$ represents the value of the functional in the absence of sources. It is obvious now that

$$\frac{\delta^2 Z[\Theta]}{\delta\bar{\Theta}(t_1)\delta\Theta(t_2)}\bigg|_{\Theta=\bar{\Theta}=0} = iG_F(t_1 - t_2)\, Z[0]\,. \tag{5.82}$$

Therefore, using Eq. (5.79) we have

$$\langle T\left(\Psi(t_1)\bar{\Psi}(t_2)\right)\rangle = -(-i)^2 \frac{1}{Z[\Theta]} \frac{\delta^2 Z[\Theta]}{\delta\bar{\Theta}(t_1)\delta\Theta(t_2)}\bigg|_{\Theta=\bar{\Theta}=0}$$

$$= -(-i)^2 \frac{1}{Z[0]}\, iG_F(t_1 - t_2)Z[0]$$

$$= i\, G_F(t_1 - t_2)\,. \tag{5.83}$$

This again shows (see Eq. (4.76)) that the time ordered two-point correlation function for the fermions in the vacuum gives the Feynman Green's function. As we have argued earlier, this is a general feature of all quantum theories.

5.5 The fermion determinant

The fermion action following from Eq. (5.44) or Eq. (5.45) is quadratic in the dynamical variables just like the action of the bosonic oscillator in Eq. (3.2). Therefore, the generating functional can be easily evaluated. In this section, we will evaluate the generating functional for the fermions in the absence of any source (which gives the determinant absorbed into the normalization constant $\overline{N}$ in Eq. (5.81)). For simplicity, let us take the dynamical Lagrangian of Eq. (5.45). Then, we can write the generating functional, in the absence of sources, to be

$$Z[0] = \tilde{N} \int \mathcal{D}\bar{\psi}\mathcal{D}\psi \; e^{iS[\psi,\bar{\psi}]}$$

$$= \tilde{N} \int \mathcal{D}\bar{\psi}\mathcal{D}\psi \; e^{i\int_{t_i}^{t_f} dt \, (i\bar{\psi}\dot{\psi} - \omega\bar{\psi}\psi)} . \tag{5.84}$$

Here we have used the anti-commuting properties of the Grassmann variables to rewrite the commutator of the fermionic variables in a simpler form. The constant $\tilde{N}$ (we use $\tilde{N}$ instead of N to avoid any possible confusion with the number of discrete time intervals to be introduced below), representing the normalization of the path integral measure is arbitrary at this point and would be appropriately chosen later. We can once again define

$$t_f - t_i = T, \tag{5.85}$$

as the time interval and translate the time coordinate to write the generating functional of Eq. (5.84) also as

$$Z[0] = \tilde{N} \int \mathcal{D}\bar{\psi}\mathcal{D}\psi \; e^{i\int_0^T dt \, (i\bar{\psi}\dot{\psi} - \omega\bar{\psi}\psi)} . \tag{5.86}$$

To evaluate the path integral, we should discretize the time interval as in Eq. (2.19). Thus, defining the intermediate time points to be

$$t_n = n\epsilon, \quad n = 1, 2, \ldots, N-1, \tag{5.87}$$

where the infinitesimal interval is defined to be

$$\epsilon = \frac{T}{N}, \tag{5.88}$$

we can write the path integral to be

$$Z[0] = \lim_{\substack{\epsilon \to 0 \\ N \to \infty}} \tilde{N} \int d\bar{\psi}_1 \cdots d\bar{\psi}_{N-1} d\psi_1 \cdots d\psi_{N-1}$$

$$\times \; e^{\; i\epsilon \sum\limits_{n=1}^{N} (i\bar{\psi}_n (\frac{\psi_n - \psi_{n-1}}{\epsilon}) - \omega\bar{\psi}_n (\frac{\psi_n + \psi_{n-1}}{2}))} . \tag{5.89}$$

Here we have used the mid-point prescription of Weyl ordering as discussed in Eqs. (2.18) and (2.22) along with the earlier observation that the variable $\bar{\psi}$ represents the momentum conjugate to ψ.

The exponent in Eq. (5.89) can be written out in detail as

$$-\left[\sum_{n=1}^{N-1} (1 + \frac{i\epsilon\omega}{2})\bar{\psi}_n\psi_n + \left(1 + \frac{i\epsilon\omega}{2}\right) \bar{\psi}_N\psi_N \right.$$

$$- \sum_{n=2}^{N-1} \left(1 - \frac{i\epsilon\omega}{2}\right) \bar{\psi}_n\psi_{n-1} - \left(1 - \frac{i\epsilon\omega}{2}\right) \bar{\psi}_1\psi_0$$

$$\left. - \left(1 - \frac{i\epsilon\omega}{2}\right) \bar{\psi}_N\psi_{N-1} \right]. \tag{5.90}$$

Thus, defining $(N-1)$ component matrices

$$\psi = \begin{pmatrix} \psi_1 \\ \psi_2 \\ \vdots \\ \psi_{N-1} \end{pmatrix},$$

$$\bar{\psi} = \begin{pmatrix} \bar{\psi}_1 \\ \bar{\psi}_2 \\ \vdots \\ \bar{\psi}_{N-1} \end{pmatrix},$$

$$J = -\left(1 - \frac{i\epsilon\omega}{2}\right) \begin{pmatrix} \psi_0 \\ 0 \\ \vdots \\ 0 \end{pmatrix},$$

$$\bar{J} = -\left(1 - \frac{i\epsilon\omega}{2}\right) \begin{pmatrix} 0 \\ \vdots \\ 0 \\ \bar{\psi}_N \end{pmatrix}, \tag{5.91}$$

we can write the path integral of Eq. (5.89) also as

$$Z[0] = \lim_{\substack{\epsilon \to 0 \\ N \to \infty}} \tilde{N} \int d\bar{\psi}d\psi \; e^{-(\bar{\psi}^T B\psi + \bar{J}^T\psi + \bar{\psi}^T J + (1+\frac{i\epsilon\omega}{2})\bar{\psi}_N\psi_N)} ,$$

$$(5.92)$$

where we have defined a $(N-1) \times (N-1)$ matrix B as

$$B = \begin{pmatrix} x & 0 & 0 & 0 & \cdots \\ y & x & 0 & 0 & \cdots \\ 0 & y & x & 0 & \cdots \\ \vdots & \vdots & \vdots & \vdots & \vdots \end{pmatrix} ,$$

$$(5.93)$$

with

$$x = \left(1 + \frac{i\epsilon\omega}{2}\right) ,$$

$$y = -\left(1 - \frac{i\epsilon\omega}{2}\right) .$$

$$(5.94)$$

The path integral in Eq. (5.92) can now be easily evaluated using Eq. (5.43) and the result is

$$Z[0] = \lim_{\substack{\epsilon \to 0 \\ N \to \infty}} \tilde{N} \; \det B \; e^{(\bar{J}^T B^{-1} J - (1+\frac{i\epsilon\omega}{2})\bar{\psi}_N\psi_N)}$$

$$= \lim_{\substack{\epsilon \to 0 \\ N \to \infty}} \tilde{N} \; \det B \; e^{(\bar{J}_{N-1}B^{-1}_{N-1,1}J_1 - (1+\frac{i\epsilon\omega}{2})\bar{\psi}_N\psi_N)} ,$$

$$(5.95)$$

where any sign factor coming from the integration over the Grassmann variables is included in $\tilde{N}$. The matrix B has a very simple structure and we can easily evaluate the determinant as well as the appropriate element of the inverse matrix which have the following forms

$$\det B = x^{N-1} = \left(1 + \frac{i\epsilon\omega}{2}\right)^{N-1},$$

$$B^{-1}_{N-1,1} = (-1)^N \frac{y^{N-2}}{x^{N-1}} = \frac{\left(1 - \frac{i\epsilon\omega}{2}\right)^{N-2}}{\left(1 + \frac{i\epsilon\omega}{2}\right)^{N-1}}. \tag{5.96}$$

In the continuum limit of $\epsilon \to 0$ and $N \to \infty$ such that $N\epsilon = T$, the path integral, therefore, has the form

$$Z[0] = \tilde{N} \, e^{\frac{i\omega T}{2}} \, e^{(e^{-i\omega T}\bar{\psi}_N\psi_0 - \bar{\psi}_N\psi_N)}$$

$$= \tilde{N} \, e^{\frac{i\omega T}{2}} \, e^{(e^{-i\omega T}\bar{\psi}_f\psi_i - \bar{\psi}_f\psi_f)}. \tag{5.97}$$

Here we have identified

$$\psi_0 = \psi_i, \quad \psi_N = \psi_f,$$

$$\bar{\psi}_0 = \bar{\psi}_i, \quad \bar{\psi}_N = \bar{\psi}_f. \tag{5.98}$$

We choose, for simplicity and for future use, the normalization of the path integral measure to be $\tilde{N} = 1$ so that the free fermion path integral takes the form

$$Z[0] = e^{\frac{i\omega T}{2}} \, e^{(e^{-i\omega T}\bar{\psi}_f\psi_i - \bar{\psi}_f\psi_f)}. \tag{5.99}$$

5.6 References

F. Berezin, *The Method of Second Quantization*, Academic Press.

B. S. DeWitt, *Supermanifolds*, Cambridge University Press.

CHAPTER 6

Supersymmetry

6.1 Supersymmetric oscillator

We have seen in Chapters 3 and 5 that a bosonic oscillator in one dimension with a natural frequency ω is described by the Hamiltonian

$$H_B = \frac{\omega}{2} \left(a_B^\dagger a_B + a_B a_B^\dagger \right) = \omega \left(a_B^\dagger a_B + \frac{1}{2} \right), \tag{6.1}$$

while a fermionic oscillator with a natural frequency ω is described by the Hamiltonian

$$H_F = \frac{\omega}{2} \left(a_F^\dagger a_F - a_F a_F^\dagger \right) = \omega \left(a_F^\dagger a_F - \frac{1}{2} \right). \tag{6.2}$$

Here we are assuming that $\hbar = 1$ (see Eqs. (5.2) and (5.3)). The creation and the annihilation operators for the bosonic oscillator satisfy the commutation relations

$$\left[a_B, a_B^\dagger \right] = 1, \tag{6.3}$$

with all others vanishing. For the fermionic oscillator, on the other hand, the creation and the annihilation operators satisfy the anti-commutation relations (see Eq. (5.6))

$$[a_F, a_F]_+ = 0 = \left[a_F^\dagger, a_F^\dagger \right]_+,$$

$$\left[a_F, a_F^\dagger \right]_+ = 1. \tag{6.4}$$

Let us note here (as we have pointed out earlier in Chapter 5) that the ground state energy for the bosonic oscillator is $\frac{\omega}{2}$ whereas that for the fermionic oscillator is $-\frac{\omega}{2}$.

Let us next consider a system consisting of a bosonic and a fermionic oscillator with the same natural frequency ω. This is known as the supersymmetric oscillator and provides the simplest example of a class of theories known as supersymmetric theories. The Hamiltonian for this system follows from Eqs. (6.1) and (6.2) to be

$$H = H_B + H_F = \frac{\omega}{2}\left(a_B^\dagger a_B + a_B a_B^\dagger + a_F^\dagger a_F - a_F a_F^\dagger\right)$$

$$= \omega\left(a_B^\dagger a_B + \frac{1}{2} + a_F^\dagger a_F - \frac{1}{2}\right)$$

$$= \omega\left(a_B^\dagger a_B + a_F^\dagger a_F\right). \tag{6.5}$$

We note from Eq. (6.5) that the constant term in the Hamiltonian for this system has cancelled out suggesting a vanishing zero point energy for this system. If we define the number operators for the bosonic and the fermionic oscillators as

$$N_B = a_B^\dagger a_B \,,$$

$$N_F = a_F^\dagger a_F \,, \tag{6.6}$$

then, we can write the Hamiltonian for the system also as

$$H = \omega(N_B + N_F) \,, \tag{6.7}$$

where N_B and N_F commute (since the bosonic creation and annihilation operators commute with those for the fermionic system).

It is clear from Eq. (6.7) that the energy eigenstates of the system will be a product of the eigenstates of the number operators N_B and N_F. Consequently, let us define

$$|n_B, n_F\rangle = |n_B\rangle \otimes |n_F\rangle \,, \tag{6.8}$$

where

$$N_B|n_B\rangle = n_B|n_B\rangle, \qquad n_B = 0, 1, 2, \ldots,$$

$$N_F|n_F\rangle = n_F|n_F\rangle, \qquad n_F = 0, 1. \tag{6.9}$$

Here we are using our earlier result in Eq. (5.12) that the eigenvalues for the fermion number operator are 0 or 1 consistent with the Pauli principle while the eigenvalues for the bosonic number operator can take any positive semidefinite integer value. From Eqs. (6.7), (6.8) and (6.9) we note that the energy eigenvalues for the supersymmetric oscillator are given by

$$H|n_B, n_F\rangle = E_{n_B, n_F}|n_B, n_F\rangle = \omega(n_B + n_F)|n_B, n_F\rangle, \tag{6.10}$$

with $n_B = 0, 1, 2\ldots$ and $n_F = 0, 1$.

We note explicitly from Eq. (6.10) that the ground state energy of the supersymmetric oscillator vanishes, namely,

$$E_{0,0} = 0, \tag{6.11}$$

and the ground state (vacuum) of the theory is assumed to satisfy

$$a_B|0, 0\rangle = 0 = a_F|0, 0\rangle. \tag{6.12}$$

The vanishing of the ground state energy is a general feature of supersymmetric theories and as we will see shortly it is a consequence of the fact that such theories are invariant under symmetry transformations (known as supersymmetry transformations) which transform bosonic variables into fermionic ones and *vice versa*. We also observe from Eq. (6.10) that except for the ground state, all other energy eigenstates of the system are doubly degenerate. Namely, the states $|n_B, 1\rangle$ and $|n_B + 1, 0\rangle$ have the same energy for any value of n_B. The degeneracy in the energy value for a bosonic ($n_F = 0$) and a fermionic ($n_F = 1$) state, as we will see, is again a consequence of the supersymmetry of the system.

Let us next consider the following two fermionic operators in the theory

$$Q = a_B^\dagger a_F \,,$$

$$\bar{Q} = a_F^\dagger a_B, \qquad \bar{Q} = Q^\dagger. \tag{6.13}$$

We can show using the commutation relations in Eqs. (6.3) and (6.4) that (the bosonic operators commute with the fermionic ones)

$$
\begin{aligned}
[Q, H] &= \left[a_B^\dagger a_F, \omega(a_B^\dagger a_B + a_F^\dagger a_F) \right] \\
&= \omega \left(a_B^\dagger \left[a_B^\dagger, a_B \right] a_F + a_B^\dagger \left[a_F, a_F^\dagger \right]_+ a_F \right) \\
&= \omega \left(-a_B^\dagger a_F + a_B^\dagger a_F \right) \\
&= 0 \,,
\end{aligned}
\tag{6.14}
$$

and similarly,

$$[\bar{Q}, H] = \left[a_F^\dagger a_B, \omega(a_B^\dagger a_B + a_F^\dagger a_F) \right] = 0 \,. \tag{6.15}$$

The two operators, Q and $\bar{Q}$, define conserved quantities (charges) of the system and would correspond to generators of infinitesimal symmetry transformations in the theory. However, since the generators are fermionic in nature, the symmetries which they would generate will correspond to fermionic symmetries. In fact, using the (anti) commutation relations, it can be checked that these charges rotate the bosonic operators into fermionic ones and *vice versa*. We also note that

$$
\begin{aligned}
[Q, \bar{Q}]_+ &= \left[a_B^\dagger a_F, a_F^\dagger a_B \right]_+ \\
&= a_B^\dagger \left[a_F, a_F^\dagger \right]_+ a_B - a_F^\dagger \left[a_B^\dagger, a_B \right] a_F \\
&= a_B^\dagger a_B + a_F^\dagger a_F \\
&= \frac{1}{\omega} H \,.
\end{aligned}
\tag{6.16}
$$

Thus, we see from Eqs. (6.14), (6.15) and (6.16) that the operators Q, $\bar{Q}$ and H define a closed algebra which involves both commutators and anti-commutators. Such an algebra is known as a graded Lie algebra and defines the infinitesimal form of the supersymmetry transformations. An immediate consequence of the supersymmetry algebra is that if the ground state of the theory is invariant under supersymmetry transformations, namely, if (see Eq. (6.12), for simplicity, we identify $|0\rangle \equiv |0,0\rangle$)

$$Q|0\rangle = 0 = \bar{Q}|0\rangle\,, \tag{6.17}$$

then it follows from Eq. (6.16) that

$$\langle 0|H|0\rangle = \omega\,\langle 0|Q\bar{Q} + \bar{Q}Q|0\rangle = 0\,. \tag{6.18}$$

Namely, the ground state energy in a supersymmetric theory vanishes as a consequence of the supersymmetry algebra. Furthermore, we note from Eq. (6.13) that $\bar{Q}$ is really the Hermitian conjugate of Q. Consequently, it follows from Eq. (6.16) that in a supersymmetric theory H is really a positive semidefinite operator and, therefore, its expectation value in any state must be positive semidefinite.

Let us next analyze the effect of Q and $\bar{Q}$ on the energy eigenstates of the system. We note from the commutation rules of the theory that

$$[Q, N_B] = \left[a_B^\dagger a_F, a_B^\dagger a_B\right]$$

$$= a_B^\dagger \left[a_B^\dagger, a_B\right] a_F$$

$$= -a_B^\dagger a_F = -Q\,,$$

$$[Q, N_F] = \left[a_B^\dagger a_F, a_F^\dagger a_F\right]$$

$$= a_B^\dagger \left[a_F, a_F^\dagger\right]_+ a_F$$

$$- a_B^\dagger a_F - Q\,, \tag{6.19}$$

and similarly

$$[\bar{Q}, N_B] = \bar{Q} \,,$$

$$[\bar{Q}, N_F] = -\bar{Q} \,. \tag{6.20}$$

In other words, we can think of Q as raising the bosonic number n_B while lowering the fermionic number n_F by one unit whereas $\bar{Q}$ does the opposite. It now follows that for

$$|n_B, n_F\rangle = \frac{\left(a_B^\dagger\right)^{n_B}}{\sqrt{n_B!}} \left(a_F^\dagger\right)^{n_F} |0\rangle \,, \tag{6.21}$$

where we recognize $n_F = 0, 1$ and $n_B = 0, 1, 2, \ldots$, we have

$$Q|n_B, n_F\rangle = \begin{cases} 0 & \text{if } n_F = 0 \,, \\ \sqrt{n_B + 1}|n_B + 1, 0\rangle & \text{if } n_F = 1 \,, \end{cases}$$

$$\bar{Q}|n_B, n_F\rangle = \begin{cases} \sqrt{n_B}|n_B - 1, 1\rangle & \text{if } n_F = 0 \,, \\ 0 & \text{if } n_F = 1 \,. \end{cases} \tag{6.22}$$

Namely, we note that acting on any state other than the ground state, the operators Q and $\bar{Q}$ take a bosonic state (with $n_F = 0$) to a fermionic state (with $n_F = 1$) and *vice versa*. This is the manifestation of supersymmetry on the states in the Hilbert space of the Hamiltonian, i.e., the bosonic and the fermionic states are paired. Furthermore, since Q and $\bar{Q}$ commute with the Hamiltonian of the system (see Eqs (6.14) and (6.15)), it now follows that such paired states (namely, $|n_B, n_F\rangle$ and $Q|n_B, n_F\rangle$ or $|n_B, n_F\rangle$ and $\bar{Q}|n_B, n_F\rangle$) will be degenerate in energy except for the ground state. Namely

$$H(Q|n_B, n_F\rangle) = Q(H|n_B, n_F\rangle) = E_{n_B, n_F}(Q|n_B, n_F\rangle) \,,$$

$$H(\bar{Q}|n_B, n_F\rangle) = \bar{Q}(H|n_B, n_F\rangle) = E_{n_B, n_F}(\bar{Q}|n_B, n_F\rangle) \,. \tag{6.23}$$

The supersymmetric oscillator is the simplest example of supersymmetric theories. The concept of supersymmetry and graded Lie algebras generalizes to other cases as well and there exist many useful realizations of these algebras in the context of field theories.

6.2 Supersymmetric quantum mechanics

Let us next study a general supersymmetric, quantum mechanical theory in one dimension. From our discussion in the last section, we note that supersymmetry necessarily involves both bosons and fermions and, therefore, let us consider a general classical Lagrangian of the form

$$L = \frac{1}{2}\dot{x}^2 - \frac{1}{2}(W(x))^2 + i\bar{\psi}\dot{\psi} - \frac{1}{2}W'(x)[\bar{\psi}, \psi]. \tag{6.24}$$

Here, for consistency with earlier discussions we have set $m = 1$ for the bosonic part of the Lagrangian. We also note here that $W(x)$ can be any chosen monomial of x at this point. It is clear from Eq. (6.24) that when

$$W(x) = \omega x, \tag{6.25}$$

the Lagrangian of Eq. (6.24) reduces to that of a supersymmetric oscillator discussed in the last section.

In general, we note that under the infinitesimal transformations

$$\delta_\epsilon x = \frac{1}{\sqrt{2}}\,\bar{\psi}\epsilon,$$

$$\delta_\epsilon \psi = -\frac{i}{\sqrt{2}}\,\dot{x}\epsilon - \frac{1}{\sqrt{2}}\,W(x)\epsilon,$$

$$\delta_\epsilon \bar{\psi} = 0, \tag{6.26}$$

and

$$\delta_{\bar{\epsilon}} x = \frac{1}{\sqrt{2}}\,\bar{\epsilon}\psi,$$

$$\delta_{\bar{\epsilon}} \psi = 0,$$

$$\delta_{\bar{\epsilon}} \bar{\psi} = \frac{i}{\sqrt{2}}\,\dot{x}\bar{\epsilon} - \frac{1}{\sqrt{2}}\,W(x)\bar{\epsilon}, \tag{6.27}$$

where ϵ and $\bar{\epsilon}$ are constant infinitesimal Grassmann parameters, the action for the Lagrangian remains unchanged. (Note that since

$\epsilon^2 = 0 = \bar{\epsilon}^2$, the qualification infinitesimal is redundant.) In other words, the transformations in Eqs. (6.26) and (6.27) define global symmetries of the system. (See Chapter 11 for a detailed discussion of symmetries.) These symmetry transformations (Eqs. (6.26) and (6.27)) mix up the bosonic and the fermionic variables of the theory and, therefore, are reminiscent of the supersymmetry transformations which we discussed earlier. In fact, one can explicitly show that the two sets of transformations in Eqs. (6.26) and (6.27) are generated respectively by the two supersymmetric charges Q and $\bar{Q}$ in the theory (which can be constructed through the Noether procedure discussed in Chapter 11.) For completeness we simply note here that in the operator language, the two supersymmetric charges can be written as (compare with Eq. (6.13))

$$Q = A^\dagger \psi = \frac{1}{\sqrt{2}} \left(p + iW(x) \right) \psi,$$

$$\bar{Q} = A\bar{\psi} = \frac{1}{\sqrt{2}} \left(p - iW(x) \right) \bar{\psi}, \qquad (6.28)$$

which, of course, reduce to Eq. (6.13) when $W(x) = \omega x$ (up to a normalization, with $\psi = a_F, \bar{\psi} = a_F^\dagger$). Note also that a realization for the fermionic anti-commutation relations in Eq. (6.4) (with $\psi = a_F, \bar{\psi} = a_F^\dagger$) is provided by the matrices (with $\sigma_\pm = (\sigma_1 \pm i\sigma_2)/2$)

$$\psi = \sigma_+, \qquad \bar{\psi} = \sigma_-, \qquad \sigma_+ = \begin{pmatrix} 0 & 1 \\ 0 & 0 \end{pmatrix}, \qquad \sigma_- = \begin{pmatrix} 0 & 0 \\ 1 & 0 \end{pmatrix}. \quad (6.29)$$

In this case, the supersymmetric Hamiltonian can be written as

$$H = \left[Q, \bar{Q} \right]_+$$

$$= \frac{1}{2} \left(p^2 + (W(x))^2 - W'(x)\sigma_3 \right), \qquad (6.30)$$

and we can identify the two diagonal elements as the supersymmetric partner Hamiltonians

$$H_- = \frac{1}{2}\left(p^2 + (W(x))^2 - W'(x)\right) = A^\dagger A,$$

$$H_+ = \frac{1}{2}\left(p^2 + (W(x))^2 + W'(x)\right) = AA^\dagger, \qquad (6.31)$$

where $W'(x)$ denotes the derivative of the superpotential $W(x)$ with respect to x.

We note here, without going into detail, that while the Lagrangian in Eq. (6.24) is supersymmetric for any monomial $W(x)$, in the case of even monomials the presence of instantons breaks supersymmetry. (Instantons are discussed in Chapter 8.) Consequently, let us consider monomials only of the form.

$$W(x) \sim x^{2n+1}, \qquad n = 0, 1, 2, \ldots . \qquad (6.32)$$

With these preparations, let us next look at the generating functional for the supersymmetric quantum mechanical theory in Eq. (6.24) (recall that $\hbar = 1$)

$$Z = N \int \mathcal{D}\bar{\psi}\mathcal{D}\psi\mathcal{D}x \; e^{iS[x,\psi,\bar{\psi}]} . \qquad (6.33)$$

As we have seen in the last section, the spectrum of a supersymmetric theory has many interesting features. Correspondingly, the generating functional for such a theory is also quite interesting. In particular, let us note from Eq. (6.24) that since the Lagrangian is quadratic in the fermionic variables, the functional integral for these variables can be done easily using our results in Chapter 5. (See Eqs. (5.43) and (5.95).) Thus, we can write the fermion part of the path integral as

$$\int \mathcal{D}\bar{\psi}\mathcal{D}\psi \; e^{i\int dt(\bar{\psi}(i\frac{d}{dt} - W'(x))\psi)} = N' \det\left(i\frac{d}{dt} - W'(x)\right) . \qquad (6.34)$$

Substituting Eq. (6.34) into Eq. (6.33), we obtain

$$Z = N \int \mathcal{D}\bar{\psi}\mathcal{D}\psi\mathcal{D}x \; e^{iS[x,\psi,\bar{\psi}]}$$

$$= N \int \mathcal{D}x \; e^{i \int dt(\frac{1}{2}\dot{x}^2 - \frac{1}{2}(W(x))^2)} \int \mathcal{D}\bar{\psi}\mathcal{D}\psi \; e^{i \int dt(\bar{\psi}(i\frac{d}{dt} - W'(x))\psi)}$$

$$= \tilde{N} \int \mathcal{D}x \; \det\left(i\frac{d}{dt} - W'(x)\right) \; e^{i \int dt(\frac{1}{2}\dot{x}^2 - \frac{1}{2}(W(x))^2)} \,. \qquad (6.35)$$

Let us next note that if we define a new bosonic variable through the relation

$$\rho = i\dot{x} - W(x) \,, \qquad\qquad\qquad\qquad (6.36)$$

then the Jacobian for this change of variables in Eq. (6.36) will be given by

$$J = \left[\det\left(i\frac{d}{dt} - W'(x)\right)\right]^{-1} \,. \qquad\qquad (6.37)$$

This is precisely the inverse of the determinant in Eq. (6.35). Furthermore, we note that

$$\int dt \, \rho^2 = \int dt \, (i\dot{x} - W(x))^2$$

$$= \int dt \left(-\dot{x}^2 - 2i\dot{x}W(x) + (W(x))^2\right)$$

$$= -\int dt \left(\dot{x}^2 - (W(x))^2\right) - 2i \int_{-\infty}^{\infty} dx \, W(x)$$

$$= -\int dt \left(\dot{x}^2 - (W(x))^2\right) \,. \qquad\qquad (6.38)$$

Here we have used the fact that for $W(x)$ of the form in Eq. (6.32), the last integral vanishes. (For even monomials, on the other hand, this does not vanish giving the contribution due to the instantons which breaks supersymmetry.) Substituting Eqs. (6.36), (6.37) and (6.38) into the generating functional in Eq. (6.35), we find

$$Z = \tilde{N} \int \mathcal{D}\rho \, e^{-\frac{i}{2} \int dt \, \rho^2} \,. \tag{6.39}$$

In other words, we see that the generating functional for a super-symmetric theory can be redefined to have the form of a free bosonic generating functional. This is known as the Nicolai map (namely, Eq. (6.36)) and generalizes to field theories in higher dimensions as well.

6.3 Shape invariance

As we have noted earlier, there are only a handful of quantum mechanical systems which can be solved analytically. The solubility of such systems now appears to be related to a special symmetry associated with these systems known as shape invariance. This symmetry is also quite useful in the evaluation of the path integrals for such systems.

Let us consider a one dimensional quantum mechanical system described by the Hamiltonian (we assume $\hbar = m = 1$)

$$H_- = \frac{p^2}{2} + V_-(x) \,. \tag{6.40}$$

If we assume the ground state of the system to have vanishing energy, then we can write the Hamiltonian in Eq. (6.40) also in the factorized form

$$H_- = A^\dagger A \,, \tag{6.41}$$

where (these have the same forms as the operators A, $A^\dagger$ in Eq. (6.28))

$$A = \frac{1}{\sqrt{2}}(p - iW(x)) \,,$$
$$A^\dagger = \frac{1}{\sqrt{2}}(p + iW(x)) \,, \tag{6.42}$$

and we identify ($W(x)$ is the superpotential)

$$V_-(x) = \frac{1}{2}(W^2(x) - W'(x)).$$ (6.43)

Such a relation is known as a Riccati relation. The ground state of the theory is assumed to satisfy (since the ground state energy vanishes)

$$A|\psi_0\rangle = 0.$$ (6.44)

We note that A and $A^\dagger$ are Hermitian conjugates of each other and that given these two operators, we can construct a second Hermitian Hamiltonian of the form

$$H_+ = AA^\dagger = \frac{p^2}{2} + V_+(x) = \frac{p^2}{2} + \frac{1}{2}(W^2(x) + W'(x)).$$ (6.45)

It now follows that if $|\psi\rangle$ is any eigenstate of the Hamiltonian H_- other than the ground state, namely, if

$$H_-|\psi\rangle = A^\dagger A|\psi\rangle = \lambda|\psi\rangle, \qquad \lambda \neq 0,$$ (6.46)

then

$$AH_-|\psi\rangle = \lambda(A|\psi\rangle),$$

$$\text{or,} \quad AA^\dagger(A|\psi\rangle) = \lambda(A|\psi\rangle),$$

$$\text{or,} \quad H_+(A|\psi\rangle) = \lambda(A|\psi\rangle).$$ (6.47)

Namely, we note that the two Hamiltonians, H_- and H_+, are almost isospectral in the sense that they share the same energy spectrum except for the ground state energy of H_-. (These are the supersymmetric partner Hamiltonians we constructed in Eq. (6.31).)

The potential, of course, depends on some parameters such as the coupling constants. However, if the potential of the theory is such that we can write

$$V_+(x, a_0) = V_-(x, a_1) + R(a_1),$$ (6.48)

with $R(a_1)$ a constant and the parameters a_0 and a_1 satisfying a known functional relationship

$$a_1 = f(a_0),\qquad(6.49)$$

then we say that the potential is shape invariant. (Namely, the potential for the partner Hamiltonian has the same form or shape as the original Hamiltonian with modified parameters and a possible shift.) In such a case, we can write using Eq. (6.48)

$$H_-(a_0) = A^\dagger(a_0)A(a_0) = \frac{p^2}{2} + V_-(x, a_0),$$

$$H_+(a_0) = A(a_0)A^\dagger(a_0) = \frac{p^2}{2} + V_+(x, a_0)$$

$$= \frac{p^2}{2} + V_-(x, a_1) + R(a_1)$$

$$= H_-(a_1) + R(a_1)$$

$$= A^\dagger(a_1)A(a_1) + R(a_1).\qquad(6.50)$$

Since the ground state energy of H_- vanishes and since we know that $H_+(a_0)$ and $H_-(a_0)$ are almost isospectral, it follows now from Eq. (6.50) that the energy value for the first excited state of $H_-(a_0)$ must be

$$E_1 = R(a_1).\qquad(6.51)$$

It is also easy to see now that for a shape invariant potential, we can construct a sequence of Hamiltonians such as

$$H^{(0)} = H_-(a_0),$$

$$H^{(1)} = H_+(a_0) = H_-(a_1) + R(a_1),$$

$$H^{(2)} = H_+(a_1) + R(a_1) = H_-(a_2) + R(a_1) + R(a_2),$$

$$\vdots\ =\ \vdots$$

$$H^{(s)} = H_+(a_{s-1}) + \sum_{k=1}^{s-1} R(a_k)$$

$$= H_-(a_s) + \sum_{k=1}^{s} R(a_k) . \tag{6.52}$$

Here we have identified

$$a_s = f^s(a_0) = f(f \ldots (f(a_0)) \ldots) . \tag{6.53}$$

Any two consecutive Hamiltonians in Eq. (6.52) will be almost isospectral and from this, with a little bit of analysis, we can determine the energy levels of $H_-(a_0)$ to be

$$E_n = \sum_{k=1}^{n} R(a_k) . \tag{6.54}$$

We emphasize here that supersymmetry allows us to construct only a pair of partner Hamiltonians which are almost isospectral, but does not determine the energy levels. However, combined with shape invariance, supersymmetry leads to a sequence of Hamiltonians where any consecutive pair is almost isospectral and, in the process, determines the energy levels.

Given the sequence of Hamiltonians in Eq. (6.52) we can write down the relation

$$A(a_s)A^\dagger(a_s) = A^\dagger(a_{s+1})A(a_{s+1}) + R(a_{s+1})$$

$$\text{or,} \quad A(a_s)H^{(s)} = H^{(s+1)}A(a_s) . \tag{6.55}$$

This defines a recursion relation between the sequence of Hamiltonians. Furthermore, for $t > 0$, defining the time evolution operator for a particular Hamiltonian $H^{(s)}$ in the sequence to be

$$U^{(s)} = e^{-itH^{(s)}} , \tag{6.56}$$

we note that Eq. (6.55) gives

$$A(a_s)U^{(s)} = A(a_s)e^{-itH^{(s)}}$$

$$= e^{-itH^{(s+1)}}A(a_s) = U^{(s+1)}A(a_s). \qquad (6.57)$$

Similarly, by taking the time derivative of Eq. (6.56), we obtain

$$\frac{\partial U^{(s)}}{\partial t} = -iH^{(s)}e^{-itH^{(s)}}$$

$$= -i\left(A^\dagger(a_s)A(a_s) + \sum_{k=1}^{s} R(a_k)\right)e^{-itH^{(s)}}$$

$$= -i\sum_{k=1}^{s} R(a_k)U^{(s)} - iA^\dagger(a_s)U^{(s+1)}A(a_s),$$

or, $$\left(\frac{\partial}{\partial t} + i\sum_{k=1}^{s} R(a_k)\right)U^{(s)} = -iA^\dagger(a_s)U^{(s+1)}A(a_s). \qquad (6.58)$$

The relations in Eqs. (6.57) and (6.58) define recursion relations for the time evolution operator and have the coordinate representation of the form

$$\left(\frac{\partial}{\partial x} + W(x, a_s)\right)U^{(s)}(x, y; t)$$

$$= -\left(\frac{\partial}{\partial y} - W(y, a_s)\right)U^{(s+1)}(x, y; t),$$

$$\left(\frac{\partial}{\partial t} + i\sum_{k=1}^{s} R(a_k)\right)U^{(s)}(x, y; t)$$

$$= -\frac{i}{2}\left(\frac{\partial}{\partial x} - W(x, a_s)\right)\left(\frac{\partial}{\partial y} - W(y, a_s)\right)U^{(s+1)}(x, y; t).$$

$$(6.59)$$

It is clear from this discussion that, for a shape invariant potential, if one of the Hamiltonians in the sequence coincides with a system which we can solve exactly, then the energy levels of the original system can be determined. Furthermore, using the recursion relations in Eq. (6.59), we can solve for the time evolution operator of the original system. This will determine the path integral for the system.

6.4 Example

Let us consider a quantum mechanical system with

$$W(x, a_0) = a_0 \tanh x, \qquad a_0 = 1.\tag{6.60}$$

From Eqs. (6.43) and (6.45), we find

$$
\begin{aligned}
V_-(x, a_0) &= \frac{1}{2}\left(W^2(x, a_0) - W'(x, a_0)\right) \\
&= \frac{1}{2}\left(a_0^2 \tanh^2 x - a_0 \operatorname{sech}^2 x\right) \\
&= \frac{1}{2}\left(a_0^2 - a_0(a_0 + 1)\operatorname{sech}^2 x\right) \\
&= \frac{1}{2} - \operatorname{sech}^2 x\,, \\
V_+(x, a_0) &= \frac{1}{2}\left(W^2(x, a_0) + W'(x, a_0)\right) \\
&= \frac{1}{2}\left(a_0^2 \tanh^2 x + a_0 \operatorname{sech}^2 x\right) \\
&= \frac{1}{2}\left(a_0^2 - a_0(a_0 - 1)\operatorname{sech}^2 x\right) \\
&= \frac{1}{2} = \frac{1}{2}\left(a_1^2 - a_1(a_1 + 1)\operatorname{sech}^2 x\right) + R(a_1)\,.\tag{6.61}
\end{aligned}
$$

In this case, therefore, we can identify

$$a_0 = 1, \quad a_1 = a_0 - 1 = 0, \quad R(a_1) = \frac{2a_1 + 1}{2} = \frac{1}{2}\,.\tag{6.62}$$

We see that the original Hamiltonian for the system, in this case, is

$$H^{(0)} = H_-(a_0) = \frac{p^2}{2} - \text{sech}^2 x + \frac{1}{2}, \tag{6.63}$$

and the next Hamiltonian in the sequence is given by (see Eq. (6.52))

$$H^{(1)} = H_+(a_0) = H_-(a_1) + R(a_1) = \frac{p^2}{2} + \frac{1}{2}. \tag{6.64}$$

This is, of course, the free particle Hamiltonian for which we know the transition amplitude to be (see Eq. (2.49))

$$U^{(1)}(x, y; t) = \frac{1}{\sqrt{2\pi i t}} e^{i(\frac{(x-y)^2}{2t} - \frac{t}{2})}. \tag{6.65}$$

Substituting Eq. (6.65) into Eq. (6.59), it is easy to see that

$$U^{(0)}(x, y; t) = \frac{1}{2} \text{sech} x \ \text{sech} y$$
$$- \frac{1}{2\pi} \int_{-\infty}^{\infty} dk \ \frac{(ik - \tanh x)(ik + \tanh y)}{1 + k^2} e^{(ik(x-y) - \frac{it}{2}(k^2+1))}.$$

This determines the transition amplitude for the original system.

We comment here that the $\text{sech}^2 x$ potentials (soliton potentials) are known to be reflectionless. The connection of this Hamiltonian with the free Hamiltonian, through shape invariance, gives an intuitive reason for this. Furthermore, this derivation also works for n-soliton potentials where $a_0 = n$, a positive integer. In this case, the parameter is reduced by 1 at every order, ultimately vanishing to give the free particle Hamiltonian.

6.5 Supersymmetry and singular potentials

Singular potentials within the context of quantum mechanics are interesting because they remind us of the necessity of regularization even in such simple systems. We know from studies in relativistic quantum field theories that a regularization must always be chosen consistent with the symmetries of the theory under study in order to be able to extract meaningful results. An improper choice of regularization can lead to incorrect conclusions about the theory. In simple quantum mechanical systems, a careful analysis of singular systems using a regularization have led to the working rule that the quantum mechanical wave function must vanish at points where the potential becomes infinite. It must be emphasized that such a condition should not be thought of as a boundary condition, rather it is a consistency condition which arises from treating the singular potential in a carefully regularized manner.

In going beyond simple quantum mechanical systems, however, such a working rule (boundary condition) should not be blindly imposed which can lead to erroneous conclusions. Singular potentials within the context of supersymmetric quantum mechanical systems provide an excellent example of this and we will discuss this in this section. This will also bring out clearly how choosing a regularization consistent with the symmetries of the system under study is important. We will study a simple model, the super-symmetric "half" oscillator, to bring out these features. Other quantum mechanical systems with a complex singularity structure can be found in the literature.

To understand the supersymmetric "half" oscillator or the supersymmetric oscillator on the half line, it is useful to recapitulate briefly the results of the "half" oscillator. Let us consider a particle moving in the potential (the zero point energy has been subtracted)

$$
V(x) = \begin{cases} \frac{1}{2}(\omega^2 x^2 - \omega), & \text{for} \quad x > 0, \\ \infty, & \text{for} \quad x < 0. \end{cases} \tag{6.66}
$$

The spectrum of this potential is quite clear intuitively. Namely, because of the infinite barrier, we expect the wave function to vanish at the origin leading to the conclusion that, of all the solutions of the oscillator on the full line, only the odd solutions (of course, on the "half" line there is no notion of even and odd) would survive in this case. While this is quite obvious, let us analyze the problem systematically for later purpose.

First, let us note that singular potentials are best studied in a regularized manner because this is the only way that appropriate boundary conditions can be determined correctly. Therefore, let us consider the particle moving in the regularized potential

$$V(x) = \begin{cases} \frac{1}{2}(\omega^2 x^2 - \omega), & \text{for} \quad x > 0 \\ \frac{c^2}{2}, & \text{for} \quad x < 0, \end{cases} \tag{6.67}$$

with the understanding that the limit $|c| \to \infty$ is to be taken only at the end. The Schrödinger equation ($\hbar = m = 1$)

$$\left(-\frac{1}{2}\frac{d^2}{dx^2} + V(x)\right)\psi(x) = \epsilon\psi(x), \tag{6.68}$$

can now be solved in the two regions. Since $|c| \to \infty$ at the end, for any finite energy solution, we have the asymptotically damped solution, for $x < 0$, (A is a constant)

$$\psi^{(II)}(x) = A\,e^{(c^2 - 2\epsilon)^{\frac{1}{2}}x}. \tag{6.69}$$

Since the system no longer has reflection symmetry, the solutions, in the region $x > 0$, cannot be classified into even and odd solutions. Rather, the normalizable (physical) solution would correspond to one which vanishes asymptotically. The solutions of the Schrödinger equation (Eq. (6.68)), in the region $x > 0$, are given by the parabolic cylinder functions and the asymptotically damped physical solution is given by (B is a constant)

$$\psi^{(I)}(x) = B\,U\left(-\left(\frac{\epsilon}{\omega} + \frac{1}{2}\right), \sqrt{2\omega}\,x\right). \tag{6.70}$$

The parabolic cylinder function, $U(a, x)$ vanishes for large values of x. For small values of x, it satisfies

$$U(a, x) \stackrel{x \to 0}{\Longrightarrow} \frac{\sqrt{\pi}}{2^{\frac{1}{4}(2a+1)}\Gamma(\frac{3}{4} + \frac{a}{2})},$$

$$U'(a, x) \stackrel{x \to 0}{\Longrightarrow} -\frac{\sqrt{\pi}}{2^{\frac{1}{4}(2a-1)}\Gamma(\frac{1}{4} + \frac{a}{2})}. \tag{6.71}$$

It is now straightforward to match the solutions in Eqs. (6.69), (6.70) and their first derivatives across the boundary at $x = 0$ and their ratio gives

$$\frac{1}{\sqrt{c^2 - 2\epsilon}} = -\frac{1}{2\sqrt{\omega}} \frac{\Gamma(-\frac{\epsilon}{2\omega})}{\Gamma(-\frac{\epsilon}{2\omega} + \frac{1}{2})}. \tag{6.72}$$

It is clear, then, that as $|c| \to \infty$, this can be satisfied only if

$$-\frac{\epsilon}{2\omega} + \frac{1}{2} \xrightarrow{|c| \to \infty} -n, \qquad n = 0, 1, 2, \ldots. \tag{6.73}$$

In other words, when the regularization is removed, the energy levels that survive are the odd ones, namely, (remember that the zero point energy is already subtracted out in (6.66) or (6.67))

$$\epsilon_n = \omega(2n + 1). \tag{6.74}$$

The corresponding physical wave functions are nontrivial only on the half line $x > 0$ and have the form

$$\psi_n(x) = B_n U\left(-\left(2n + \frac{3}{2}\right), \sqrt{2\omega}\, x\right) = \tilde{B}_n\, e^{-\frac{1}{2}\omega x^2}\, H_{2n+1}(\sqrt{\omega}\, x). \tag{6.75}$$

Namely, only the odd Hermite polynomials survive leading to the fact that the wave function vanishes at $x = 0$. Thus, we see that the correct boundary condition naturally arises from regularizing the singular potential and studying the problem systematically.

We now turn to the analysis of the supersymmetric oscillator on the half line. One can define a superpotential

$$W(x) = \begin{cases} \omega x, & \text{for } x > 0, \\ \infty, & \text{for } x < 0, \end{cases} \tag{6.76}$$

which would, naively, lead to the pair of potentials

$$V_{\mp}(x) = \frac{1}{2}(W^2(x) \mp W'(x))$$
$$= \begin{cases} \frac{1}{2}(\omega^2 x^2 \mp \omega), & \text{for } x > 0, \\ \infty, & \text{for } x < 0. \end{cases} \tag{6.77}$$

Since, this involves singular potentials, we can study it, as before, by regularizing the singular potentials as

$$
V_-(x) = \begin{cases} \frac{1}{2}(\omega^2 x^2 - \omega), & \text{for} \quad x > 0, \\ \frac{c_-^2}{2}, & \text{for} \quad x < 0, \end{cases}
$$

$$
V_+(x) = \begin{cases} \frac{1}{2}(\omega^2 x^2 + \omega), & \text{for} \quad x > 0, \\ \frac{c_+^2}{2}, & \text{for} \quad x < 0, \end{cases} \tag{6.78}
$$

with the understanding that $|c_\pm| \to \infty$ at the end.

The earlier analysis can now be repeated for the pair of potentials in Eq. (6.78). It is straightforward and without going into details, let us simply note the results in this case

$$
\epsilon_{-,n} = \omega(2n + 1), \quad \psi_{-,n}(x) = B_{-,n}\, e^{-\frac{1}{2}\omega x^2}\, H_{2n+1}(\sqrt{\omega}\, x),
$$

$$
\epsilon_{+,n} = 2\omega(n + 1), \quad \psi_{+,n}(x) = B_{+,n}\, e^{-\frac{1}{2}\omega x^2}\, H_{2n+1}(\sqrt{\omega}\, x). \tag{6.79}
$$

Here $n = 0, 1, 2, \ldots$. There are several things to note from this analysis. First, only the odd Hermite polynomials survive as physical solutions since the wave function has to vanish at the origin. This boundary condition arises from a systematic study involving a regularized potential. Second, the energy levels for the supersymmetric pair of Hamiltonians are no longer degenerate (one is odd, the other even). Furthermore, the state with $\epsilon = 0$ no longer belongs to the Hilbert space (since it corresponds to an even Hermite polynomial solution). This leads to the conventional conclusion that supersymmetry is broken in such a case and let us note, in particular, that in such a case, it would appear that the superpartner states do not belong to the physical Hilbert space (Namely, in this case, the supercharge is an odd operator and hence connects even and odd Hermite polynomials. However, the boundary condition selects out only odd Hermite polynomials as belonging to the physical Hilbert space.).

There is absolutely no doubt that the results of the calculation show that supersymmetry is broken in this case. The question that needs to be addressed is whether it is a dynamical property of the system or an artifact of the regularization (and, hence the boundary condition) used. The answer is quite obvious, namely, that supersymmetry is broken mainly because the regularization (and, therefore, the boundary condition) breaks supersymmetry. In other words, for any value of the regularizing parameters, $c_\pm$ (even if $|c_+| = |c_-|$), the pair of potentials in Eq. (6.78) do not define a supersymmetric system and hence the regularization itself breaks supersymmetry. Consequently, the breaking of supersymmetry that results when the regularization is removed cannot be trusted as a dynamical effect.

6.5.1 Regularized superpotential

The proper way to understand the problem is to note that for a supersymmetric system, it is not the potential that is fundamental. Rather, it is the superpotential which gives the pair of supersymmetric potentials through Riccati type relations. It is natural, therefore, to regularize the superpotential which would automatically lead to a pair of regularized potentials which would be supersymmetric for any value of the regularization parameter. Namely, such a regularization will respect supersymmetry and, with such a regularization, it is, then, meaningful to ask if supersymmetry is broken when the regularization parameter is removed at the end. With this in mind, let us look at the regularized superpotential

$$W(x) = \omega x \theta(x) - c\theta(-x).\tag{6.80}$$

Here c is the regularization parameter and we are supposed to take $|c| \to \infty$ at the end. Note that the existence of a normalizable ground state for H_- selects out $c > 0$ (otherwise, the regularization would have broken supersymmetry through instanton effects as we have mentioned earlier).

The regularized superpotential now leads to the pair of regularized supersymmetric potentials

$$V_-(x) = \frac{1}{2} \left[(\omega^2 x^2 - \omega)\theta(x) + c^2\theta(-x) - c\delta(x) \right],$$

$$V_+(x) = \frac{1}{2} \left[(\omega^2 x^2 + \omega)\theta(x) + c^2\theta(-x) + c\delta(x) \right],\tag{6.81}$$

which are supersymmetric for any $c > 0$. Let us note that the difference here from the earlier case where the potentials were directly regularized (see Eq. (6.78)) lies only in the presence of the $\delta(x)$ terms in the potentials. Consequently, the earlier solutions in the regions $x > 0$ and $x < 0$ continue to hold. However, the matching conditions are now different because of the delta function terms. Carefully matching the wave function and the discontinuity of the first derivative across $x = 0$ for each of the wavefunctions and taking their ratio, we obtain the two conditions

$$\frac{1}{(c^2 - 2\epsilon_-)^{1/2} - c} = -\frac{1}{2\sqrt{\omega}} \frac{\Gamma(-\frac{\epsilon_-}{2\omega})}{\Gamma(-\frac{\epsilon_-}{2\omega} + \frac{1}{2})}, \tag{6.82}$$

$$\frac{1}{(c^2 - 2\epsilon_+)^{1/2} + c} = -\frac{1}{2\sqrt{\omega}} \frac{\Gamma(-\frac{\epsilon_+}{2\omega} + \frac{1}{2})}{\Gamma(-\frac{\epsilon_+}{2\omega} + 1)}. \tag{6.83}$$

It is now clear that, as $c \to \infty$, (6.82) and (6.83) give respectively

$$\epsilon_{-,n} = 2\omega n, \quad n = 0, 1, 2, \dots,$$

$$\epsilon_{+,n} = 2\omega(n + 1). \tag{6.84}$$

The corresponding wave functions, in this case, have the forms

$$\psi_{-,n}(x) = B_{-,n} \, e^{-\frac{1}{2}\omega x^2} \, H_{2n}(\sqrt{\omega}\, x),$$

$$\psi_{+,n}(x) = B_{+,n} \, e^{-\frac{1}{2}\omega x^2} \, H_{2n+1}(\sqrt{\omega}\, x). \tag{6.85}$$

This is indeed quite interesting for it shows that the spectrum of H_- contains the ground state with vanishing energy. Furthermore, all the other states of H_- and H_+ are degenerate in energy corresponding to even and odd Hermite polynomials as one would expect from superpartner states. (Basically, the delta function term in V_- softens the divergence at $x = 0$.) Consequently, it is quite clear that if the supersymmetric "half" oscillator is defined carefully by regularizing the superpotential, then, supersymmetry is manifest in the limit of removing the regularization. This should be contrasted with the general belief that supersymmetry is broken in this system (which is a consequence of using boundary conditions or, equivalently, of regularizing the potentials in a manner which violates supersymmetry).

6.5.2 Alternate regularization

Of course, we should worry at this point as to how regularization independent our conclusion really is. Namely, our results appear to follow from the matching conditions in the presence of singular delta potential terms and, consequently, it is worth investigating whether our conclusions would continue to hold with an alternate regularization of the superpotential which would not introduce such singular terms to the potentials. With this in mind, let us choose a regularized superpotential of the form

$$W(x) = \omega x \theta(x) + \lambda x \theta(-x) \,. \tag{6.86}$$

Here λ is the regularization parameter and we are to take the limit $|\lambda| \to \infty$ at the end. Once again, we note that, although both signs of λ appear to be allowed, existence of a normalizable ground state would select $\lambda > 0$.

This regularized superpotential would now lead to the pair of supersymmetric potentials of the form

$$V_-(x) = \frac{1}{2} \left[(\omega^2 x^2 - \omega)\theta(x) + (\lambda^2 x^2 - \lambda)\theta(-x) \right] ,$$

$$V_+(x) = \frac{1}{2} \left[(\omega^2 x^2 + \omega)\theta(x) + (\lambda^2 x^2 + \lambda)\theta(-x) \right] . \tag{6.87}$$

There are no singular delta potential terms with this regularization. In fact, the regularization merely introduces a supersymmetric pair of oscillators for $x < 0$ whose frequency is to be taken to infinity at the end.

Since there is a harmonic oscillator potential for both $x > 0$ and $x < 0$, the solutions are straightforward. They are the parabolic cylinder functions which we have mentioned earlier. Now matching the wave function and its first derivative at $x = 0$ for each of the Hamiltonians and taking the ratio, we obtain

$$\frac{1}{\sqrt{\lambda}} \frac{\Gamma(-\frac{\epsilon_-}{2\lambda})}{\Gamma(-\frac{\epsilon_-}{2\lambda} + \frac{1}{2})} = \frac{1}{\sqrt{\omega}} \frac{\Gamma(-\frac{\epsilon_-}{2\omega})}{\Gamma(-\frac{\epsilon_-}{2\omega} + \frac{1}{2})} , \tag{6.88}$$

$$\frac{1}{\sqrt{\lambda}} \frac{\Gamma(-\frac{\epsilon_+}{2\lambda} + \frac{1}{2})}{\Gamma(-\frac{\epsilon_+}{2\lambda} + 1)} = \frac{1}{\sqrt{\omega}} \frac{\Gamma(-\frac{\epsilon_+}{2\omega} + \frac{1}{2})}{\Gamma(-\frac{\epsilon_+}{2\omega} + 1)} . \tag{6.89}$$

It is clear now that, as $\lambda \to \infty$, Eqs. (6.88) and (6.89) give respectively

$$\epsilon_{-,n} = 2\omega n, \quad n = 0, 1, 2, \ldots,$$
$$\epsilon_{+,n} = 2\omega(n + 1). \tag{6.90}$$

The corresponding wave functions are given by

$$\psi_{-,n}(x) = B_{-,n}\, e^{-\frac{1}{2}\omega x^2}\, H_{2n}(\sqrt{\omega}\, x),$$
$$\psi_{+,n}(x) = B_{+,n}\, e^{-\frac{1}{2}\omega x^2}\, H_{2n+1}(\sqrt{\omega}\, x). \tag{6.91}$$

These are, of course, the same energy levels and wave functions as obtained in Eqs. (6.84) and (6.85) respectively showing again that supersymmetry is manifest. Furthermore, this shows that this conclusion is independent of the regularization used as long as the regularization preserves supersymmetry which can be achieved by properly regularizing the superpotential. This analysis can be carried out in a straight forward manner to more complicated superpotentials and the conclusions hold without any change.

6.6 References

A. Das and W. J. Huang, Phys. Rev. **D41**, 3241 (1990).

A. Das, A. Kharev and V. S. Mathur, Phys. Lett. **181**, 299 (1986).

A. Das and S. Pernice, Nucl. Phys. **B505**, 123 (1997).

A. Das and S. Pernice, Nucl. Phys. **561**, 357 (1999).

L. Gendenshtein, JETP Lett. **38**, 356 (1983).

H. Nicolai, Nucl. Phys. **B170**, 419 (1980).

E. Witten, Nucl. Phys. **B188**, 513 (1981).

E. Witten, Nucl. Phys. **B202**, 253 (1982).

Semi-classical methods

7.1 WKB approximation

As we know, most quantum mechanical systems cannot be solved analytically. In such a case, we use perturbation theory (which we have already discussed) and perturbation theory brings out many interesting properties of the system. However, by definition perturbation theory cannot provide information about nonperturbative aspects of the theory. For example, the Born approximations used in scattering theory give more accurate estimates of the scattering amplitudes as we go to higher orders of perturbation, but we cannot obtain information about the bound states of the system from this analysis. Similarly, even though we may be able to obtain the energy levels and the eigenstates for the motion of a particle in a potential well by using perturbation theory, we will never learn about barrier penetration from such an analysis. These are inherently nonperturbative phenomena.

It is, therefore, useful to develop an approximation scheme which brings out some of these nonperturbative characteristics. The WKB method provides such an approximation scheme. The basic idea behind this is quite simple. Let us assume that we have a particle moving in a complicated potential $V(x)$. Then, the stationary states of the system will satisfy the time-independent Schrödinger equation given by

$$\left(-\frac{\hbar^2}{2m}\frac{d^2}{dx^2} + V(x)\right)\psi(x) = E\psi(x).$$ (7.1)

Here E is a constant representing the energy of the state. We know that if the potential were a constant, namely, if

$$V(x) = V = \text{constant}, \tag{7.2}$$

then, the solutions of Eq. (7.1) will be plane waves (for $E > V$) of the form

$$\psi(x) = A\, e^{\pm \frac{i}{\hbar} p x}, \tag{7.3}$$

where

$$p = \sqrt{2m(E - V)}. \tag{7.4}$$

When the potential changes with the coordinate, but changes slowly, then it is easy to convince ourselves that within a region where the potential does not change appreciably, the solutions of Eq. (7.1) can still be written as plane waves of the form of Eq. (7.3) with

$$p(x) = \sqrt{2m(E - V(x))}. \tag{7.5}$$

It is clear, therefore, that we can try a general solution to the time-independent Schrödinger equation Eq. (7.1) of the form

$$\psi(x) = N A(x)\, e^{i B(x)}, \tag{7.6}$$

where N is a normalization constant and furthermore, noting that we can write (for non-negative $A(x)$)

$$A(x) = e^{\ln A(x)}, \tag{7.7}$$

we conclude that the general solution of Eq. (7.1) can be represented as a phase where the phase, in general, is complex. Since the Schrödinger operator depends on $\hbar$, the phase clearly will also be a function of $\hbar$. With all these information, let us write the general solution of the time-independent Schrödinger equation to have the form

$$\psi(x) = N\, e^{\frac{i}{\hbar} \phi(x)}. \tag{7.8}$$

If we substitute this wave function back into the Schrödinger equation
(Eq. (7.1)), we obtain

$$\frac{\mathrm{d}^2\psi(x)}{\mathrm{d}x^2} + \frac{2m}{\hbar^2}(E - V(x))\psi(x) = 0,$$

or, $\left(-\frac{1}{\hbar^2}(\phi'(x))^2 + \frac{i}{\hbar}\phi''(x) + \frac{2m}{\hbar^2}(E - V(x))\right)e^{\frac{i}{\hbar}\phi(x)} = 0,$

or, $\left(\frac{1}{\hbar^2}\left(-(\phi'(x))^2 + 2m(E - V(x))\right) + \frac{i}{\hbar}\phi''(x)\right) = 0.$ (7.9)

So far, everything has been exact. Let us next assume a power
series expansion for $\phi(x)$ of the form (such an expansion is meaningful
since Eq. (7.9) has the structure of a Laurent series in $\hbar$)

$$\phi(x) = \phi_0(x) + \hbar\phi_1(x) + \hbar^2\phi_2(x) + \cdots. \qquad (7.10)$$

(We comment here parenthetically that an expansion in powers of $\hbar$
is very interesting in relativistic quantum field theories and leads to
the loop expansion.) It is clear, then, that $\phi_0(x)$ will correspond to
the classical phase since that is what will survive in the limit $\hbar \to 0$.
Other terms in the series, therefore, represent quantum corrections to
the classical phase. Substituting the power series back into Eq. (7.9),
we obtain

$$\frac{1}{\hbar^2}\left[-(\phi_0'(x) + \hbar\phi_1'(x) + \cdots)^2 + 2m(E - V(x))\right]$$

$$+ \frac{i}{\hbar}\left(\phi_0''(x) + \hbar\phi_1''(x) + \cdots\right) = 0,$$

or, $\frac{1}{\hbar^2}\left(-(\phi_0'(x))^2 + 2m(E - V(x))\right)$

$$+ \frac{1}{\hbar}\left(i\phi_0''(x) - 2\phi_0'(x)\phi_1'(x)\right) + O(\hbar^0) = 0. \qquad (7.11)$$

For this to be true, the coefficients of the individual powers of $\hbar$ in
Eq. (7.11) must be zero. Equating the coefficient of the $\frac{1}{\hbar^2}$ term to
zero, we obtain

$$- (\phi_0'(x))^2 + 2m(E - V(x)) = 0,$$

or, $\quad (\phi_0'(x))^2 = 2m(E - V(x)) = p^2(x),$

or, $\quad \phi_0'(x) = \pm p(x),$

or, $\quad \phi_0(x) = \pm \int_{x_0}^{x} dx' \, p(x').$ \hfill (7.12)

Here x_0 is a reference coordinate and $p(x)$ denotes the momentum of the particle at the point x defined by Eq. (7.5), corresponding to motion with energy E. Furthermore, let us note that even though both the signs of the solution for ϕ_0' are allowed in Eq. (7.12), consistency of the subsequent relations will pick out only the positive sign. (The right and the left moving waves would, then, correspond to $x > x_0$ or $x < x_0$ respectively.) This, therefore, determines the classical phase to be

$$\phi_0(x) = \int_{x_0}^{x} dx' \, p(x').$$ \hfill (7.13)

If we keep only the leading order term in the expansion of $\phi(x)$, then the wave function would have the form

$$\psi(x) = N \, e^{\frac{i}{\hbar}\phi_0(x)} = N \, e^{\frac{i}{\hbar} \int_{x_0}^{x} dx' \, p(x')}.$$ \hfill (7.14)

The time-dependent stationary wave function, in this case, would be given by

$$\psi(x, t) = e^{-\frac{i}{\hbar} E t} \psi(x)$$

$$= N \, e^{-\frac{i}{\hbar} E t + \frac{i}{\hbar} \int_{x_0}^{x} dx' \, p(x')}$$

$$= N \, e^{\frac{i}{\hbar} \int_{0}^{t} dt' \, (p\dot{x} - E)}$$

$$= N \, e^{\frac{i}{\hbar} S[x_{\mathrm{cl}}]}.$$ \hfill (7.15)

This is exactly what we would expect in the classical limit.

Let us next determine the first order correction to the classical phase. Setting the coefficient of the $\frac{1}{\hbar}$ term in Eq. (7.11) to zero, we obtain

$$i\phi_0''(x) - 2\phi_0'(x)\phi_1'(x) = 0,$$

$$\text{or,} \quad \phi_1'(x) = \frac{i}{2}\frac{\phi_0''(x)}{\phi_0'(x)} = \frac{i}{2}\left(\ln \phi_0'(x)\right)',$$

$$\text{or,} \quad \phi_1(x) = \frac{i}{2}\ln \phi_0'(x) = \frac{i}{2}\ln p(x). \tag{7.16}$$

The constant of integration in Eq. (7.16) can be absorbed into the normalization constant N of the wave function. We note here that this selects out the positive root for $\phi_0'(x)$ in Eq. (7.12). (Remember that a scalar quantum mechanical wave function is assumed to be single valued.)

Thus, keeping only the first order correction to the classical phase, we can write the wave function to be

$$\psi(x) \simeq N\, e^{\frac{i}{\hbar}(\phi_0(x) + \hbar\phi_1(x))}$$

$$= N\, e^{\frac{i}{\hbar}\left(\int^x dx'\, p(x') + \frac{i\hbar}{2}\ln p(x)\right)}$$

$$= \frac{N}{\sqrt{p(x)}}\, e^{\frac{i}{\hbar}\int^x dx'\, p(x')}. \tag{7.17}$$

This is known as the WKB approximation for the quantum mechanical wave function of the system. This approximation, clearly, breaks down for small $p(x)$ and in particular when

$$p(x) = 0. \tag{7.18}$$

Namely, at the classical turning points, the classical momentum vanishes and, consequently, the WKB approximation breaks down and we must reexamine the Schrödinger equation and its solutions more carefully in these regions.

From the form of the WKB wave function in Eq. (7.17), we note that

$$\psi^*(x)\psi(x) \propto \frac{1}{p(x)} \,. \tag{7.19}$$

In other words, the probability density, in this case, is inversely proportional to the momentum or the velocity of the particle. This is, of course, what we would expect from classical considerations alone. Namely, classically, we would qualitatively expect that the system is more likely to be found at points where its velocity is smaller (where it is likely to spend more time). Thus, the WKB approximation gives us a quantum mechanical wave function which retains some of the classical properties. It is for this reason that the WKB approximation is often also called the semi-classical approximation.

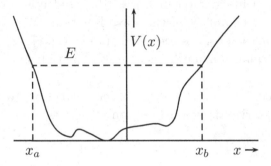

Figure 7.1: A general potential with two classical turning points at x_a and x_b.

Let us consider a particle moving in a potential of the form shown in Fig. 7.1. Then, the normalization constant for the WKB wave function can be determined approximately in the following way. Since the wave function damps outside the potential well, we can write the normalization condition to be approximately

$$\int_{-\infty}^{\infty} dx\, \psi^*(x)\psi(x) \simeq \int_{x_a}^{x_b} dx\, \psi^*(x)\psi(x) = |N|^2 \int_{x_a}^{x_b} \frac{dx}{p(x)} = 1 \,. \tag{7.20}$$

Recalling that the classical period of oscillation is related to the integral above as (since $dx = dt\, v(x)$, as $v(x) \to 0$ at the turning points dx also vanishes and the integral $\int \frac{dx}{v(x)}$ remains finite)

$$\int\limits_{x_a}^{x_b} \frac{\mathrm{d}x}{p(x)} = \frac{1}{m} \int\limits_{x_a}^{x_b} \frac{\mathrm{d}x}{v(x)} = \frac{1}{m} \int\limits_{t_a}^{t_b} \mathrm{d}t = \frac{T_{\mathrm{cl}}}{2m}, \tag{7.21}$$

where T_{cl} denotes the classical period of motion, we obtain

$$|N|^2 \frac{T_{\mathrm{cl}}}{2m} = 1,$$

$$\text{or,} \quad N = N^* = \sqrt{\frac{2m}{T_{\mathrm{cl}}}}. \tag{7.22}$$

Therefore, we can write the normalized WKB wave function to be

$$\psi(x) \simeq \sqrt{\frac{2m}{T_{\mathrm{cl}}\, p(x)}}\, e^{\frac{i}{\hbar} \int\limits^{x} \mathrm{d}x'\, p(x')} = \sqrt{\frac{\omega}{\pi v(x)}}\, e^{\frac{i}{\hbar} \int\limits^{x} \mathrm{d}x'\, p(x')}, \tag{7.23}$$

where $\omega = \frac{2\pi}{T_{\mathrm{cl}}}$ denotes the classical angular frequency of motion.

7.2 Saddle point method

Let us consider an integral of the form

$$I = \int\limits_{-\infty}^{\infty} \mathrm{d}x\, e^{\frac{1}{a} f(x)}, \tag{7.24}$$

where a is a small constant parameter. Furthermore, let us assume that the function $f(x)$ (which is in general a complicated function) has an extremum at $x = x_0$ which is a maximum. In other words, we are assuming that

$$f'(x)\big|_{x=x_0} = 0,$$

$$f''(x)\big|_{x=x_0} < 0. \tag{7.25}$$

We can now expand the function around this extremum as (the linear or the first derivative term vanishes because of the first relation in Eq. (7.25))

$$f(x) = f(x_0) + \frac{1}{2}(x - x_0)^2 f''(x_0) + O((x - x_0)^3).$$ (7.26)

Substituting this back into the integral in Eq. (7.24), we obtain

$$I = \int\limits_{-\infty}^{\infty} dx \, e^{\frac{1}{a}(f(x_0)+\frac{1}{2}(x-x_0)^2 f''(x_0)+O((x-x_0)^3))}$$

$$= e^{\frac{1}{a}f(x_0)} \int\limits_{-\infty}^{\infty} dx \, e^{\frac{1}{a}(-\frac{1}{2}(x-x_0)^2|f''(x_0)|+O((x-x_0)^3))}$$

$$= e^{\frac{1}{a}f(x_0)} \int\limits_{-\infty}^{\infty} dy \, \sqrt{a} \, e^{(-\frac{1}{2}y^2|f''(x_0)|+O(\sqrt{a}y^3))}$$

$$\simeq e^{\frac{1}{a}f(x_0)} \sqrt{a} \, \sqrt{\frac{2\pi}{|f''(x_0)|}}$$

$$= \sqrt{\frac{2\pi a}{|f''(x_0)|}} \, e^{\frac{1}{a}f(x_0)}.$$ (7.27)

It is easy to see that the terms we neglected are of higher order in a and, therefore, this is the most dominant contribution to the integral when a is small.

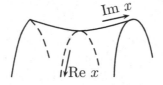

Figure 7.2: Saddle point of a function in the complex plane.

Note from Eq. (7.26) that if we consider the function in the complex x-plane, then along the imaginary axis it has a minimum at the extremal point as shown in Fig. 7.2. Therefore, the extremal point of the function $f(x)$ is really a saddle point in the complex x-plane, hence the name, the saddle point method for such a calculation. Let us also note that the direction of our integration has been along the real axis (which is the direction of steepest descent) and, therefore, this method of evaluating the integral is also referred to as the method of steepest descent. It is worth emphasizing here that if the function in the exponent has several extrema, then the value of the integral will approximately equal the sum of the contributions from each of the extrema.

This method is quite useful in obtaining approximate values for complicated integrals. As an example, let us analyze the behavior of the Gamma function for large values of its argument. The Gamma function has the integral representation given by

$$\Gamma(n+1) = n! = \int_0^\infty dx\, x^n\, e^{-x}$$

$$= \int_0^\infty dx\, e^{(-x + n \ln x)}$$

$$= \int_0^\infty dx\, e^{n(\ln x - \frac{x}{n})} \, . \tag{7.28}$$

Let us next assume that n is very large. This would, then, correspond to $\frac{1}{a}$ in our previous discussion in Eq. (7.24). In the present case,

$$f(x) = \ln x - \frac{x}{n},$$

$$f'(x) = \frac{1}{x} - \frac{1}{n} \, . \tag{7.29}$$

Requiring the first derivative to vanish gives

$$x_0 = n,$$

$$f(x_0) = \ln x_0 - \frac{x_0}{n} = (\ln n - 1) \, . \tag{7.30}$$

Furthermore,

$$f''(x) = -\frac{1}{x^2},$$

(7.31)

and, therefore, we have

$$f''(x_0) = -\frac{1}{n^2} < 0.$$

(7.32)

Therefore, $x_0 = n$ is a maximum of the function and, in this case, is the only extremum. As a result, for large n, we can write

$$\Gamma(n+1) \simeq e^{nf(x_0)} \int_0^\infty dx\, e^{n(-\frac{1}{2n^2}(x-x_0)^2)}$$

$$= e^{nf(x_0)} \int_0^\infty dx\, e^{-\frac{1}{2n}(x-x_0)^2}$$

$$\simeq e^{nf(x_0)} \int_{-\infty}^\infty dx\, e^{-\frac{1}{2n}(x-x_0)^2}$$

$$= \sqrt{2\pi n}\, e^{nf(x_0)}$$

$$= \sqrt{2\pi n}\, e^{n(\ln n - 1)},$$

$$\text{or,} \quad n! \simeq \sqrt{2\pi n}\left(\frac{n}{e}\right)^n, \quad \text{for large } n.$$

(7.33)

(Since $x_0 = n$ is large and positive, we have extended the integration to the negative axis in the intermediate step because the contribution from this region is negligible.) The result in Eq. (7.33) is known as Stirling's approximation for the Gamma function which holds when n is large. Sometimes, this is also written in the form

$$\ln n! \simeq n(\ln n - 1), \quad \text{for large } n.$$

(7.34)

7.3 Semi-classical methods in path integrals

The saddle point method or the method of steepest descent can be applied to path integrals as well. Note that so far we have only evaluated path integrals which involve quadratic actions. But any realistic theory will involve interactions which are inherently nonlinear. The path integral cannot always be evaluated exactly for such systems and the method of steepest descent gives rise to a very useful approximation in such a case.

Let us consider a general action $S[x]$ which is not necessarily quadratic. The transition amplitude associated with this action, as we have seen, is given by (see Eq. (2.29))

$$\langle x_f, t_f | x_i, t_i \rangle = N \int \mathcal{D}x \, e^{\frac{i}{\hbar} S[x]} \,. \tag{7.35}$$

Since $\hbar$ is a small parameter, this integral is similar to the one in Eq. (7.24), but is not exactly in the same form. Namely, this integral is oscillatory. However, we know that, in such a case, we can rotate to the Euclidean space as we have discussed earlier and the integrand becomes well behaved. We will continue to use the real time description keeping in mind the fact that in all our discussions, we are assuming that the actual calculations are always done in the Euclidean space and then the results are rotated back to Minkowski space.

Let us recall that the classical trajectory satisfies the Euler-Lagrange equation (see Eq. (1.31) or (3.10))

$$\frac{\delta S[x]}{\delta x(t)}\bigg|_{x=x_{\text{cl}}} = 0 \,. \tag{7.36}$$

Therefore, it defines an extremum of the exponent in the path integral. Furthermore, the action is a minimum for the classical trajectory (when gravitation is not involved). Thus, we can expand the action around the classical trajectory as we have done earlier in the case of the harmonic oscillator. Namely, let us define

$$x(t) = x_{\text{cl}}(t) + \eta(t) \,. \tag{7.37}$$

Then, we have

$$S[x] = S[x_{\mathrm{cl}} + \eta]$$

$$= S[x_{\mathrm{cl}}] + \frac{1}{2} \iint \mathrm{d}t_1 \mathrm{d}t_2 \, \eta(t_1) \frac{\delta^2 S[x_{\mathrm{cl}}]}{\delta x_{\mathrm{cl}}(t_1)\delta x_{\mathrm{cl}}(t_2)} \eta(t_2) + O(\eta^3). \tag{7.38}$$

Substituting this back into the transition amplitude in Eq. (7.35), we obtain

$$\langle x_f, t_f | x_i, t_i \rangle$$

$$= N \int \mathcal{D}\eta \, e^{\frac{i}{\hbar}\left(S[x_{\mathrm{cl}}] + \frac{1}{2} \iint \mathrm{d}t_1 \mathrm{d}t_2 \eta(t_1) \frac{\delta^2 S[x_{\mathrm{cl}}]}{\delta x_{\mathrm{cl}}(t_1)\delta x_{\mathrm{cl}}(t_2)} \eta(t_2) + O(\eta^3)\right)}$$

$$\simeq N \, e^{\frac{i}{\hbar}S[x_{\mathrm{cl}}]} \int \mathcal{D}\eta \, e^{\frac{i}{2\hbar} \iint \mathrm{d}t_1 \mathrm{d}t_2 \eta(t_1) \frac{\delta^2 S[x_{\mathrm{cl}}]}{\delta x_{\mathrm{cl}}(t_1)\delta x_{\mathrm{cl}}(t_2)} \eta(t_2)}$$

$$= \frac{N}{\sqrt{\det\left(\frac{1}{\hbar} \frac{\delta^2 S[x_{\mathrm{cl}}]}{\delta x_{\mathrm{cl}}(t_1)\delta x_{\mathrm{cl}}(t_2)}\right)}} e^{\frac{i}{\hbar}S[x_{\mathrm{cl}}]}, \tag{7.39}$$

where we have used the result of Eq. (4.3). (We are also keeping the $\hbar$ term explicitly to bring out the quantum nature of the calculations.) Clearly, the saddle point method breaks down if

$$\det \frac{1}{\hbar} \frac{\delta^2 S[x_{\mathrm{cl}}]}{\delta x_{\mathrm{cl}}(t_1)\delta x_{\mathrm{cl}}(t_2)} = 0. \tag{7.40}$$

Normally, this happens when there is some symmetry or its spontaneous breakdown present in the theory. When this happens, the path integral has to be evaluated more carefully using the method of collective coordinates as we will discuss in the next chapter.

We note here that the form of the transition amplitude in the saddle point approximation in Eq. (7.39) is surprisingly similar in form to the WKB wave function in Eq. (7.23). (Recall from Eq. (1.38) that the transition amplitude is the Schrödinger wave function for a delta function source.) In fact, the phases are identical (for example, see Eq. (7.15)). It is the multiplying factors that we readily do not see to be comparable. We will describe below, without going into too much detail, how we can, in fact, relate the two multiplying factors as well. Let us note that a general action has the form

$$S[x] = \int\limits_{t_i}^{t_f} dt \left(\frac{1}{2} m \dot{x}^2 - V(x) \right) . \tag{7.41}$$

For simplicity, we will choose $t_i = -\frac{T}{2}$ and $t_f = \frac{T}{2}$ $(T = t_f - t_i)$ which will also be useful later. In this case, we note that

$$\frac{\delta^2 S[x_{cl}]}{\delta x_{cl}(t_1) \delta x_{cl}(t_2)} = - \left(m \frac{d^2}{dt_1^2} + V''(x_{cl}) \right) \delta(t_1 - t_2) . \tag{7.42}$$

Therefore, we see that

$$\det \left(\frac{1}{\hbar} \frac{\delta^2 S[x_{cl}]}{\delta x_{cl}(t_1) \delta x_{cl}(t_2)} \right) \propto \det \left(\frac{1}{\hbar} \left(m \frac{d^2}{dt^2} + V''(x_{cl}) \right) \right) . \tag{7.43}$$

Thus, we are interested in evaluating determinants of the form

$$\det \left(\frac{1}{\hbar} \left(m \frac{d^2}{dt^2} + W(x) \right) \right) , \tag{7.44}$$

in the space of functions where $\eta(\frac{T}{2}) = 0 = \eta(-\frac{T}{2})$. We have already evaluated such determinants earlier in connection with the harmonic oscillator (see Eq. (4.13)). Let us recall here some results which hold for determinants of operators containing general bounded potentials. With a little bit of analysis, it is possible to show that if (λ will be an eigenvalue of interest if $\psi_W^{(\lambda)}(\frac{T}{2}) = 0$ as well)

$$\frac{1}{\hbar} \left(m \frac{d^2}{dt^2} + W(x) \right) \psi_W^{(\lambda)} = \lambda \psi_W^{(\lambda)} , \tag{7.45}$$

with the initial value conditions $\psi_W^{(\lambda)}(-\frac{T}{2}) = 0$ and $\dot{\psi}_W^{(\lambda)}(-\frac{T}{2}) = 1$, then (for a general λ)

$$\frac{\det(\frac{1}{\hbar}(m\frac{d^2}{dt^2} + W_1(x)) - \lambda)}{\det(\frac{1}{\hbar}(m\frac{d^2}{dt^2} + W_2(x)) - \lambda)} = \frac{\psi_{W_1}^{(\lambda)}(\frac{T}{2})}{\psi_{W_2}^{(\lambda)}(\frac{T}{2})}, \tag{7.46}$$

where $W_1(x)$ and $W_2(x)$ are two bounded potentials. We note that for λ coinciding with an eigenvalue of one of the operators, the left hand side will have a zero or a pole. But so will the right hand side for the same value of λ because in such a case $\psi^{(\lambda)}$ would correspond to an energy eigenfunction satisfying the vanishing boundary conditions at both the end points. Furthermore, for $\lambda \to \infty$, both sides equal 1 (basically, $W(x)$ becomes irrelevant). Since both the left and the right hand sides of the above equation are meromorphic functions of λ with identical zeroes and poles and coincide for the value $\lambda \to \infty$, they must be equal. It follows from this result that

$$\frac{\det(\frac{1}{\hbar}(m\frac{d^2}{dt^2} + W_1(x)))}{\psi_{W_1}^{(0)}(\frac{T}{2})} = \frac{\det(\frac{1}{\hbar}(m\frac{d^2}{dt^2} + W_2(x)))}{\psi_{W_2}^{(0)}(\frac{T}{2})} = \text{constant}. \tag{7.47}$$

That is, this ratio is independent of the particular form of the potential $W(x)$ and can be used to define the normalization constant in the path integral. We will define a particular normalization later. But, for the moment, let us use this result to write the transition amplitude in Eq. (7.39) in the form

$$\langle x_f, t_f | x_i, t_i \rangle = \frac{N}{\sqrt{\det\left(\frac{1}{\hbar}\left(m\frac{d^2}{dt^2} + V''(x_{cl})\right)\right)}} e^{\frac{i}{\hbar}S[x_{cl}]}$$

$$= \frac{N}{\sqrt{\psi_{V''}^{(0)}(\frac{T}{2})}} e^{\frac{i}{\hbar}S[x_{cl}]}. \tag{7.48}$$

Let us note that by definition (see Eq. (7.45)), $\psi_{V''}^{(0)}(t)$ satisfies the equation (it is a solution with $\lambda = 0$ with the given initial conditions, but may not satisfy $\psi_{V''}^{(0)}(\frac{T}{2}) = 0$ unless the operator has a zero eigenvalue (mode))

$$\left(m\frac{\mathrm{d}^2}{\mathrm{d}t^2} + V''(x_{\mathrm{cl}})\right)\psi_{V''}^{(0)}(t) = 0\,. \tag{7.49}$$

We note here that the classical equations (Euler-Lagrange equations) following from our action have the form (see Eq. (1.31))

$$m\frac{\mathrm{d}^2 x_{\mathrm{cl}}}{\mathrm{d}t^2} + V'(x_{\mathrm{cl}}) = 0\,. \tag{7.50}$$

It follows from this that

$$\frac{\mathrm{d}}{\mathrm{d}t}\left(m\frac{\mathrm{d}^2 x_{\mathrm{cl}}}{\mathrm{d}t^2} + V'(x_{\mathrm{cl}})\right) = 0,$$

$$\text{or,}\quad m\frac{\mathrm{d}^2}{\mathrm{d}t^2}\left(\frac{\mathrm{d}x_{\mathrm{cl}}}{\mathrm{d}t}\right) + V''(x_{\mathrm{cl}})\frac{\mathrm{d}x_{\mathrm{cl}}}{\mathrm{d}t} = 0,$$

$$\text{or,}\quad \left(m\frac{\mathrm{d}^2}{\mathrm{d}t^2} + V''(x_{\mathrm{cl}})\right)\frac{\mathrm{d}x_{\mathrm{cl}}}{\mathrm{d}t} = 0\,. \tag{7.51}$$

Comparing with Eq. (7.49), we readily identify that

$$\psi_{V''}^{(0)}(t) \propto \frac{\mathrm{d}x_{\mathrm{cl}}}{\mathrm{d}t} \propto p(x_{\mathrm{cl}})\,. \tag{7.52}$$

Consequently, we recognize that we can write the transition amplitude in Eq. (7.48) also in the form

$$\left\langle x_f, \frac{T}{2} \Big| x_i, -\frac{T}{2}\right\rangle = \frac{N}{\sqrt{p(x_f)}}\, e^{\frac{i}{\hbar}S[x_{\mathrm{cl}}]}\,. \tag{7.53}$$

The correspondence with the WKB wave function in Eq. (7.23) is now complete and we recognize that the method of steepest descent in the case of path integrals merely gives the WKB approximation.

7.4 Double well potential

As an application of the WKB method, let us consider a particle moving in one dimension in an anharmonic potential of the form

$$V(x) = \frac{g^2}{8}(x^2 - a^2)^2 \,, \tag{7.54}$$

where g and a are real constants. Consequently, the action has the form

$$S[x] = \int dt \left(\frac{1}{2}m\dot{x}^2 - \frac{g^2}{8}(x^2 - a^2)^2 \right) \,. \tag{7.55}$$

The potential in Eq. (7.54), when plotted, has the shape of a double well as shown in Fig. 7.3. This is a very interesting potential and shows up in all branches of physics in different forms. Note that it is an even potential with a local maximum at the origin ($x = 0$). The two minima of the potential are symmetrically located at

$$x = \pm a \,. \tag{7.56}$$

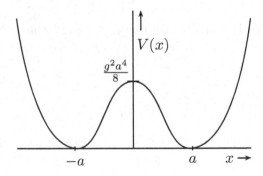

Figure 7.3: Double well potential.

The height of the potential at the origin is given by

$$V(x = 0) = \frac{g^2 a^4}{8} \,. \tag{7.57}$$

Let us also define here for later use

$$V''(x = \pm a) = g^2 a^2 = m\omega^2 \, , \tag{7.58}$$

where we can identify ω with the natural angular frequency of harmonic oscillations near the minima.

We note from Eq. (7.57) that for an infinitely large coupling, the potential separates into two symmetrical wells with an infinite barrier. The motion of the particle is easy to analyze in this case. Each well has quantized levels of energy and if the particle is in one of the wells, then it stays there forever. Namely, there will be no tunneling from one well to the other. Furthermore, from the symmetry of the problem at hand (namely, $x \leftrightarrow -x$), we conclude that both the wells in the present case will have degenerate energy levels. Thus, if $\psi_0(x)$ represents the ground state wave function in the well in the positive x-axis with energy E_0, then $\psi_0(-x)$ will describe the ground state wave function in the left well with the same energy. In fact, any linear combination of the two wave functions and, in particular, the combinations

$$\psi_1(x) = \frac{1}{\sqrt{2}}(\psi_0(x) + \psi_0(-x)) \, ,$$

$$\psi_2(x) = \frac{1}{\sqrt{2}}(\psi_0(x) - \psi_0(-x)) \, , \tag{7.59}$$

will also be degenerate in energy. When the coupling constant g is finite, then the potential barrier between the two wells is also finite. Consequently, the particle initially confined to one of the wells can tunnel into the other well and the states of the two wells will mix. The reflection (parity) symmetry of the system (Hamiltonian) still requires that the eigenstates of the Hamiltonian are only even or odd linear combinations as described in Eq. (7.59). They will, however, not be degenerate in energy any longer because of tunneling and we wish to calculate the splitting in the energy levels due to tunneling using the WKB approximation.

Let us note that $\psi_1(x)$ in Eq. (7.59) is a symmetric wave function whereas $\psi_2(x)$ is anti-symmetric. Consequently, it is obvious that $\psi_1(x)$ would represent the ground state wave function of the system (after taking tunneling into account). Let us write down the time-independent Schrödinger equations that various wave functions satisfy

$$\frac{d^2\psi_0(x)}{dx^2} + \frac{2m}{\hbar^2}(E_0 - V(x))\psi_0(x) = 0, \qquad x \geq 0, \qquad (7.60)$$

$$\frac{d^2\psi_1(x)}{dx^2} + \frac{2m}{\hbar^2}(E_1 - V(x))\psi_1(x) = 0, \qquad (7.61)$$

$$\frac{d^2\psi_2(x)}{dx^2} + \frac{2m}{\hbar^2}(E_2 - V(x))\psi_2(x) = 0. \qquad (7.62)$$

Furthermore, let us assume, for simplicity, that the wave functions are all real. If $\psi_0(x)$ denotes the wave function in the well that is on the right (with $x \geq 0$), then, we can easily see that its value will be vanishingly small in the well on the left (with $x \leq 0$). Thus, we can normalize the wave function as

$$\int_0^\infty dx\, \psi_0^2(x) \simeq 1. \qquad (7.63)$$

Let us similarly note that if $\psi_0(-x)$ denotes the wave function in the left well, then it would have a vanishingly small value in the right well. Consequently, a product such as $\psi_0(x)\psi_0(-x)$ will be negligible everywhere. Therefore, we see from Eq. (7.59) that

$$\psi_1(x)\psi_0(x) \simeq \frac{1}{\sqrt{2}}\,\psi_0^2(x), \quad x \geq 0, \qquad (7.64)$$

so that

$$\int_0^\infty dx\, \psi_1(x)\psi_0(x) \simeq \frac{1}{\sqrt{2}}\int_0^\infty dx\, \psi_0^2(x) \simeq \frac{1}{\sqrt{2}}. \qquad (7.65)$$

Let us next multiply Eq. (7.60) by $\psi_1(x)$ and Eq. (7.61) by $\psi_0(x)$ and subtract one from the other. This gives

$$\psi_1(x)\frac{d^2\psi_0(x)}{dx^2} - \psi_0(x)\frac{d^2\psi_1(x)}{dx^2} + \frac{2m}{\hbar^2}(E_0 - E_1)\psi_1(x)\psi_0(x) = 0\,.$$
$$(7.66)$$

Integrating this equation and using Eq. (7.65), we obtain

$$\int_0^\infty dx \left(\psi_1(x)\frac{d^2\psi_0(x)}{dx^2} - \psi_0(x)\frac{d^2\psi_1(x)}{dx^2} \right)$$

$$+ \frac{2m}{\hbar^2}(E_0 - E_1)\int_0^\infty dx\,\psi_1(x)\psi_0(x) = 0,$$

or, $\quad \dfrac{2m}{\hbar^2}(E_0 - E_1)\dfrac{1}{\sqrt{2}} = -\left(\psi_1(x)\psi_0'(x) - \psi_0(x)\psi_1'(x)\right)\big|_0^\infty$

$$= \psi_1(0)\psi_0'(0) - \psi_0(0)\psi_1'(0),$$

or, $\quad E_0 - E_1 = \dfrac{\hbar^2}{\sqrt{2}m}\left(\psi_1(0)\psi_0'(0) - \psi_0(0)\psi_1'(0)\right)\,.$ $\quad(7.67)$

Here we have used Eq. (7.65) as well as the fact that the quantum mechanical bound state wave functions and their derivatives vanish at spatial infinity. Let us also note that since

$$\psi_1(x) = \frac{1}{\sqrt{2}}\left(\psi_0(x) + \psi_0(-x)\right),$$

$$\psi_1(0) = \sqrt{2}\,\psi_0(0)\,,$$

$$\psi_1'(0) = 0\,. \qquad\qquad (7.68)$$

Substituting this back into Eq. (7.67), we obtain

$$E_0 - E_1 = \frac{\hbar^2}{\sqrt{2}m}\,\sqrt{2}\psi_0(0)\psi_0'(0) = \frac{\hbar^2}{m}\psi_0(0)\psi_0'(0)\,. \qquad (7.69)$$

It is at this point that we would like to use the WKB approximation. We recall from Eq. (7.23) that we can write the WKB wave function as

$$\psi_0(0) = \sqrt{\frac{\omega}{\pi v(0)}}\, e^{-\frac{1}{\hbar}\int\limits_0^a dx\, |p(x)|},$$

$$\psi_0'(0) = \frac{mv(0)}{\hbar}\,\psi_0(0)\,. \tag{7.70}$$

Here, we note that

$$v(0) = \sqrt{\frac{2}{m}(V(0) - E_0)}, \quad V(0) \gg E_0\,. \tag{7.71}$$

Putting these back, we obtain

$$
\begin{aligned}
E_0 - E_1 &= \frac{\hbar^2}{m}\frac{mv(0)}{\hbar}(\psi_0(0))^2 \\
&= \hbar v(0)\frac{\omega}{\pi v(0)}\, e^{-\frac{2}{\hbar}\int\limits_0^a dx\,|p(x)|} \\
&= \frac{\hbar\omega}{\pi}\, e^{-\frac{2}{\hbar}\int\limits_0^a dx\,|p(x)|}\,.
\end{aligned}
\tag{7.72}
$$

This is the splitting in the energy level of the true ground state from that of the infinite well. We can similarly show that

$$E_2 - E_0 = \frac{\hbar\omega}{\pi}\, e^{-\frac{2}{\hbar}\int\limits_0^a dx\,|p(x)|}, \tag{7.73}$$

and, consequently,

$$E_2 - E_1 = \frac{2\hbar\omega}{\pi}\, e^{-\frac{2}{\hbar}\int\limits_0^a dx\,|p(x)|}, \tag{7.74}$$

which gives the splitting between the two degenerate levels in this approximation.

The damping exponential in Eq. (7.74) reflects the effects of tunneling. In fact, note that because of the reflection symmetry in the problem, we can write

$$e^{-\frac{2}{\hbar}\int_0^a dx \, |p(x)|} = e^{-\frac{1}{\hbar}\int_{-a}^a dx \, |p(x)|}, \tag{7.75}$$

and this gives the coefficient for tunneling from the minimum at $x = -a$ to the one at $x = a$. This, of course, assumes that the particle under consideration has vanishing energy. In reality, however, we know that the quantum mechanical ground state energy is nonzero in general. In fact, if we approximate the potential near each of the minima by a harmonic oscillator potential, then we can identify the ground state energy of the system with

$$E_0 = \frac{\hbar\omega}{2}. \tag{7.76}$$

This would, then, imply that the turning points for motion in both the wells, in this case, are given by (see Fig. 7.4)

$$\frac{1}{2} m\omega^2 (x - a)^2 = E_0 = \frac{\hbar\omega}{2},$$

$$\text{or,} \quad x - a = \pm\sqrt{\frac{\hbar}{m\omega}},$$

$$\text{or,} \quad x_{\mathrm{I}} = a \pm \sqrt{\frac{\hbar}{m\omega}}, \tag{7.77}$$

and, similarly,

$$\frac{1}{2} m\omega^2 (x + a)^2 = E_0 = \frac{\hbar\omega}{2},$$

$$\text{or,} \quad x + a = \pm\sqrt{\frac{\hbar}{m\omega}},$$

$$\text{or,} \quad x_{\mathrm{II}} = -a \pm \sqrt{\frac{\hbar}{m\omega}}. \tag{7.78}$$

The tunneling from one well to the other in this case, therefore, would correspond to tunneling from $-a + \sqrt{\frac{\hbar}{m\omega}}$ to $a - \sqrt{\frac{\hbar}{m\omega}}$. Correspondingly, for a more accurate estimation of the splitting in the energy levels, we can replace the exponential in Eq. (7.74) by

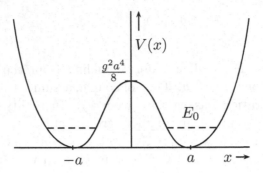

Figure 7.4: Turning points for the double well potential with ground state energy E_0.

$$e^{-\frac{2}{\hbar}\int_0^a dx\, |p(x)|} \rightarrow e^{-\frac{2}{\hbar}\int_0^{a-\sqrt{\frac{\hbar}{m\omega}}} dx\, |p(x)|} . \tag{7.79}$$

Furthermore, recalling that (see Eq. (7.5))

$$|p(x)| = \sqrt{2m(V(x) - E_0)}, \tag{7.80}$$

we can evaluate the exponent in a straightforward manner as

$$\int_0^{a-\sqrt{\frac{\hbar}{m\omega}}} dx\, |p(x)|$$

$$= \int_0^{a-\sqrt{\frac{\hbar}{m\omega}}} dx\, \sqrt{2m\left(\frac{g^2}{8}(x^2 - a^2)^2 - \frac{\hbar\omega}{2}\right)}$$

$$= \int_0^{a-\sqrt{\frac{\hbar}{m\omega}}} dx\, \frac{m\omega(a^2 - x^2)}{2a}\left(1 - \frac{4a^2\hbar}{m\omega(a^2 - x^2)^2}\right)^{\frac{1}{2}}$$

$$\simeq \int_0^{a-\sqrt{\frac{\hbar}{m\omega}}} dx\, \left(\frac{m\omega(a^2 - x^2)}{2a} - \frac{a\hbar}{(a^2 - x^2)}\right). \tag{7.81}$$

Here we have used the fact that $\hbar$ is a small parameter. This integral can be trivially done and has the value

$$
\int_0^{a-\sqrt{\frac{\hbar}{m\omega}}} dx\, |p(x)|
$$

$$
\simeq \frac{m\omega a^2}{3} - \frac{\hbar}{2} + \frac{\hbar}{2}\ln\sqrt{\frac{\hbar}{m\omega a^2}} - \frac{\hbar}{2}\ln\left(2 - \sqrt{\frac{\hbar}{m\omega a^2}}\right) + O(\hbar^{\frac{3}{2}})
$$

$$
= \frac{m\omega a^2}{3} + \frac{\hbar}{2}\left(\ln e^{-1} + \ln\sqrt{\frac{\hbar}{4m\omega a^2}} - \ln\left(1 - \sqrt{\frac{\hbar}{4m\omega a^2}}\right)\right)
$$

$$
\qquad + O(\hbar^{\frac{3}{2}}),
$$

$$
= \frac{1}{2}S_0 + \frac{\hbar}{2}\ln\sqrt{\frac{\hbar}{4m\omega a^2 e^2}} + O(\hbar^{\frac{3}{2}}). \tag{7.82}
$$

Here we have defined

$$
S_0 = \frac{2m\omega a^2}{3}. \tag{7.83}
$$

Substituting Eqs. (7.79) and (7.83) into Eq. (7.74), we obtain the splitting in the energy levels to be

$$
E_2 - E_1 \simeq \frac{2\hbar\omega}{\pi}\sqrt{\frac{4m\omega a^2 e^2}{\hbar}}\, e^{-\frac{1}{\hbar}S_0}
$$

$$
= \frac{4e}{\pi}\sqrt{m\hbar}\,\omega^{\frac{3}{2}}a\, e^{-\frac{1}{\hbar}S_0}. \tag{7.84}
$$

In the next chapter, we will calculate this energy splitting using the path integrals and compare the two results.

7.5 References

S. **Coleman**, *The Uses of Instantons*, Erice Lectures, 1977.

L. D. **Landau and L. M. Lifshitz**, *Nonrelativistic Quantum Mechanics*, Pergamon Press.

A. B. **Migdal and V. Krainov**, *Approximation Methods in Quantum Mechanics*, Benjamin Publishing.

B. **Sakita**, *Quantum Theory of Many Variable Systems and Fields*, World Scientific Publishing.

CHAPTER 8

Path integral for the double well

8.1 Instantons

Let us next evaluate the path integral for the double well potential. We recall from Eq. (2.29) that the transition amplitude is defined as

$$\left\langle x_f, \frac{T}{2} | x_i, -\frac{T}{2} \right\rangle = \left\langle x_f | e^{-\frac{i}{\hbar}HT} | x_i \right\rangle = N \int \mathcal{D}x \, e^{\frac{i}{\hbar}S[x]}, \qquad (8.1)$$

where the action for the double well potential (see Eq. (7.55)) is given by

$$S[x] = \int\limits_{-\frac{T}{2}}^{\frac{T}{2}} dt \, \left(\frac{1}{2}m\dot{x}^2 - V(x) \right)$$

$$= \int\limits_{-\frac{T}{2}}^{\frac{T}{2}} dt \, \left(\frac{1}{2}m\dot{x}^2 - \frac{g^2}{8}(x^2 - a^2)^2 \right). \qquad (8.2)$$

As we have seen earlier in Chapter 4, the best way to evaluate the path integral is to go to the Euclidean space. Thus, by rotating to imaginary time

$$t \rightarrow -it, \qquad (8.3)$$

and using Eq. (1.45) we obtain

$$\langle x_f | e^{-\frac{1}{\hbar}HT} | x_i \rangle = N \int \mathcal{D}x \, e^{-\frac{1}{\hbar}S_E[x]}, \qquad (8.4)$$

159

where the Euclidean action has the form

$$S_E[x] = \int_{-\frac{T}{2}}^{\frac{T}{2}} dt \left(\frac{1}{2} m\dot{x}^2 + V(x) \right). \tag{8.5}$$

In the semi-classical approximation, we can evaluate the path integral by the saddle point method. The classical equation, which is obtained from the extremum of the action in Eq. (8.5), has the form

$$\frac{\delta S_E[x]}{\delta x(t)} = 0,$$

or, $\quad m\ddot{x} - V'(x) = 0$,

or, $\quad m\ddot{x} - \frac{g^2}{2} x(x^2 - a^2) = 0$, \tag{8.6}

where we have used

$$V(x) = \frac{g^2}{8}(x^2 - a^2)^2. \tag{8.7}$$

The Euclidean equations, therefore, correspond to a particle moving in an inverted potential $(-V(x))$, otherwise also known as the double humped potential shown in Fig. 8.1. The Euclidean energy associated with such a motion is correspondingly given by

$$E = \frac{1}{2} m\dot{x}^2 - V(x). \tag{8.8}$$

Two solutions of the Euclidean classical equation of motion in Eq. (8.6) are obvious. Namely,

$$x(t) = \pm a, \tag{8.9}$$

satisfy the classical equation with $E = 0$ and with vanishing Euclidean action $(S_E = 0)$. In other words, the particle sits at rest on

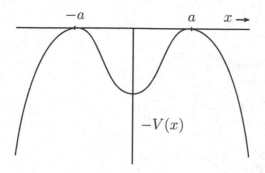

Figure 8.1: Inverted double well potential in Euclidean space.

top of one of the hills in such a case. Quantum mechanically, this would correspond to the case where the particle executes small oscillations at the bottom of either of the wells in the Minkowski space and these small oscillations can be approximated by a harmonic oscillator motion. (The other obvious solution to the Euclidean equation of motion Eq. (8.6) corresponds to

$$x(t) = 0, \tag{8.10}$$

which, in fact, has a lower energy value ($E = -\frac{g^2 a^4}{8}$) as can be checked from Eq. (8.8). However, the action (see Eq. (8.5)) corresponding to this solution is infinite in the limit $T \to \infty$ and, consequently, does not contribute to the path integral.) On the other hand, in the large T limit, (namely, when $T \to \infty$) in which we are ultimately interested, there are other nontrivial solutions to the Euclidean equation of motion in Eq. (8.6) which play an important role in evaluating the path integral. Let

$$x_{cl}(t) = \pm a \tanh \frac{\omega(t - t_c)}{2}, \tag{8.11}$$

where t_c is a constant and we have identified as in Eq. (7.58)

$$m\omega^2 = V''(\pm a) = g^2 a^2, \quad \text{or,} \quad \frac{\omega}{a} = \frac{g}{\sqrt{m}}. \tag{8.12}$$

Then, we see that

$$\dot{x}_{cl} = \pm \frac{a\omega}{2} \text{sech}^2 \frac{\omega(t - t_c)}{2}$$

$$= \pm \frac{a\omega}{2} \left(1 - \frac{x_{cl}^2}{a^2}\right) = \mp \frac{\omega}{2a}(x_{cl}^2 - a^2)$$

$$= \mp \frac{g}{2\sqrt{m}}(x_{cl}^2 - a^2) = \mp \sqrt{\frac{2V(x_{cl})}{m}}, \tag{8.13}$$

$$\ddot{x}_{cl} = \mp \frac{\omega}{2a} 2x_{cl}\dot{x}_{cl}$$

$$= \mp \frac{\omega}{a} x_{cl} \left(\mp \frac{\omega}{2a}(x_{cl}^2 - a^2)\right)$$

$$= \frac{\omega^2}{2a^2} x_{cl}(x_{cl}^2 - a^2) = \frac{g^2}{2m} x_{cl}(x_{cl}^2 - a^2)$$

$$= \frac{1}{m} V'(x_{cl}). \tag{8.14}$$

Consequently, it follows that

$$m\ddot{x}_{cl} - V'(x_{cl}) = 0, \tag{8.15}$$

and we conclude that the solutions in Eq. (8.11) represent nontrivial solutions of the Euclidean equation of motion in Eq. (8.6).

We note from Eq. (8.11) that, for these solutions,

$$x_{cl}(t \to -\infty) = \mp a,$$

$$x_{cl}(t \to \infty) = \pm a. \tag{8.16}$$

These solutions, therefore, correspond to the particle starting out on top of one of the hills at $t \to -\infty$ and then moving over to the top of the other hill at $t \to \infty$ as shown in Fig. 8.2. Let us also note (see Eq. (8.13)) that for such solutions

$$\frac{1}{2}m\dot{x}_{cl}^2 = \frac{1}{2} m \times \frac{2}{m} V(x_{cl}) = V(x_{cl}). \tag{8.17}$$

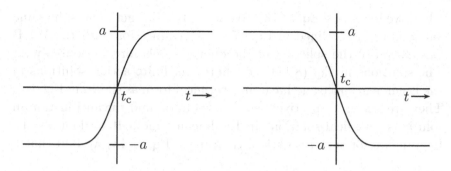

Figure 8.2: Nontrivial classical solutions of the Euclidean equation of motion.

Therefore, for such classical motion, the Euclidean energy defined in Eq. (8.8) has the value

$$E = \frac{1}{2} m \dot{x}_{\text{cl}}^2 - V(x_{\text{cl}}) = 0 \,. \tag{8.18}$$

In other words, these solutions are degenerate in energy with the two trivial solutions in Eq. (8.9). The value of the action corresponding to such a classical motion can be easily calculated to be

$$
\begin{aligned}
S_E[x_{\text{cl}}] &= \int\limits_{-\infty}^{\infty} \mathrm{d}t \, \left(\frac{1}{2} m \dot{x}_{\text{cl}}^2 + V(x_{\text{cl}}) \right) \\
&= m \int\limits_{-\infty}^{\infty} \mathrm{d}t \, \dot{x}_{\text{cl}}^2 = m \int\limits_{\mp a}^{\pm a} \mathrm{d}x_{\text{cl}} \, \dot{x}_{\text{cl}} \\
&= m \int\limits_{\mp a}^{\pm a} \mathrm{d}x_{\text{cl}} \, \left(\mp \frac{\omega}{2a} (x_{\text{cl}}^2 - a^2) \right) \\
&= \mp \frac{m\omega}{2a} \left(\frac{1}{3} x_{\text{cl}}^3 - x_{\text{cl}} a^2 \right) \Big|_{\mp a}^{\pm a} = \frac{m\omega}{2a} \frac{4a^3}{3} \\
&= \frac{2}{3} m\omega a^2 = S_0 \,,
\end{aligned} \tag{8.19}
$$

where we have used Eq. (8.13). We note that this action has the same value as the S_0 defined in Eq. (7.83) in connection with the WKB calculation of the splitting of the energy levels for the double well. The solutions in Eq. (8.11) are, therefore, finite action solutions in the Euclidean space and have the graphical form as shown in Fig. 8.2. They are known respectively as the instanton and the anti-instanton solutions (classical solutions in Euclidean time). If we look at the Lagrangian for such a solution, then using Eq. (8.13) we find that

$$
\begin{aligned}
L_E &= \frac{1}{2} m \dot{x}_{\text{cl}}^2 + V(x_{\text{cl}}) = m \dot{x}_{\text{cl}}^2 \\
&= m \left(\frac{a^2 \omega^2}{4} \text{sech}^4 \frac{\omega(t - t_c)}{2} \right) \\
&= \frac{m a^2 \omega^2}{4} \text{sech}^4 \frac{\omega(t - t_c)}{2} .
\end{aligned}
\tag{8.20}
$$

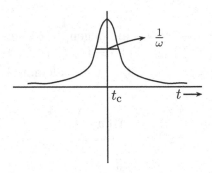

Figure 8.3: The value of the Euclidean Lagrangian for the classical motion as a function of t.

In other words, the Lagrangian is fairly localized around $t = t_c$ with a width of about

$$
\Delta t \sim \frac{1}{\omega} = \frac{\sqrt{m}}{ga} .
\tag{8.21}
$$

It is in this sense that one says that instantons (anti-instantons) are localized solutions in Euclidean time with a size of about $\frac{1}{\omega}$. The constant t_c, which signifies the time when the solution reaches the valley of the Euclidean potential, is really arbitrary. This is a direct reflection of the time translation invariance in the theory.

Just as we can have a one instanton or one anti-instanton solution, we can also have multi-instanton solutions in such a theory. However, before going into that, let us calculate the contribution of the one instanton or the one anti-instanton trajectory to the transition amplitude. From our earlier discussion in Eq. (7.39) of the saddle point method in connection with path integrals, we conclude that (O.I. stands for "one instanton")

$$\langle a|e^{-\frac{1}{\hbar}HT}|-a\rangle_{\text{O.I.}}$$

$$= N \int \mathcal{D}x \, e^{-\frac{1}{\hbar}S_E[x]}$$

$$\simeq N \, e^{-\frac{1}{\hbar}S_E[x_{\text{cl}}]} \int \mathcal{D}\eta \, e^{-\frac{1}{2\hbar} \iint dt_1 dt_2 \, \eta(t_1)\frac{\delta^2 S_E[x_{\text{cl}}]}{\delta x_{\text{cl}}(t_1)\delta x_{\text{cl}}(t_2)}\eta(t_2)}$$

$$= \frac{N}{\sqrt{\det\left(\frac{1}{\hbar}\left(-m\frac{d^2}{dt^2} + V''(x_{\text{cl}})\right)\right)}} e^{-\frac{1}{\hbar}S_0}. \qquad (8.22)$$

Here we have defined as before (the t_c dependence has been put into the solutions explicitly)

$$x(t) = x_{\text{cl}}(t - t_c) + \eta(t - t_c), \qquad (8.23)$$

and, for the one instanton case, we have already seen in Eq. (8.11) that,

$$x_{\text{cl}}(t - t_c) = a \tanh\frac{\omega(t - t_c)}{2}. \qquad (8.24)$$

Let us next analyze the determinant in Eq. (8.22) for the one instanton case in a bit more detail. We know from Eq. (8.13) that

$$\frac{dx_{\text{cl}}}{dt} = \frac{a\omega}{2} \text{sech}^2\frac{\omega(t - t_c)}{2}, \qquad (8.25)$$

and it satisfies the zero eigenvalue equation (see Eq. (7.51) rotated to imaginary time)

$$\left(-m\frac{d^2}{dt^2} + V''(x_{cl})\right)\frac{dx_{cl}}{dt} = 0, \tag{8.26}$$

which can be checked easily using Eq. (8.25) as well as the fact that

$$V''(x_{cl}) = \frac{g^2}{2}\left(3x_{cl}^2 - a^2\right)$$

$$= \frac{m\omega^2}{2}\left(2\tanh^2\frac{\omega(t-t_c)}{2} - \text{sech}^2\frac{\omega(t-t_c)}{2}\right). \tag{8.27}$$

In fact, using Eq. (8.19), we can define the normalized zero eigenvalue solution of Eq. (8.26) as ($\int_{-\infty}^{\infty} dt\,\psi_0^2(t) = 1$)

$$\psi_0(t-t_c) = \left(\frac{S_0}{m}\right)^{-\frac{1}{2}}\frac{dx_{cl}}{dt} = \left(\frac{S_0}{m}\right)^{-\frac{1}{2}}\frac{a\omega}{2}\text{sech}^2\frac{\omega(t-t_c)}{2}. \tag{8.28}$$

As we had seen earlier in Eq. (7.47), the determinant in Eq. (8.22) can be obtained from this solution simply as

$$\det\left(\frac{1}{\hbar}\left(-m\frac{d^2}{dt^2} + V''(x_{cl})\right)\right) \propto \psi_0\left(\frac{T}{2} - t_c\right), \quad T \to \infty. \tag{8.29}$$

But, from the form of the solution in Eq. (8.28), it is clear that

$$\lim_{T\to\infty} \psi_0\left(\frac{T}{2} - t_c\right) \to 0. \tag{8.30}$$

In other words, in this case, the determinant identically vanishes. The reason for this is obvious, namely, in the present case $\psi_0(t)$ happens to be an exact eigenstate of the operator $(-m\frac{d^2}{dt^2} + V''(x_{cl}))$ with zero eigenvalue. (This means that $\psi_0(\pm\frac{T}{2} - t_c) = 0$ for $T \to \infty$.) This is what one means in saying that there is a zero mode in this theory.

8.2 Zero modes

As we have argued before, a zero mode is present in the theory whenever there is a symmetry operative in the system. To see this, let us recall that the determinant in Eq. (8.22) arose from integrating out the Gaussian fluctuations. Therefore, the term that we need to re-examine is

$$
\int \mathcal{D}\eta \, e^{-\frac{1}{2\hbar} \iint dt_1 dt_2 \, \eta(t_1) \frac{\delta^2 S_E[x_{cl}]}{\delta x_{cl}(t_1)\delta x_{cl}(t_2)} \eta(t_2)} .
\tag{8.31}
$$

Note that, in this case, since $\psi_0(t)$ represents a zero mode of the operator $\frac{\delta^2 S_E[x_{cl}]}{\delta x_{cl}(t_1)\delta x_{cl}(t_2)}$, if we make a change of the integration variable as

$$
\eta(t) \to \eta(t) + \delta\eta(t), \quad \delta\eta(t) = \epsilon\psi_0(t) ,
\tag{8.32}
$$

where ϵ is an infinitesimal constant parameter, then the Gaussian does not change. In other words, the transformation in Eq. (8.32) defines a symmetry of the quadratic action. Another way to visualize the trouble is to note that if we were to expand the fluctuations around the classical trajectory in a complete basis of the (normalized) eigenstates of the operator $\frac{\delta^2 S_E[x_{cl}]}{\delta x_{cl}(t_1)\delta x_{cl}(t_2)}$, then we can write

$$
\eta(t) = \sum_{n \geq 0} c_n \psi_n(t),
$$

$$
\mathcal{D}\eta(t) = \prod_{n \geq 0} dc_n .
\tag{8.33}
$$

Substituting this expansion into Eq. (8.31), we obtain

$$
\int \mathcal{D}\eta \, e^{-\frac{1}{2\hbar} \iint dt_1 dt_2 \, \eta(t_1) \frac{\delta^2 S_E[x_{cl}]}{\delta x_{cl}(t_1)\delta x_{cl}(t_2)} \eta(t_2)}
$$

$$
= \int \prod_{n \geq 0} dc_n \, e^{-\frac{1}{2\hbar} \sum_{n>0} \lambda_n c_n^2}
$$

$$
- \int dc_0 \int \prod_{n>0} dc_n \, e^{-\frac{1}{2\hbar} \sum_{n>0} \lambda_n c_n^2} ,
\tag{8.34}
$$

where λ_n denote the eigenvalues associated with the eigenstates ψ_n. Here we note that the zero mode drops out of the exponent and, consequently, there is no Gaussian damping for the dc_0 integration.

In such a case, we have to evaluate the integral more carefully. To understand further the origin of the problem as well as to find a possible way out of this difficulty, let us examine a simple two dimensional integral. Let

$$
I = \int\!\!\int_{-\infty}^{\infty} dx_1 \, dx_2 \, e^{\frac{1}{a} f(\mathbf{x})}
$$

$$
= \int\!\!\int_{-\infty}^{\infty} dx_1 \, dx_2 \, e^{\frac{1}{a}\left(\frac{1}{2}\mathbf{x}^2 - g^2(\mathbf{x}^2)^2\right)} .
\tag{8.35}
$$

Here a is a small parameter, $g > 0$ and we have defined

$$
f(\mathbf{x}) = \frac{1}{2}\mathbf{x}^2 - g^2(\mathbf{x}^2)^2 ,
$$

$$
\mathbf{x}^2 = x_1^2 + x_2^2 .
\tag{8.36}
$$

This example is, in fact, quite analogous to the instanton calculation which we are interested in. The classical equation (extremum), in this case, leads to the maximum

$$
\frac{\partial f}{\partial \mathbf{x}} = \mathbf{x}\left(1 - 4g^2\mathbf{x}^2\right) = 0 ,
$$

$$
\text{or,} \quad |\mathbf{x}_{\mathrm{cl}}| = \frac{1}{2g} .
\tag{8.37}
$$

It is easy to see that the other solution, namely, $\mathbf{x}_{\mathrm{cl}} = 0$, in this case, corresponds to a local minimum. The most general classical solution following from Eq. (8.37) can, therefore, be written as

$$
x_{\mathrm{cl},1} = \frac{1}{2g} \cos\theta ,
$$

$$
x_{\mathrm{cl},2} = \frac{1}{2g} \sin\theta ,
\tag{8.38}
$$

where θ is an arbitrary constant angular parameter. This is very much like the arbitrary parameter t_c which arises in the case of the instantons. In the present case, the presence of the angular parameter θ is a reflection of the rotational invariance of the function $f(\mathbf{x})$ in Eq. (8.36).

Expanding around the classical solution in Eq. (8.38), namely, choosing

$$x_\alpha = x_{\mathrm{cl},\alpha} + \eta_\alpha, \quad \alpha = 1, 2, \tag{8.39}$$

we have

$$f(\mathbf{x}) = f(\mathbf{x}_{\mathrm{cl}} + \boldsymbol{\eta}) \simeq f(\mathbf{x}_{\mathrm{cl}}) + \frac{1}{2}\eta_\alpha \frac{\partial^2 f(\mathbf{x}_{\mathrm{cl}})}{\partial x_{\mathrm{cl},\alpha} \partial x_{\mathrm{cl},\beta}} \eta_\beta + O(\eta^3). \tag{8.40}$$

From the form of $f(\mathbf{x})$ in Eq. (8.36), we note that

$$\frac{\partial^2 f(\mathbf{x}_{\mathrm{cl}})}{\partial x_{\alpha,\mathrm{cl}} \partial x_{\beta,\mathrm{cl}}} = \delta_{\alpha\beta}(1 - 4g^2 \mathbf{x}_{\mathrm{cl}}^2) - 8g^2 x_{\mathrm{cl},\alpha} x_{\mathrm{cl},\beta}$$

$$= -8g^2 x_{\mathrm{cl},\alpha} x_{\mathrm{cl},\beta}. \tag{8.41}$$

In the matrix form, therefore, we can write

$$\frac{\partial^2 f(\mathbf{x}_{\mathrm{cl}})}{\partial x_{\mathrm{cl},\alpha} \partial x_{\mathrm{cl},\beta}} = -2 \left(\begin{array}{cc} \cos^2\theta & \cos\theta\sin\theta \\ \cos\theta\sin\theta & \sin^2\theta \end{array} \right). \tag{8.42}$$

Thus, if we use the saddle point method naively, we would obtain the value of the integral in Eq. (8.35) to be

$$I \simeq e^{\frac{1}{a} f(\mathbf{x}_{\mathrm{cl}})} \int \mathrm{d}\eta_\alpha \, e^{\frac{1}{2a}\eta_\alpha \frac{\partial^2 f(\mathbf{x}_{\mathrm{cl}})}{\partial x_{\mathrm{cl},\alpha} \partial x_{\mathrm{cl},\beta}} \eta_\beta}. \tag{8.43}$$

Let us note that the matrix $\frac{\partial^2 f(\mathbf{x}_{\mathrm{cl}})}{\partial x_{\mathrm{cl},\alpha} \partial x_{\mathrm{cl},\beta}}$ in Eq. (8.42) has two eigenvalues, $\lambda = 0, -2$. As a result, the Gaussian integral in Eq. (8.43) does not exist. This is very much like the instanton calculation that we have done. In fact, it is easy to check from Eq. (8.42) that the eigenstate of the matrix with zero eigenvalue has the form (note the similarity of this form with the zero mode in Eq. (8.28))

$$x_0 = \begin{pmatrix} \sin\theta \\ -\cos\theta \end{pmatrix} = -2g\,\frac{d}{d\theta}\begin{pmatrix} x_{cl,1} \\ x_{cl,2} \end{pmatrix}. \tag{8.44}$$

Consequently, under a transformation of the variables of integration of the form

$$\delta\begin{pmatrix} \eta_1 \\ \eta_2 \end{pmatrix} = \epsilon x_0 = \epsilon\begin{pmatrix} \sin\theta \\ -\cos\theta \end{pmatrix} = -2g\epsilon\frac{d}{d\theta}\begin{pmatrix} x_{cl,1} \\ x_{cl,2} \end{pmatrix}, \tag{8.45}$$

which we recognize as an infinitesimal rotation, the quadratic exponent in Eq. (8.43) does not change. In fact, writing out the exponent explicitly, we have

$$I \simeq e^{\frac{1}{a}f(\mathbf{x}_{cl})}\iint\limits_{-\infty}^{\infty} d\eta_1 d\eta_2\, e^{-\frac{1}{a}(\eta_1\cos\theta + \eta_2\sin\theta)^2}. \tag{8.46}$$

Furthermore, redefining the variables as

$$\tilde{\eta}_1 = \eta_1\cos\theta + \eta_2\sin\theta\,,$$
$$\tilde{\eta}_2 = -\eta_1\sin\theta + \eta_2\cos\theta\,, \tag{8.47}$$

we note that we can write the integral in Eq. (8.46) also as

$$I \simeq e^{\frac{1}{a}f(\mathbf{x}_{cl})}\iint\limits_{-\infty}^{\infty} d\tilde{\eta}_1 d\tilde{\eta}_2\, e^{-\frac{1}{a}\tilde{\eta}_1^2}$$

$$= e^{\frac{1}{a}f(\mathbf{x}_{cl})}\int\limits_{-\infty}^{\infty} d\tilde{\eta}_2\int\limits_{-\infty}^{\infty} d\tilde{\eta}_1\, e^{-\frac{1}{a}\tilde{\eta}_1^2}. \tag{8.48}$$

The analogy with the instanton case is now complete. There is no damping for the $d\tilde{\eta}_2$ integration. That is the origin of the divergence and it is a consequence of rotational invariance in this case. In this simple example, the solution to the problem is obvious. Namely, since the function $f(\mathbf{x})$ in Eq. (8.36) is rotationally invariant, it is appropriate to use circular (polar) coordinates. The angular integral, which will have no zero mode in this case, can be trivially performed giving a finite result, after which the saddle point approximation can be applied to the radial integral. This method generalizes readily to other systems with more degrees of freedom through what is known as the method of collective coordinates. This is what we will discuss next and we will use this method to evaluate the instanton integral.

8.3 The instanton integral

In the case of the instanton, we have already seen in Eq. (8.34) that

$$\int \mathcal{D}\eta \, e^{-\frac{1}{2\hbar} \iint dt_1 dt_2 \, \eta(t_1) \frac{\delta^2 S_E[x_{cl}]}{\delta x_{cl}(t_1) \delta x_{cl}(t_2)} \eta(t_2)}$$

$$= \int dc_0 \int \prod_{n>0} dc_n \, e^{-\frac{1}{2\hbar} \sum_{n>0} \lambda_n c_n^2}, \qquad (8.49)$$

which is divergent. The divergence, in the present case, is a consequence of time translation invariance, namely, the position of the center of the instanton can be arbitrary. So, following our earlier discussion of the rotationally symmetric example, we would like to replace the dc_0 integration by an integration over the position of the center of the instanton. Let us discuss very briefly how this is done. Let us recall that expanding around the instanton trajectory yields (here, as before, we have indicated explicitly the t_c dependence of the solution)

$$x(t) = x_{cl}(t - t_c) + \eta(t - t_c),$$

$$\text{or,} \quad x(t + t_c) = x_{cl}(t) + \eta(t) = x_{cl}(t) + \sum_{n \geq 0} c_n \psi_n(t). \qquad (8.50)$$

(Since the trajectory is independent of the center of the instanton trajectory, the fluctuations must balance out the t_c dependence. Furthermore, $\eta(t - t_c)$ should really be written as $\eta(t - t_c, t_c)$ to allow for additional t_c dependence.) Multiplying Eq. (8.50) with $\psi_0(t)$ and integrating over time, we obtain

$$
\int_{-\frac{T}{2}}^{\frac{T}{2}} \mathrm{d}t\, x(t + t_c)\psi_0(t)
$$

$$
= \int_{-\frac{T}{2}}^{\frac{T}{2}} \mathrm{d}t\, \left(x_{\mathrm{cl}}(t) + \sum_{n \geq 0} c_n\, \psi_n(t) \right) \psi_0(t)
$$

$$
= \int_{-\frac{T}{2}}^{\frac{T}{2}} \mathrm{d}t\, \left(\left(\frac{S_0}{m} \right)^{-\frac{1}{2}} x_{\mathrm{cl}}(t) \frac{\mathrm{d}x_{\mathrm{cl}}(t)}{\mathrm{d}t} + c_0 \psi_0^2(t) \right)
$$

$$
= \frac{1}{2} \left(\frac{S_0}{m} \right)^{-\frac{1}{2}} x_{\mathrm{cl}}^2(t) \Big|_{-\frac{T}{2}}^{\frac{T}{2}} + c_0
$$

$$
= c_0\,. \tag{8.51}
$$

The first term vanishes because it has the same value at both the limits and the higher eigenfunctions, in the second term are orthogonal to $\psi_0(t)$ giving vanishing contribution. (It is worth emphasizing here that we are only interested in large T limits when all these results hold.) This simple analysis shows that

$$
c_0 = c_0(t_c)\,. \tag{8.52}
$$

Therefore, we can easily change the c_0-integration to an integration over t_c. To obtain the Jacobian of this transformation to the leading order, let us consider an infinitesimal change in the path in Eq. (8.50) arising from a change in the coefficient of the zero mode. Namely, let

$$
\delta\eta(t) = \delta c_0\, \psi_0(t)\,, \tag{8.53}
$$

where we assume that δc_0 is infinitesimal. In this case,

$$\delta x(t + t_c) = \delta\eta(t) = \delta c_0\,\psi_0(t) = \delta c_0 \left(\frac{S_0}{m}\right)^{-\frac{1}{2}} \frac{\mathrm{d}x_{\mathrm{cl}}(t)}{\mathrm{d}t}, \qquad (8.54)$$

where we have used Eq. (8.28). However, we also note that this is precisely the change in the path that we would have obtained to the leading order had we translated the center of the instanton as

$$t_c \to t_c + \delta t_c = t_c + \delta c_0 \left(\frac{S_0}{m}\right)^{-\frac{1}{2}}. \qquad (8.55)$$

Namely, in this case,

$$\begin{aligned}
\delta x(t + t_c) &= x(t + t_c + \delta t_c) - x(t + t_c) \\
&\simeq \delta t_c \frac{\mathrm{d}x(t + t_c)}{\mathrm{d}t} = \delta c_0 \left(\frac{S_0}{m}\right)^{-\frac{1}{2}} \frac{\mathrm{d}x(t + t_c)}{\mathrm{d}t} \\
&\simeq \delta c_0 \left(\frac{S_0}{m}\right)^{-\frac{1}{2}} \frac{\mathrm{d}x_{\mathrm{cl}}(t)}{\mathrm{d}t}.
\end{aligned} \qquad (8.56)$$

(Namely, we are neglecting contributions coming from $\frac{\mathrm{d}\eta(t)}{\mathrm{d}t}$ which are higher order.) Thus, from Eq. (8.55), we note that to leading order, we can determine the Jacobian of the transformation from the integration variable c_0 to t_c to be

$$\frac{\mathrm{d}c_0(t_c)}{\mathrm{d}t_c} \simeq \left(\frac{S_0}{m}\right)^{\frac{1}{2}}. \qquad (8.57)$$

A more direct way to arrive at this result is to note from Eq. (8.51) that since

$$c_0(t_c) = \int_{-\frac{T}{2}}^{\frac{T}{2}} dt\, x(t + t_c)\psi_0(t)$$

$$\frac{dc_0}{dt_c} = \int_{-\frac{T}{2}}^{\frac{T}{2}} dt\, \frac{dx(t + t_c)}{dt_c}\, \psi_0(t)$$

$$= \int_{-\frac{T}{2}}^{\frac{T}{2}} dt\, \frac{dx(t + t_c)}{dt}\, \psi_0(t)$$

$$= \int_{-\frac{T}{2}}^{\frac{T}{2}} dt\, \left(\frac{dx_{cl}(t)}{dt} + \sum_{n\geq 0} c_n \frac{d\psi_n(t)}{dt} \right)\psi_0(t)$$

$$= \left(\frac{S_0}{m}\right)^{\frac{1}{2}} \int_{-\frac{T}{2}}^{\frac{T}{2}} dt\, \psi_0^2(t) + \sum_{n>0} c_n \int_{-\frac{T}{2}}^{\frac{T}{2}} dt\, \frac{d\psi_n(t)}{dt}\, \psi_0(t), \qquad (8.58)$$

where we have used Eq. (8.28). The $n = 0$ term drops out in the second term because for $n = 0$, the integrand is a total derivative of $\psi_0^2(t)$ which vanishes at both the limits. It is clear, therefore, that to leading order (since ψ_0 is normalized to unity),

$$\frac{dc_0(t_c)}{dt_c} = \left(\frac{S_0}{m}\right)^{\frac{1}{2}} + O(\hbar). \qquad (8.59)$$

Namely, we are using here the fact that the higher moments of a Gaussian of the kind that we are dealing with in Eq. (8.49) are higher orders in $\hbar$.

Thus, we are ready to do the determinant calculation now. We substitute Eq. (8.57) or (8.59) into Eq. (8.49) to obtain

$$\int \mathcal{D}\eta\, e^{-\frac{1}{2\hbar} \iint dt_1 dt_2\, \eta(t_1) \frac{\delta^2 S_E[x_{cl}]}{\delta x_{cl}(t_1)\delta x_{cl}(t_2)} \eta(t_2)}$$

$$= \left(\frac{S_0}{m}\right)^{\frac{1}{2}} \int_{-\frac{T}{2}}^{\frac{T}{2}} dt_c \int \prod_{n>0} dc_n\, e^{-\frac{1}{2\hbar}\sum_{n>0} \lambda_n c_n^2}$$

$$= \left(\frac{S_0}{m}\right)^{\frac{1}{2}} \frac{\int_{-\frac{T}{2}}^{\frac{T}{2}} dt_c}{\sqrt{\det'(\frac{1}{\hbar}(-m\frac{d^2}{dt^2} + V''(x_{cl})))}}. \tag{8.60}$$

Here $\det'$ stands for the value of the determinant of the operator without the zero mode. Let us also note here that even though the dt_c integral in Eq. (8.60) can be done trivially, we will leave it as it is for later purposes. Thus, from Eqs. (8.22) and (8.60) we obtain the form of the transition amplitude in the presence of an instanton to be

$$\langle a| e^{-\frac{1}{\hbar}HT}|-a\rangle_{\text{O.I.}} = \frac{N\left(\frac{S_0}{m}\right)^{\frac{1}{2}} e^{-\frac{1}{\hbar}S_0}}{\sqrt{\det'(\frac{1}{\hbar}(-m\frac{d^2}{dt^2} + V''(x_{cl})))}} \int_{-\frac{T}{2}}^{\frac{T}{2}} dt_c. \tag{8.61}$$

8.4 Evaluating the determinant

To evaluate $\det'$ in Eq. (8.61), let us study the ratio of the determinants

$$\Delta(E) = \frac{\det(\frac{1}{\hbar}(-m\frac{d^2}{dt^2} + V''(x_{cl})) - E)}{\det(\frac{1}{\hbar}(-m\frac{d^2}{dt^2} + m\omega^2) - E)} = \frac{\prod_n (E_n - E)}{\prod_n (E_n^{(0)} - E)}, \tag{8.62}$$

where the determinant in the denominator corresponds to that of a free harmonic oscillator (in Euclidean space) which we have already evaluated (and which does not have a zero mode). Here $E_n, E_n^{(0)}$

denote respectively the eigenvalues of the operators (whose determinants we are calculating when $E = 0$) in the numerator and the denominator. It is easy to see that both sides of Eq. (8.62) have the same analytic structure and, therefore must be equal. We note that

$$\Delta(E = 0) = 0, \tag{8.63}$$

since there is a zero eigenvalue for the determinant in the numerator and further,

$$\Delta(E = \infty) = 1. \tag{8.64}$$

If we eliminate the zero mode in Eq. (8.62) by dividing it out, then we obtain

$$-\frac{1}{E} \Delta(E) \mid_{E=0} = \frac{\prod_{n>0} E_n}{\prod_n E_n^{(0)}},$$

or, $$-\frac{\partial \Delta(E)}{\partial E} \mid_{E=0} = \frac{\det'(\frac{1}{\hbar}(-m\frac{\mathrm{d}^2}{\mathrm{d}t^2} + V''(x_{\mathrm{cl}})))}{\det(\frac{1}{\hbar}(-m\frac{\mathrm{d}^2}{\mathrm{d}t^2} + m\omega^2))}. \tag{8.65}$$

Clearly, if we can evaluate the left hand side of Eq. (8.65), then we would have evaluated det$'$ since we already know the value of the determinant for the harmonic oscillator.

To evaluate this, let us consider the scattering problem for the one dimensional Schrödinger equation

$$\frac{1}{\hbar}\left(-m\frac{\mathrm{d}^2}{\mathrm{d}t^2} + V''(x_{\mathrm{cl}})\right)\psi = E\psi,$$

or, $$\left(-m\frac{\mathrm{d}^2}{\mathrm{d}t^2} + V''(x_{\mathrm{cl}})\right)\psi = \hbar E\psi. \tag{8.66}$$

If we define the asymptotic solutions (Jost functions) as

$$\lim_{t\to\infty} f_+(t, E) \to e^{-ikt},$$

$$\lim_{t\to-\infty} f_-(t, E) \to e^{ikt}, \tag{8.67}$$

where we have used the asymptotic form of $V''(x_{cl})$ (see, for example, Eq. (8.27)) and have identified

$$\lim_{t \to \pm\infty} V''(x_{cl}) \to m\omega^2,$$

$$k^2 = \frac{\hbar E}{m} - \omega^2. \tag{8.68}$$

The Jost functions are two linearly independent (asymptotic) solutions (basically they correspond to the asymptotic incoming and outgoing solutions) of the Schrödinger equation in Eq. (8.66) and consequently, any general solution can be written as a linear combination of the two. In particular, we can write (the coefficients of expansions can be related to scattering amplitudes)

$$\lim_{t \to -\infty} f_+(t, E) \to A_+(E) e^{ikt} + B_+(E) e^{-ikt},$$

$$\lim_{t \to \infty} f_-(t, E) \to B_-(E) e^{ikt} + A_-(E) e^{-ikt}. \tag{8.69}$$

The linear independence of the Jost functions can be easily seen by calculating the Wronskian which is defined as

$$W(f_+(t, E), f_-(t, E))$$

$$= f_+(t, E) \frac{\partial f_-(t, E)}{\partial t} - \frac{\partial f_+(t, E)}{\partial t} f_-(t, E). \tag{8.70}$$

Using Eqs. (8.67) and (8.69), for $t \to -\infty$ this leads to $2ikB_+(E)$ while in the limit $t \to \infty$ this takes the value $2ikB_-(E)$. Furthermore, since the Wronskian is time independent, it follows that

$$W(f_+(t, E), f_-(t, E)) = 2ik\, B_+(E) = 2ik\, B_-(E). \tag{8.71}$$

This, in fact, shows that the two coefficients $B_+(E)$ and $B_-(E)$ are identical. With a bit more analysis, they can also be shown to be equal to $\Delta(E)$, namely,

$$B_+(E) = B_-(E) = \Delta(E). \tag{8.72}$$

Let us also recall from Eq. (8.28) that the zero mode of Eq. (8.66) has the asymptotic form

$$
\lim_{t \to \pm\infty} \psi_0(t) = \lim_{t \to \pm\infty} \left(\frac{S_0}{m} \right)^{-\frac{1}{2}} \frac{a\omega}{2} \operatorname{sech}^2 \frac{\omega t}{2}
$$

$$
\to 2a\omega \left(\frac{S_0}{m} \right)^{-\frac{1}{2}} e^{\mp\omega t} = K\, e^{\mp\omega t}, \tag{8.73}
$$

where we have defined

$$
K = 2a\omega \left(\frac{S_0}{m} \right)^{-\frac{1}{2}}. \tag{8.74}
$$

Thus, from Eqs. (8.67), (8.68), (8.69) and (8.73), we note that we can identify $(k(E = 0) = -i\omega)$

$$
\lim_{|t| \to \infty} f_\pm(t, E = 0) \to e^{-\omega|t|} = \lim_{|t| \to \infty} \frac{1}{K} \psi_0(t). \tag{8.75}
$$

Comparing Eq. (8.75) with the asymptotic form of the Jost functions in Eqs. (8.67) and (8.69), we conclude that

$$
\begin{aligned}
A_+(E = 0) &= 1, & A_-(E = 0) &= 1, \\
B_+(E = 0) &= 0, & B_-(E = 0) &= 0.
\end{aligned} \tag{8.76}
$$

Consequently, we obtain

$$
\Delta(E = 0) = B_+(E = 0) = 0, \tag{8.77}
$$

a result which we already know.

The asymptotic equations which the Jost functions satisfy (see Eqs. (8.66) and (8.68)) are

$$
-m\frac{\partial^2 f_+(t, E)}{\partial t^2} + m\omega^2 f_+(t, E) = \hbar E f_+(t, E),
$$

$$
-m\frac{\partial^2 f_-(t, E')}{\partial t^2} + m\omega^2 f_-(t, E') = \hbar E' f_-(t, E'). \tag{8.78}
$$

Multiplying the first of these equations by $f_-(t, E')$ and the second by $f_+(t, E)$ and subtracting one from the other, we obtain

$$m\left(f_+(t, E)\frac{\partial^2 f_-(t, E')}{\partial t^2} - f_-(t, E')\frac{\partial^2 f_+(t, E)}{\partial t^2}\right)$$

$$= \hbar(E - E')f_+(t, E)f_-(t, E'),$$

or, $\quad \dfrac{\partial}{\partial t}\left(f_+(t, E)\dfrac{\partial f_-(t, E')}{\partial t} - f_-(t, E')\dfrac{\partial f_+(t, E)}{\partial t}\right)$

$$= \frac{\hbar}{m}(E - E')f_+(t, E)f_-(t, E'),$$

or, $\quad \dfrac{\partial}{\partial t}W(f_+(t, 0), f_-(t, E)) = -\dfrac{\hbar}{m}Ef_+(t, 0)f_-(t, E),$

or, $\quad \dfrac{\partial^2}{\partial E \partial t}W(f_+(t, 0), f_-(t, E))\Big|_{E=0} = -\dfrac{\hbar}{m}f_+(t, 0)f_-(t, 0)$

$$= -\frac{\hbar}{mK^2}\psi_0^2(t). \tag{8.79}$$

Integrating this equation between $\left(-\frac{T}{2}, \frac{T}{2}\right)$ with $T \to \infty$, we obtain

$$\lim_{E \to 0}\frac{\partial}{\partial E}W(f_+(t, 0), f_-(t, E))\Big|_{t=-\infty}^{t=\infty} = -\frac{\hbar}{mK^2}. \tag{8.80}$$

On the other hand, from the asymptotic form of the Jost functions in Eq. (8.69), we see that

$$W(f_+(t, 0), f_-(t, E))|_{t=-\infty}^{t=\infty}$$

$$= \left(e^{-\omega t}\left(ik\, B_-(E)e^{ikt} - ik\, A_-(E)e^{-ikt}\right)\right.$$

$$\left. + \omega e^{-\omega t}(B_-(E)e^{ikt} + A_-(E)e^{-ikt})\right)_{t=\infty}$$

$$- \left(e^{\omega t}ik\, e^{ikt} - \omega e^{\omega t}e^{ikt}\right)_{t=-\infty}$$

$$= \left(e^{-\omega t}\left((\omega + ik)e^{ikt}B_-(E) + (\omega - ik)e^{-ikt}A_-(E)\right)\right)_{t=\infty}$$

$$+ \left((\omega - ik)e^{(\omega+ik)t}\right)_{t=-\infty}, \tag{8.81}$$

from which we determine

$$
\lim_{E \to 0} \frac{\partial}{\partial E} W(f_+(t,0), f_-(t,E)) \Big|_{-\infty}^{\infty}
$$

$$
= 2\omega \frac{\partial B_-(E)}{\partial E} \Big|_{E=0} = 2\omega \frac{\partial \Delta(E)}{\partial E} \Big|_{E=0}. \tag{8.82}
$$

Here we have used the identification in Eq. (8.72) as well as the results of Eq. (8.76). Comparing Eqs. (8.80) and (8.82), then, we obtain

$$
2\omega \frac{\partial \Delta(E)}{\partial E} \Big|_{E=0} = -\frac{\hbar}{mK^2},
$$

$$
\text{or,} \quad -\frac{\partial \Delta(E)}{\partial E} \Big|_{E=0} = \frac{\hbar}{2m\omega K^2}. \tag{8.83}
$$

Therefore, we determine the ratio in Eq. (8.65) to be

$$
\frac{\det'(\frac{1}{\hbar}(-m\frac{d^2}{dt^2} + V''(x_{cl})))}{\det(\frac{1}{\hbar}(-m\frac{d^2}{dt^2} + m\omega^2))} = -\frac{\partial \Delta(E)}{\partial E} \Big|_{E=0} = \frac{\hbar}{2m\omega K^2}. \tag{8.84}
$$

The one instanton contribution in Eq. (8.61) can now be explicitly determined and takes the form

$$
\langle a|e^{-\frac{1}{\hbar}HT}| - a\rangle_{\text{O.I.}}
$$

$$
= \frac{N}{\sqrt{\det(\frac{1}{\hbar}(-m\frac{d^2}{dt^2} + \omega^2))}} \sqrt{\frac{\det(\frac{1}{\hbar}(-m\frac{d^2}{dt^2} + \omega^2))}{\det'(\frac{1}{\hbar}(-m\frac{d^2}{dt^2} + V''(x_{cl})))}}
$$

$$
\times \left(\frac{S_0}{m}\right)^{\frac{1}{2}} e^{-\frac{1}{\hbar}S_0} \int_{-\frac{T}{2}}^{\frac{T}{2}} dt_c
$$

$$
= \left(\frac{m\omega}{\pi\hbar}\right)^{\frac{1}{2}} e^{-\frac{\omega T}{2}} \sqrt{\frac{2m\omega}{\hbar}} K \left(\frac{S_0}{m}\right)^{\frac{1}{2}} e^{-\frac{1}{\hbar}S_0} \int_{-\frac{T}{2}}^{\frac{T}{2}} dt_c
$$

$$= \left(\frac{m\omega}{\pi\hbar}\right)^{\frac{1}{2}} e^{-\frac{\omega T}{2}} \sqrt{\frac{2m\omega}{\hbar}} \times 2a\omega \left(\frac{S_0}{m}\right)^{-\frac{1}{2}} \left(\frac{S_0}{m}\right)^{\frac{1}{2}} e^{-\frac{1}{\hbar}S_0} \int\limits_{-\frac{T}{2}}^{\frac{T}{2}} dt_c$$

$$= \left(\frac{m\omega}{\pi\hbar}\right)^{\frac{1}{2}} e^{-\frac{\omega T}{2}} r \int\limits_{-\frac{T}{2}}^{\frac{T}{2}} dt_c \,, \tag{8.85}$$

where we have used Eq. (8.84) as well as the value of the path integral for the harmonic oscillator given in Eq. (3.32) (rotated to Euclidean space with large Euclidean time which also determines the normalization constant). We have also defined a new quantity r whose value is given by

$$r = 2\sqrt{\frac{2m}{\hbar}} \, \omega^{\frac{3}{2}} a \, e^{-\frac{1}{\hbar}S_0} \,. \tag{8.86}$$

The transition amplitude, in this case, separates into a product of two factors — the first corresponding to the contribution of a simple harmonic oscillator arising from the trivial solution of the Euclidean equation of motion while the second giving the true contribution due to an instanton. We can, similarly, calculate the transition amplitude in the presence of an anti-instanton and it can be shown to be identical to the result obtained in Eq. (8.85).

8.5 Multi-instanton contributions

As we had discussed earlier, a string of widely separated instantons and anti-instantons also satisfies the Euclidean classical equation given in Eq. (8.6). The instanton density is small for weak coupling and, in such a case, the contribution of these multi-instanton solutions to the transition amplitude can be evaluated under an approximation commonly known as the dilute gas approximation. A typical example of a "multi-instanton" solution has the form shown in Fig. 8.4.

Let us consider a n-"instanton" solution for which the centers of the instantons at $t_1, t_2, \cdots, t_n$ satisfy

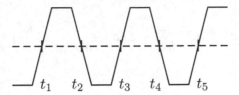

Figure 8.4: Graphical representation for a "multi-instanton" solution.

$$-\frac{T}{2} \leq t_n \leq t_{n-1} \cdots \leq t_1 \leq \frac{T}{2}. \tag{8.87}$$

In such a case, the integral over the centers of the instantons gives

$$\int\limits_{-\frac{T}{2}}^{\frac{T}{2}} dt_1 \int\limits_{-\frac{T}{2}}^{t_1} dt_2 \cdots \int\limits_{-\frac{T}{2}}^{t_{n-1}} dt_n = \frac{T^n}{n!}. \tag{8.88}$$

Furthermore, since the instantons and the anti-instantons are assumed to be noninteracting in this dilute gas approximation, their contributions to the transition amplitude will simply be multiplicative. Thus, a n-"instanton" solution will contribute an amount (see Eqs. (8.85) and (8.88))

$$\left(\frac{m\omega}{\pi\hbar}\right)^{\frac{1}{2}} e^{-\frac{\omega T}{2}} r^n \frac{T^n}{n!} = \left(\frac{m\omega}{\pi\hbar}\right)^{\frac{1}{2}} e^{-\frac{\omega T}{2}} \frac{(rT)^n}{n!}. \tag{8.89}$$

(Basically, the entire region can be divided into n-"instanton" regions and in each region we can divide the corresponding instanton determinant by a harmonic oscillator determinant leading to a factor of r for each region. The overall harmonic oscillator determinant for the entire region, necessary to compensate for the product of individual harmonic oscillator determinants, will give rise to the first two factors.)

We have to recognize here that only an even number of instantons and anti-instantons (in total) can contribute to the transition amplitude of the form

$$\langle a|e^{-\frac{1}{\hbar}HT}|a\rangle, \qquad \langle -a|e^{-\frac{1}{\hbar}HT}|-a\rangle. \tag{8.90}$$

Similarly, only an odd number of instantons and anti-instantons (in total) can contribute to transition amplitudes of the form

$$\langle a|e^{-\frac{1}{\hbar}HT}|-a\rangle, \qquad \langle -a|e^{-\frac{1}{\hbar}HT}|a\rangle. \tag{8.91}$$

Adding all such "instanton" contributions, we see from Eq. (8.89) that we will have

$$
\begin{aligned}
\langle -a|e^{-\frac{1}{\hbar}HT}|-a\rangle &= \sum_n \left(\frac{m\omega}{\pi\hbar}\right)^{\frac{1}{2}} e^{-\frac{\omega T}{2}} \frac{(rT)^{2n}}{(2n)!} \\
&= \left(\frac{m\omega}{\pi\hbar}\right)^{\frac{1}{2}} e^{-\frac{\omega T}{2}} \cosh(rT) \\
&= \left(\frac{m\omega}{\pi\hbar}\right)^{\frac{1}{2}} e^{-\frac{\omega T}{2}} \frac{(e^{rT} + e^{-rT})}{2} \\
&= \frac{1}{2}\left(\frac{m\omega}{\pi\hbar}\right)^{\frac{1}{2}} \left(e^{-(\frac{\omega}{2}-r)T} + e^{-(\frac{\omega}{2}+r)T}\right). \tag{8.92}
\end{aligned}
$$

Similarly, we can show that for odd numbers of "instanton" contributions, we will obtain

$$
\begin{aligned}
\langle a|e^{-\frac{1}{\hbar}HT}|-a\rangle &= \left(\frac{m\omega}{\pi\hbar}\right)^{\frac{1}{2}} e^{-\frac{\omega T}{2}} \sinh(rT) \\
&= \frac{1}{2}\left(\frac{m\omega}{\pi\hbar}\right)^{\frac{1}{2}} \left(e^{-(\frac{\omega}{2}-r)T} - e^{-(\frac{\omega}{2}+r)T}\right). \tag{8.93}
\end{aligned}
$$

If we identify the two low lying states of the quantum Hamiltonian as $|\pm\rangle$ with energy eigenvalues $E_\pm$ respectively (these are what we had called E_2, E_1 respectively in the previous chapter), then we note that by inserting a complete set of energy states we will obtain for large T

$$\langle -a|e^{-\frac{1}{\hbar}HT}|-a\rangle$$

$$\simeq \langle -a|e^{-\frac{1}{\hbar}HT}|-\rangle\langle -|-a\rangle + \langle -a|e^{-\frac{1}{\hbar}HT}|+\rangle\langle +|-a\rangle \qquad (8.94)$$

$$= e^{-\frac{1}{\hbar}E_- T}\langle -a|-\rangle\langle -|-a\rangle + e^{-\frac{1}{\hbar}E_+ T}\langle -a|+\rangle\langle +|-a\rangle .$$

Comparing the exponents of these two terms with those in Eq. (8.92), we obtain

$$E_\pm = \hbar\left(\frac{\omega}{2} \pm r\right). \qquad (8.95)$$

Therefore, the splitting between the two energy levels of the quantum theory is determined to be

$$\Delta E = E_+ - E_- = 2\hbar r$$

$$= 2\hbar \times 2\sqrt{\frac{2m}{\hbar}}\, \omega^{\frac{3}{2}} a\, e^{-\frac{1}{\hbar}S_0}$$

$$= 4\sqrt{2m\hbar}\, \omega^{\frac{3}{2}} a\, e^{-\frac{1}{\hbar}S_0} . \qquad (8.96)$$

This splitting of energy levels calculated in the path integral formalism can now be compared with the result obtained through the WKB approximation in Eq. (7.84) (remember that $\frac{e}{\pi} \simeq 0.88$ while $\sqrt{2} \simeq 1.4$).

8.6 References

S. Coleman, *The Uses of Instantons*, Erice Lectures, 1977.

B. Sakita, *Quantum Theory of Many Variable Systems and Fields*, World Scientific Publishing.

M. Shifman *et al.*, Sov. Phys. Usp. **25**, 195 (1982).

J. Zinn-Justin, *Quantum Field Theory and Critical Phenomena*, Oxford Univ. Press.

CHAPTER 9

Path integral for relativistic theories

9.1 Systems with many degrees of freedom

Thus far, we have discussed only one particle systems. However, the method of path integrals generalizes naturally to systems with many particles or systems with many degrees of freedom. Let us consider a system with n-degrees of freedom characterized by the coordinates $x^\alpha(t)$, $\alpha = 1, 2, \cdots, n$. These coordinates, for example, can denote the coordinates of n-particles in one dimension or the coordinates of a single particle in n-dimensions. If $S[x]$ denotes the appropriate action for the system (namely, if it describes the dynamics of the system), then the transition amplitude in Eq. (2.29) can be easily shown to generalize to

$$\langle x_f^\alpha, t_f | x_i^\alpha, t_i \rangle = N \int \mathcal{D}x^\alpha \, e^{\frac{i}{\hbar} S[x^\alpha]} \,. \tag{9.1}$$

The action generically has the form

$$S[x^\alpha] = \int_{t_i}^{t_f} dt \, L(x^\alpha, \dot{x}^\alpha) \,, \tag{9.2}$$

and the integration in Eq. (9.1) is supposed to be over all paths starting at x_i^α at $t = t_i$ and ending at x_f^α at $t = t_f$. We can also introduce appropriate sources, in this case, through the linear interaction

$$S[x^\alpha, J^\alpha] = S[x^\alpha] + \int_{t_i}^{t_f} dt \, J^\alpha(t) x^\alpha(t) \,, \tag{9.3}$$

to define the transition amplitude in the presence of these sources as (see Eq. (4.49))

$$\langle x_f^\alpha, t_f | x_i^\alpha, t_i \rangle^J = N \int \mathcal{D}x^\alpha \, e^{\frac{i}{\hbar} S[x^\alpha, J^\alpha]} . \tag{9.4}$$

As we have seen earlier in Eqs. (4.47) and (4.62), this allows us to derive various transition amplitudes or matrix elements in a simple manner. We can also define, as before, the vacuum to vacuum transition amplitude in the limit of infinite time interval as (see Section 4.4)

$$Z[J] = \langle 0|0 \rangle^J = N \int \mathcal{D}x^\alpha \, e^{\frac{i}{\hbar} S[x^\alpha, J^\alpha]} , \tag{9.5}$$

where in the infinite time interval limit, the action in Eq. (9.5) has the form

$$S[x^\alpha, J^\alpha] = \int\limits_{-\infty}^{\infty} dt \left(L(x^\alpha, \dot{x}^\alpha) + J^\alpha(t) x^\alpha(t) \right) , \tag{9.6}$$

and the integration over the paths in Eq. (9.5) has no end point restriction in the sense that the initial and the final coordinates of the paths can be chosen arbitrarily.

The path integrals can also be extended to continuum field theories once we recognize that these theories describe physical systems with an infinite number of degrees of freedom. Thus, if $\phi(x, t)$ is the basic variable of a $1 + 1$ dimensional field theory, then the vacuum to vacuum transition amplitude in the presence of an external source can be written as

$$Z[J] = \langle 0|0 \rangle^J = N \int \mathcal{D}\phi \, e^{\frac{i}{\hbar} S[\phi, J]} , \tag{9.7}$$

where

$$S[\phi, J] = S[\phi] + \int\limits_{-\infty}^{\infty} \!\! \int dt \, dx \, J(x, t) \phi(x, t) . \tag{9.8}$$

(Incidentally, in all these discussions, we are assuming that the relation between the Lagrangian and the Hamiltonian of the system is the canonical one which would lead to path integrals of the form in Eq. (9.5) or (9.7). If this is not the case, then one should take as the starting point, the path integral in the phase space as obtained in Eq. (2.23).) Before going into the discussion about the functional integration in the present case, it is worth emphasizing what we have discussed earlier, namely, it is the time ordered Green's functions in the vacuum which play the most important role in a field theory because the scattering matrix or the S-matrix can be obtained from them. This is why it is the vacuum functional which is the quantity of fundamental significance in these studies. The second point to note is that we have left the specific form of $S[\phi]$ arbitrary. Depending on the particular form of the action, we will be dealing with different kinds of field theories — both non-relativistic and relativistic.

Returning now to the question of the functional integration, let us recall that in the $0 + 1$ dimensional case, we defined the path integral by dividing the time interval into infinitesimal segments (see Eqs. (2.19) and (2.21)). Here, in addition, we have to divide the space interval into infinitesimal segments as shown in Fig. 9.1. Thus, we assume that

$$-\frac{L}{2} \leq x \leq \frac{L}{2}, \tag{9.9}$$

with the understanding that we will take the limit $L \to \infty$ at the end. We divide the finite length interval into $\bar{N}$ equal segments of infinitesimal length ϵ such that

$$\bar{N}\epsilon = L. \tag{9.10}$$

With this, the space-time manifold is divided into infinitesimal cells which we can label with an index "i". We denote the average value of the field variable in the i-th cell of infinitesimal area δA_i by ϕ_i

$$\phi_i = \frac{1}{\delta A_i} \int_{\delta A_i} \mathrm{d}t\, \mathrm{d}x\, \phi(t, x), \tag{9.11}$$

$$-\frac{L}{2} \qquad\qquad\qquad \frac{L}{2}$$

Figure 9.1: Division of space interval into infinitesimal segments.

we can define the path integral measure as in the case of quantum mechanics (see Eq. (2.30), for example)

$$\int \mathcal{D}\phi = \int \prod_i \mathrm{d}\phi_i \,, \tag{9.12}$$

where the limiting procedures are to be understood. However, unlike the case of $0 + 1$ dimensional quantum mechanics, which we have extensively discussed, the path integral, in the present case, does not exist in the sense that the integrations defined in Eq. (9.12) lead to divergences in the continuum limit. However, if we absorb the divergence into the normalization constant N in Eq. (9.5) or (9.7), then the Green's functions can still be defined uniquely since they are defined as ratios for which the divergent constants simply drop out.

For field theories in higher dimensions where the basic variables are $\phi(\mathbf{x}, t)$, we can define the vacuum generating functional exactly in an analogous manner. Namely, we have

$$Z[J] = \langle 0|0\rangle^J = N \int \mathcal{D}\phi \, e^{\frac{i}{\hbar} S[\phi, J]} \,, \tag{9.13}$$

where

$$S[\phi, J] = S[\phi] + \int \mathrm{d}^n x \, J(\mathbf{x}, t)\phi(\mathbf{x}, t) \,, \tag{9.14}$$

and the integration in Eq. (9.14) is over the entire space-time manifold in higher dimension with n denoting the dimensionality of the space-time manifold (this would correspond to a $((n - 1) + 1)$ dimensional field theory). The functional integral in Eq. (9.13), in such a case, is defined by taking a hypercube in n dimensions (the space-time manifold with suitable limiting procedure) divided into a large number of infinitesimal hypercubes and then identifying the functional integral with a product of ordinary integrals of the field values at each of the infinitesimal hypercubes (in an averaged sense over the volume of the hypercube as defined in Eq. (9.11)).

9.2 Relativistic scalar field theory

With all these preliminaries, let us choose a specific form for the action in Eq. (9.14). Namely, let us choose a relativistic scalar field theory in $3 + 1$ dimensions described by the Lagrangian density

$$\mathcal{L}(\phi, \partial_\mu \phi) = \frac{1}{2} \partial_\mu \phi \partial^\mu \phi - \frac{m^2}{2} \phi^2 - \frac{\lambda}{4!} \phi^4, \qquad (9.15)$$

with $\lambda > 0$ (this condition merely corresponds to the fact that we would like the potential to be bounded from below in which case the quantum theory will have a meaningful ground state) so that

$$S[\phi] = \int d^4 x \, \mathcal{L}(\phi, \partial_\mu \phi), \qquad (9.16)$$

and

$$S[\phi, J] = S[\phi] + \int d^4 x \, J(\mathbf{x}, t) \phi(\mathbf{x}, t) = \int d^4 x \, \mathcal{L}^J(\phi, \partial_\mu \phi). \quad (9.17)$$

It is worth noting here that the Lagrangian can be obtained from the Lagrangian density in Eq. (9.15) by integrating over the space variables as

$$L = \int d^3 x \, \mathcal{L}(\phi, \partial_\mu \phi). \qquad (9.18)$$

This theory is quite similar (see Eq. (4.77)) to the anharmonic oscillator which we discussed earlier except that it is a relativistic field theory invariant under global Poincaré transformations. This is a self-interacting theory which can describe charge neutral spin zero particles of mass m. The Euler-Lagrange equation following from the action in Eq. (9.17) takes the form

$$-\frac{\delta S[\phi, J]}{\delta \phi(x)} = \partial_\mu \frac{\partial \mathcal{L}^J}{\partial \partial_\mu \phi} - \frac{\partial \mathcal{L}^J}{\partial \phi}$$

$$= \partial_\mu \partial^\mu \phi + m^2 \phi + \frac{\lambda}{3!} \phi^3 - J(x) = 0. \qquad (9.19)$$

Here we have chosen to represent the space-time variables in a compact notation of x for simplicity. Commonly, this theory described by the action in Eq. (9.17) or by the dynamical equations in Eq. (9.19) is also known as the ϕ^4-theory.

In the absence of self-interaction, namely, when $\lambda = 0$, the action in Eq. (9.17) is at most quadratic in the field variables and hence the generating functional can be evaluated in much the same way as in the case of quantum mechanical systems (see Chapters 2, 3 and in particular, Chapter 4). However, let us first define the Feynman Green's function associated with this theory. The equation satisfied by the Green's function is given by

$$\left(\partial_\mu \partial^\mu + m^2\right) G(x - x') = -\delta^4(x - x'). \qquad (9.20)$$

Defining the Fourier transforms as

$$G(x - x') = \int \frac{\mathrm{d}^4 k}{(2\pi)^4} G(k) e^{-ik\cdot(x-x')},$$

$$\delta^4(x - x') = \int \frac{\mathrm{d}^4 k}{(2\pi)^4} e^{-ik\cdot(x-x')}, \qquad (9.21)$$

and substituting these back into the differential equation Eq. (9.20) we obtain

$$\frac{1}{(2\pi)^4} \left(-k^2 + m^2\right) G(k) = -\frac{1}{(2\pi)^4}$$

$$\text{or,} \quad G(k) = \frac{1}{k^2 - m^2}. \qquad (9.22)$$

Here we are using the scalar product for the four vectors with the metrics introduced in Eqs. (1.3) and (1.4) and $k^2 = k_\mu k^\mu$ represents the invariant length square of the conjugate four vector k_μ. The Feynman Green's function is, then defined following Eq. (3.80) as

$$G_F(x - x') = \lim_{\epsilon \to 0^+} \int \frac{\mathrm{d}^4 k}{(2\pi)^4} \frac{1}{k^2 - m^2 + i\epsilon} e^{-ik \cdot (x - x')}. \qquad (9.23)$$

We can also think of the Feynman Green's function as satisfying the differential equation (see Eq. (3.82))

$$\lim_{\epsilon \to 0^+} \left(\partial_\mu \partial^\mu + m^2 - i\epsilon \right) G_F(x - x') = -\delta^4(x - x'). \qquad (9.24)$$

It is clear from (9.23) that the Feynman Green's function in this theory is a symmetric function

$$G_F(x - x') = G_F(x' - x). \qquad (9.25)$$

Going back to the generating functional in Eq. (9.14), we note that for the present case, if $\lambda = 0$, then we can define

$$Z_0[J] = N \int \mathcal{D}\phi \, e^{\frac{i}{\hbar} S_0[x, J]}$$

$$= N \int \mathcal{D}\phi \, e^{\frac{i}{\hbar} \int \mathrm{d}^4 x \, \left(\frac{1}{2} \partial_\mu \phi \partial^\mu \phi - \frac{m^2}{2} \phi^2 + J\phi \right)}$$

$$= N \int \mathcal{D}\phi \, e^{-\frac{i}{\hbar} \int \mathrm{d}^4 x \, \left(\frac{1}{2} \phi (\partial_\mu \partial^\mu + m^2) \phi - J\phi \right)}, \qquad (9.26)$$

with $S_0[x, J]$ representing the action Eq. (9.17) with $\lambda = 0$. The field $\phi(\mathbf{x}, t)$ is assumed to satisfy the asymptotic condition

$$\lim_{|\mathbf{x}| \to \infty} \phi(\mathbf{x}, t) \to 0. \qquad (9.27)$$

Let us note here once again that the integral in Eq. (9.26) should be properly evaluated by rotating to Euclidean space as discussed in Section 4.1. Alternatively, we can also define

$$Z_0[J] = \lim_{\epsilon \to 0^+} N \int \mathcal{D}\phi \, e^{-\frac{i}{\hbar} \int d^4x \left(\frac{1}{2} \phi(\partial_\mu \partial^\mu + m^2 - i\epsilon)\phi - J\phi \right)} . \qquad (9.28)$$

If we now redefine the variable of integration to be

$$\tilde{\phi}(x) = \phi(x) + \int d^4x' \, G_F(x - x') J(x') , \qquad (9.29)$$

with G_F defined in Eq. (9.23), then, we obtain

$$\lim_{\epsilon \to 0^+} \int d^4x \, \frac{1}{2} \tilde{\phi}(\partial_\mu \partial^\mu + m^2 - i\epsilon)\tilde{\phi}$$

$$= \lim_{\epsilon \to 0^+} \int d^4x \, \frac{1}{2} \left(\phi(x) + \int d^4x' \, G_F(x - x') J(x') \right)$$

$$\times \; (\partial_\mu \partial^\mu + m^2 - i\epsilon) \left(\phi(x) + \int d^4x'' \, G_F(x - x'') J(x'') \right)$$

$$= \lim_{\epsilon \to 0^+} \int d^4x \left[\frac{1}{2}\phi(x)(\partial_\mu \partial^\mu + m^2 - i\epsilon)\phi(x) - J(x)\phi(x) \right.$$

$$\left. - \frac{1}{2} \int d^4x' \, J(x)G_F(x - x')J(x') \right] , \qquad (9.30)$$

where we have used Eq. (9.24). Substituting Eq. (9.30) back into the generating functional in Eq. (9.28), we obtain (note that the Jacobian for the change of variable in Eq. (9.29) is unity)

$$Z_0[J] = \lim_{\epsilon \to 0^+} N \int \mathcal{D}\phi \, e^{-\frac{i}{\hbar} \int d^4x (\frac{1}{2}\phi(x)(\partial_\mu \partial^\mu + m^2 - i\epsilon)\phi(x) - J(x)\phi(x))}$$

$$= \lim_{\epsilon \to 0^+} e^{-\frac{i}{2\hbar} \iint d^4x \, d^4x' \, J(x)G_F(x-x')J(x')}$$

$$\times \; N \int \mathcal{D}\tilde{\phi} \, e^{-\frac{i}{\hbar} \int d^4x \, \frac{1}{2}\tilde{\phi}(x)(\partial_\mu \partial^\mu + m^2 - i\epsilon)\tilde{\phi}(x)}$$

$$= N \left[\det(\partial_\mu \partial^\mu + m^2) \right]^{-\frac{1}{2}}$$

$$\times \; e^{-\frac{i}{2\hbar} \iint d^4x \, d^4x' \, J(x)G_F(x-x')J(x')}$$

$$= Z_0[0] \, e^{-\frac{i}{2\hbar} \iint d^4x \, d^4x' \, J(x)G_F(x-x')J(x')} . \qquad (9.31)$$

Here we have used a generalization of the result in Eq. (4.3) for a field theory.

As in the case of the harmonic oscillator, we note that when $\lambda = 0$ (we are calculating the normalized vacuum expectation values below),

$$
\begin{aligned}
\langle 0|\phi(x)|0\rangle &= \frac{(-i\hbar)}{Z_0[J]} \frac{\delta Z_0[J]}{\delta J(x)}\bigg|_{J=0} \\
&= \frac{(-i\hbar)}{Z_0[J]} \left(-\frac{i}{\hbar} \int \mathrm{d}^4 x' \, G_F(x - x') J(x') \right) Z_0[J]\bigg|_{J=0} \\
&= 0,
\end{aligned}
\tag{9.32}
$$

and

$$
\begin{aligned}
\langle 0|T(\phi(x)\phi(y))|0\rangle &= \frac{(-i\hbar)^2}{Z_0[J]} \frac{\delta^2 Z_0[J]}{\delta J(x)\delta J(y)}\bigg|_{J=0} \\
&= \frac{(-i\hbar)^2}{Z_0[J]} \left(-\frac{i}{\hbar} G_F(x - y) \right) Z_0[J]\bigg|_{J=0} \\
&= i\hbar \, G_F(x - y).
\end{aligned}
\tag{9.33}
$$

Namely, we obtain once again the familiar result that the Feynman propagator is nothing other than the time ordered two point correlation function in the vacuum (see Eq. (4.76)).

Just as the path integral for the anharmonic oscillator cannot be evaluated in a closed form, the ϕ^4-theory does not also have a closed form expression for the generating functional. However, we can evaluate it perturbatively at least when the coupling λ is weak (small). We note that we can write (as we had also noted earlier in Eq. (4.83) in the case of the anharmonic oscillator)

$$
\phi(x) \to \frac{\delta}{\delta J(x)},
\tag{9.34}
$$

when acting on the action $S_0[\phi, J]$. Therefore, we can rewrite the generating functional of Eq. (9.13), in the present case, also as

$$Z[J] = N \int \mathcal{D}\phi \, e^{\frac{i}{\hbar} \int d^4x \left(\frac{1}{2}\partial_\mu \phi \partial^\mu \phi - \frac{m^2}{2}\phi^2 - \frac{\lambda}{4!}\phi^4 + J\phi\right)}$$

$$= N \int \mathcal{D}\phi \left(e^{-\frac{i\lambda}{4!\hbar} \int d^4x \, \phi^4(x)} \right) e^{\frac{i}{\hbar} S_0[\phi,J]}$$

$$= N \int \mathcal{D}\phi \left(e^{-\frac{i\lambda}{4!\hbar} \int d^4x \, \left(-i\hbar \frac{\delta}{\delta J(x)}\right)^4} \right) e^{\frac{i}{\hbar} S_0[\phi,J]}$$

$$= \left(e^{-\frac{i\lambda}{4!\hbar} \int d^4x \, \left(-i\hbar \frac{\delta}{\delta J(x)}\right)^4} \right) N \int \mathcal{D}\phi \, e^{\frac{i}{\hbar} S_0[\phi,J]}$$

$$= \left(e^{-\frac{i\lambda\hbar^3}{4!} \int d^4x \, \left(\frac{\delta}{\delta J(x)}\right)^4} \right) Z_0[J]. \tag{9.35}$$

Once again, we note that this is very analogous to the result obtained in Eq. (4.84) for the anharmonic oscillator.

A power series expansion in λ for the generating functional in Eq. (9.35) follows by Taylor expanding the exponential involving the interaction terms. Thus, we have

$$Z[J] = \left[1 - \frac{i\lambda\hbar^3}{4!} \int d^4x \frac{\delta^4}{\delta J^4(x)} + \frac{1}{2!} \left(-\frac{i\lambda\hbar^3}{4!} \right)^2 \right.$$

$$\times \left(\int d^4x \frac{\delta^4}{\delta J^4(x)} \right) \left(\int d^4y \frac{\delta^4}{\delta J^4(y)} \right) + \cdots \left] Z_0[J] \right.$$

$$= Z_0[0] \left[1 - \frac{i\lambda\hbar^3}{4!} \int d^4x \frac{\delta^4}{\delta J^4(x)} + \frac{1}{2!} \left(-\frac{i\lambda\hbar^3}{4!} \right)^2 \right.$$

$$\times \left(\int d^4x \frac{\delta^4}{\delta J^4(x)} \right) \left(\int d^4y \frac{\delta^4}{\delta J^4(y)} \right) + \cdots \right]$$

$$\times e^{-\frac{i}{2\hbar} \int\int d^4x_1 \, d^4x_2 \, J(x_1) G_F(x_1 - x_2) J(x_2)}. \tag{9.36}$$

To obtain a feeling for how the actual calculations are carried out, let us derive some of the Green's functions to low orders in the coupling constant for the ϕ^4-theory. Let us recall from Eq. (4.62) that, by definition,

$$\langle 0|T(\phi(x_1)\phi(x_2)\cdots\phi(x_n))|0\rangle$$

$$= \frac{(-i\hbar)^n}{Z[J]} \left.\frac{\delta^n Z[J]}{\delta J(x_1)\delta J(x_2)\cdots\delta J(x_n)}\right|_{J=0}. \qquad (9.37)$$

Furthermore, from the symmetry of $Z[J]$ in Eqs. (9.31) and (9.35) (namely, from the fact that it is invariant under $J \leftrightarrow -J$), we conclude that the vacuum expectation value of the time ordered product of an odd number of fields must vanish in this theory. In other words,

$$\langle 0|T(\phi(x_1)\cdots\phi(x_{2n+1}))|0\rangle$$

$$= \frac{(-i\hbar)^{2n+1}}{Z[J]} \left.\frac{\delta^{2n+1} Z[J]}{\delta J(x_1)\cdots\delta J(x_{2n+1})}\right|_{J=0} = 0. \qquad (9.38)$$

Consequently, only the even order Green's functions will be nontrivial in this theory and let us calculate only the 2-point and the 4-point functions up to order λ. Keeping terms up to order λ, we note from Eq. (9.36) that the generating functional takes the form

$$Z[J] \simeq Z_0[0]\left(1 - \frac{i\lambda\hbar^3}{4!}\int d^4x \frac{\delta^4}{\delta J^4(x)}\right)$$

$$\times\ e^{-\frac{i}{2\hbar}\iint d^4x_1 d^4x_2 J(x_1)G_F(x_1-x_2)J(x_2)}. \qquad (9.39)$$

To evaluate this, let us note that

$$\frac{\delta^2}{\delta J^2(x)} e^{-\frac{i}{2\hbar}\iint d^4x_1 d^4x_2\, J(x_1)G_F(x_1-x_2)J(x_2)}$$

$$= \frac{\delta}{\delta J(x)}\left[-\frac{i}{\hbar}\left(\int d^4x_3 G_F(x-x_3)J(x_3)\right)\right.$$

$$\left.\times\ e^{-\frac{i}{2\hbar}\iint d^4x_1 d^4x_2 J(x_1)G_F(x_1-x_2)J(x_2)}\right]$$

$$
= \left[-\frac{i}{\hbar} G_F(0) \right.
$$

$$
\left. - \frac{1}{\hbar^2} \left(\int \mathrm{d}^4 x_3 G_F(x - x_3) J(x_3) \right) \left(\int \mathrm{d}^4 x_4 G_F(x - x_4) J(x_4) \right) \right]
$$

$$
\times \; e^{-\frac{i}{2\hbar} \iint \mathrm{d}^4 x_1 \mathrm{d}^4 x_2 J(x_1) G_F(x_1 - x_2) J(x_2)} . \tag{9.40}
$$

With some algebraic manipulations this, then, leads to the result

$$
\int \mathrm{d}^4 x \frac{\delta^4}{\delta J^4(x)} e^{-\frac{i}{2\hbar} \iint \mathrm{d}^4 x_1 \mathrm{d}^4 x_2 J(x_1) G_F(x_1 - x_2) J(x_2)}
$$

$$
= \int \mathrm{d}^4 x \frac{\delta^2}{\delta J^2(x)} \left(\frac{\delta^2}{\delta J^2(x)} e^{-\frac{i}{2\hbar} \iint \mathrm{d}^4 x_1 \mathrm{d}^4 x_2 J(x_1) G_F(x_1 - x_2) J(x_2)} \right)
$$

$$
= \int \mathrm{d}^4 x \left[-\frac{3}{\hbar^2} G_F(0) G_F(0) \right.
$$

$$
+ \frac{6i}{\hbar^3} G_F(0) \left(\int \mathrm{d}^4 x_3 G_F(x - x_3) J(x_3) \right)
$$

$$
\times \left(\int \mathrm{d}^4 x_4 G_F(x - x_4) J(x_4) \right)
$$

$$
+ \frac{1}{\hbar^4} \left(\int \mathrm{d}^4 x_3 G_F(x - x_3) J(x_3) \right) \left(\int \mathrm{d}^4 x_4 G_F(x - x_4) J(x_4) \right)
$$

$$
\left. \times \left(\int \mathrm{d}^4 x_5 G_F(x - x_5) J(x_5) \right) \left(\int \mathrm{d}^4 x_6 G_F(x - x_6) J(x_6) \right) \right]
$$

$$
\times \; e^{-\frac{i}{2\hbar} \iint \mathrm{d}^4 x_1 \mathrm{d}^4 x_2 J(x_1) G_F(x_1 - x_2) J(x_2)} . \tag{9.41}
$$

Putting this back into Eq. (9.39), we obtain the generating functional to linear power in λ to be

$$
Z[J] = Z_0[0] \left[1 + \frac{i \lambda \hbar}{8} G_F(0) G_F(0) \int \mathrm{d}^4 x \right.
$$

$$
+ \frac{\lambda}{4} G_F(0) \int \mathrm{d}^4 x \left(\int \mathrm{d}^4 x_3 G_F(x - x_3) J(x_3) \right)
$$

$$
\times \left(\int \mathrm{d}^4 x_4 G_F(x - x_4) J(x_4) \right)
$$

$$-\frac{i\lambda}{4!\hbar}\int d^4x \left(\int d^4x_3 G_F(x-x_3)J(x_3)\right)$$

$$\times \left(\int d^4x_4 G_F(x-x_4)J(x_4)\right)$$

$$\times \left(\int d^4x_5 G_F(x-x_5)J(x_5)\right)\left(\int d^4x_6 G_F(x-x_6)J(x_6)\right)\bigg]$$

$$\times e^{-\frac{i}{2\hbar}\int\int d^4x_1 d^4x_2 J(x_1)G_F(x_1-x_2)J(x_2)} .\tag{9.42}$$

It now follows that

$$Z[0] = Z_0[0]\left(1 + \frac{i\lambda\hbar}{8}G_F(0)G_F(0)\int d^4x\right).\tag{9.43}$$

This is clearly divergent (both the space-time volume as well as the Green's function at vanishing arguments are divergent) and as we have argued earlier, the divergence can be absorbed into the normalization constant.

We recall that by definition, the time ordered two point function (in vacuum) is given by

$$\langle 0|T(\phi(x_1)\phi(x_2))|0\rangle = \frac{(-i\hbar)^2}{Z[J]}\frac{\delta^2 Z[J]}{\delta J(x_1)\delta J(x_2)}\bigg|_{J=0},\tag{9.44}$$

and from Eq. (9.42) we note that to linear order in λ we have

$$\frac{\delta^2 Z[J]}{\delta J(x_1)\delta J(x_2)}\bigg|_{J=0}$$

$$= Z_0[0]\left[\left(1 + \frac{i\lambda\hbar}{8}G_F(0)G_F(0)\int d^4x\right)\left(-\frac{i}{\hbar}G_F(x_1-x_2)\right)\right.$$

$$\left. + \frac{\lambda}{2}G_F(0)\int d^4x\, G_F(x-x_1)G_F(x-x_2)\right].\tag{9.45}$$

Therefore, to linear order in λ, we obtain

$$\langle 0|T(\phi(x_1)\phi(x_2))|0\rangle = \frac{(-i\hbar)^2}{Z[J]} \frac{\delta^2 Z[J]}{\delta J(x_1)\delta J(x_2)}\bigg|_{J=0}$$

$$= -\frac{\hbar^2}{Z_0[0]\,(1+\frac{i\lambda\hbar}{8}G_F(0)G_F(0)\int \mathrm{d}^4x)}$$

$$\times Z_0[0]\left[\left(1+\frac{i\lambda\hbar}{8}G_F(0)G_F(0)\int \mathrm{d}^4x\right)\left(-\frac{i}{\hbar}G_F(x_1-x_2)\right)\right.$$

$$\left. +\frac{\lambda}{2}G_F(0)\int \mathrm{d}^4x\, G_F(x-x_1)G_F(x-x_2)\right]$$

$$\simeq -\hbar^2\left[-\frac{i}{\hbar}G_F(x_1-x_2)\right.$$

$$\left. +\frac{\lambda}{2}G_F(0)\int \mathrm{d}^4x G_F(x-x_1)G_F(x-x_2)\right]$$

$$= i\hbar G_F(x_1-x_2) - \frac{\lambda\hbar^2}{2}G_F(0)\int \mathrm{d}^4x G_F(x-x_1)G_F(x-x_2).$$

$$(9.46)$$

Here in the second term, we have only kept the leading order term (which is unity) coming from the expansion of the denominator since we are interested in the 2-point function up to order λ. We note that the first term is, of course, the Feynman propagator for the free theory defined in Eq. (9.23). The second term, on the other hand, is a first order quantum correction to the propagator. It is worth pointing out here that $G_F(0)$ is a divergent quantity as we can readily check from the form of the Green's function in Eq. (9.23). Thus, we find that the first order correction to the propagator in this theory is divergent. This is, indeed, a general feature of quantum field theories, namely, the quantum corrections in a field theory lead to divergences which are then taken care of by a procedure known as renormalization.

Next, let us calculate the 4-point function up to order λ. To leading order, we note from Eq. (9.42) that

$$\frac{\delta^4 Z[J]}{\delta J(x_1)\delta J(x_2)\delta J(x_3)\delta J(x_4)}\bigg|_{J=0}$$

$$= Z_0[0] \left[-\frac{1}{\hbar^2} \left(1 + \frac{i\lambda\hbar}{8} G_F(0)G_F(0) \int d^4x \right) \right.$$

$$\times \left(G_F(x_1 - x_2)G_F(x_3 - x_4) \right.$$

$$+ G_F(x_1 - x_3)G_F(x_2 - x_4) + G_F(x_1 - x_4)G_F(x_2 - x_3) \Big)$$

$$- \frac{i\lambda}{2\hbar} G_F(0) \int d^4x \Big\{ G_F(x_1 - x_2)G_F(x - x_3)G_F(x - x_4)$$

$$+ G_F(x_1 - x_3)G_F(x - x_2)G_F(x - x_4)$$

$$+ G_F(x_1 - x_4)G_F(x - x_2)G_F(x - x_3)$$

$$+ G_F(x_2 - x_3)G_F(x - x_1)G_F(x - x_4)$$

$$+ G_F(x_2 - x_4)G_F(x - x_1)G_F(x - x_3)$$

$$+ G_F(x_3 - x_4)G_F(x - x_1)G_F(x - x_2) \Big\}$$

$$\left. - \frac{i\lambda}{\hbar} \int d^4x G_F(x - x_1)G_F(x - x_2)G_F(x - x_3)G_F(x - x_4) \right].$$

$$(9.47)$$

Substituting this back into the definition of the 4-point function in Eq. (9.37)and keeping terms only up to order λ, we obtain

$$\langle 0|T(\phi(x_1)\phi(x_2)\phi(x_3)\phi(x_4))|0\rangle$$

$$= \frac{(-i\hbar)^4}{Z[J]} \frac{\delta^4 Z[J]}{\delta J(x_1)\delta J(x_2)\delta J(x_3)\delta J(x_4)} \Big|_{J=0}$$

$$= -\hbar^2 \Big(G_F(x_1 - x_2)G_F(x_3 - x_4) + G_F(x_1 - x_3)G_F(x_2 - x_4)$$

$$+ G_F(x_1 - x_4)G_F(x_2 - x_3) \Big)$$

$$- \frac{i\lambda\hbar^3}{2} G_F(0) \int d^4x \Big(G_F(x_1 - x_2)G_F(x - x_3)G_F(x - x_4)$$

$$+ G_F(x_1 - x_3)G_F(x - x_2)G_F(x - x_4)$$

$$+ G_F(x_1 - x_4)G_F(x - x_2)G_F(x - x_3)$$

$$+ G_F(x_2 - x_3)G_F(x - x_1)G_F(x - x_4)$$

$$+ \; G_F(x_2 - x_4)G_F(x - x_1)G_F(x - x_3)$$

$$+ \; G_F(x_3 - x_4)G_F(x - x_1)G_F(x - x_2)\Big)$$

$$- i\lambda\hbar^3 \int d^4x \, G_F(x - x_1)G_F(x - x_2)G_F(x - x_3)G_F(x - x_4).$$

$$(9.48)$$

9.3 Feynman rules

These lowest order calculations are sufficient to convince anyone interested in the subject that a systematic procedure needs to be developed to keep track of the perturbative expansion. The Feynman rules do precisely this. Let us note that the basic elements in our ϕ^4-theory are the Feynman propagator for the free theory and the interaction. Let us represent them diagrammatically as

$$x_1 \; \rule{2cm}{0.4pt} \; x_2 = i\hbar \, G_F(x_1 - x_2),$$

$$
\begin{aligned}
\begin{matrix} x_1 & & x_4 \\ & \times & \\ x_2 & & x_3 \end{matrix} &= V(x_1, x_2, x_3, x_4) \\[2mm]
&= \frac{i}{\hbar} \left. \frac{\delta^4 S[\phi]}{\delta\phi(x_1)\delta\phi(x_2)\delta\phi(x_3)\delta\phi(x_4)} \right|_{\phi=0} \\[2mm]
&= -\frac{i\lambda}{\hbar} \int d^4x \, \delta(x - x_1)\delta(x - x_2)\delta(x - x_3)\delta(x - x_4).
\end{aligned}
$$

$$(9.49)$$

The interaction vertex is understood to be the part of the graph without the external lines or the propagators (the delta functions make this explicit). It is clear that given these basic elements, we can construct various nontrivial graphs by joining the vertex to the propagators. Let us further use the rule that in evaluating such graphs, we integrate over the intermediate points (coordinates) where a vertex connects with the propagators. With these rules, then, we can obtain the value for the following simple diagram to be

$$
\begin{array}{c}
\includegraphics{loop}\\
\underline{\quad\quad y_3\ \bigcirc\ y_4 \quad\quad}\\
x_1 \quad\quad y_1\ y_2 \quad\quad x_2
\end{array}
$$

$$
= \int d^4y_1 d^4y_2 d^4y_3 d^4y_4\, i\hbar G_F(x_1 - y_1) i\hbar G_F(y_2 - x_2)
$$

$$
\times\ i\hbar G_F(y_3 - y_4) V(y_1, y_2, y_3, y_4)
$$

$$
= (i\hbar)^3 \int d^4y d^4y_1 d^4y_2 d^4y_3 d^4y_4 G_F(x_1 - y_1) G_F(y_2 - x_2)
$$

$$
\times\ G_F(y_3 - y_4) \left(-\frac{i\lambda}{\hbar}\delta(y - y_1)\delta(y - y_2)\delta(y - y_3)\delta(y - y_4) \right)
$$

$$
= -\lambda\hbar^2 G_F(0) \int d^4y\, G_F(x_1 - y) G_F(y - x_2)\,. \tag{9.50}
$$

There is one final rule. Namely, if the internal part of a diagram has a symmetry, then the true value of the diagram is obtained by dividing with this symmetry factor. For the present case, the internal bubble diagram in Eq. (9.50) is invariant under a rotation by 180° (namely, under $y_3 \leftrightarrow y_4$). The symmetry factor, in this case, is $2^1 = 2$. (The symmetry factor for a nontrivial Feynman diagram is the most difficult to determine by naive inspection of the diagram and should be obtained through a careful evaluation of the functional derivation (see, for example, Eq. (9.46)), or when necessary going back to Wick's expansion of quantum field theory.) Dividing by the symmetry factor, we obtain the value of this diagram to be

$$
x_1 \ \underline{\quad\bigcirc\quad}\ x_2 = -\frac{\lambda\hbar^2}{2} G_F(0) \int d^4x\, G_F(x_1 - x) G_F(x - x_2)\,, \tag{9.51}
$$

which we recognize to be the first order (linear in λ) correction to the propagator in Eq. (9.46) (note that for the scalar theory $G_F(x - y) = G_F(y - x)$).

Each such diagram that can be constructed from the basic elements in Eq. (9.49) is known as a Feynman diagram of the theory and corresponds to a particular term in the perturbative expansion of the amplitudes (Green's functions). Thus, for example, we now immediately recognize from Eqs. (9.46) and (9.48) that up to order λ, we can write

$$\langle 0|T(\phi(x_1)\phi(x_2))|0\rangle$$

$$= x_1 \rule{2em}{0.4pt} x_2 \; + \; x_1 \underset{}{\overset{\text{\Large Q}}{\rule{2em}{0.4pt}}} x_2 \; ,$$

$$\langle 0|T(\phi(x_1)\phi(x_2)\phi(x_3)\phi(x_4))|0\rangle$$

$$= \begin{matrix} x_1 \rule{2em}{0.4pt} x_2 \\ x_3 \rule{2em}{0.4pt} x_4 \end{matrix} + \begin{matrix} x_1 \rule{2em}{0.4pt} x_3 \\ x_2 \rule{2em}{0.4pt} x_4 \end{matrix} + \begin{matrix} x_1 \rule{2em}{0.4pt} x_4 \\ x_2 \rule{2em}{0.4pt} x_3 \end{matrix}$$

$$+ \begin{matrix} x_3 \overset{\text{\Large Q}}{\rule{2em}{0.4pt}} x_4 \\ x_1 \rule{2em}{0.4pt} x_2 \end{matrix} + \begin{matrix} x_2 \overset{\text{\Large Q}}{\rule{2em}{0.4pt}} x_4 \\ x_1 \rule{2em}{0.4pt} x_3 \end{matrix} + \begin{matrix} x_2 \overset{\text{\Large Q}}{\rule{2em}{0.4pt}} x_3 \\ x_1 \rule{2em}{0.4pt} x_4 \end{matrix}$$

$$+ \begin{matrix} x_1 \overset{\text{\Large Q}}{\rule{2em}{0.4pt}} x_4 \\ x_2 \rule{2em}{0.4pt} x_3 \end{matrix} + \begin{matrix} x_1 \overset{\text{\Large Q}}{\rule{2em}{0.4pt}} x_3 \\ x_2 \rule{2em}{0.4pt} x_4 \end{matrix} + \begin{matrix} x_1 \overset{\text{\Large Q}}{\rule{2em}{0.4pt}} x_2 \\ x_3 \rule{2em}{0.4pt} x_4 \end{matrix}$$

$$+ \begin{matrix} x_1 \; \diagdown\!\!\!\!\diagup \; x_4 \\ x_2 \; \diagup\!\!\!\!\diagdown \; x_3 \end{matrix} \; .$$

$$(9.52)$$

9.4 Connected diagrams

The Feynman diagrams, as can be seen from Eq. (9.52), clearly consist of two classes of diagrams, one where each part of the diagram is connected to the rest of the diagram and another where parts of the diagram are disconnected from the rest. The Feynman diagrams, which consist of parts that are not connected to one another, are known as disconnected Feynman diagrams. It is clear from the above simple examples that the generating functional $Z[J]$ generates Green's functions which contain disconnected diagrams as well as connected ones. As we have discussed earlier (see the discussion following Eq. (4.63)), the logarithm of $Z[J]$ generates Green's functions which describe only the connected matrix elements (also known as the connected Green's functions) and these give rise to the physical scattering matrix elements. Namely,

$$W[J] = -i\hbar \ln Z[J], \qquad (9.53)$$

generates connected Green's functions. We note from Eqs. (9.53), (9.37) and (9.38) that

$$\left.\frac{\delta W[J]}{\delta J(x_1)}\right|_{J=0} = \frac{(-i\hbar)}{Z[J]} \left.\frac{\delta Z[J]}{\delta J(x_1)}\right|_{J=0} = \langle 0|\phi(x_1)|0\rangle. \tag{9.54}$$

Similarly, for the two point function, we obtain

$$-i\hbar \left.\frac{\delta^2 W[J]}{\delta J(x_1)\delta J(x_2)}\right|_{J=0}$$

$$= (-i\hbar)^2 \left[\frac{1}{Z[J]}\frac{\delta^2 Z[J]}{\delta J(x_1)\delta J(x_2)} - \frac{1}{Z^2[J]}\frac{\delta Z[J]}{\delta J(x_1)}\frac{\delta Z[J]}{\delta J(x_2)}\right]_{J=0}$$

$$= \langle 0|T(\phi(x_1)\phi(x_2))|0\rangle - \langle 0|\phi(x_1)|0\rangle\,\langle 0|\phi(x_2)|0\rangle$$

$$= \langle 0|T(\phi(x_1)\phi(x_2))|0\rangle_c. \tag{9.55}$$

In a similar manner, with some algebra, it can be shown that

$$(-i\hbar)^2 \left.\frac{\delta^3 W[J]}{\delta J(x_1)\delta J(x_2)\delta J(x_3)}\right|_{J=0}$$

$$= \langle 0|T(\phi(x_1)\phi(x_2)\phi(x_3))|0\rangle - \langle 0|\phi(x_1)|0\rangle\,\langle 0|T(\phi(x_2)\phi(x_3))|0\rangle$$

$$- \langle 0|\phi(x_2)|0\rangle\langle 0|T(\phi(x_3)\phi(x_1))|0\rangle$$

$$- \langle 0|\phi(x_3)|0\rangle\,\langle 0|T(\phi(x_1)\phi(x_2))|0\rangle$$

$$+ 2\langle 0|\phi(x_1)|0\rangle\,\langle 0|\phi(x_2)|0\rangle\,\langle 0|\phi(x_3)|0\rangle$$

$$= \langle 0|T(\phi(x_1)\phi(x_2)\phi(x_3)|0\rangle_c, \tag{9.56}$$

and so on. In general, the connected n-point Green's function is given by

$$\langle 0|T(\phi(x_1)\phi(x_2)\cdots\phi(x_n))|0\rangle_c$$

$$= (-i\hbar)^{n-1} \left.\frac{\delta^n W[J]}{\delta J(x_1)\delta J(x_2)\cdots\delta J(x_n)}\right|_{J=0}. \tag{9.57}$$

We can, of course, check explicitly that the connected Green's functions involve only connected Feynman diagrams as follows. Note that for the ϕ^4-theory, as we have discussed earlier in Eq. (9.38)

$$\langle 0|\phi(x)|0\rangle = 0. \tag{9.58}$$

Consequently, up to order λ we note from Eq. (9.55) that

$$\langle 0|T(\phi(x_1)\phi(x_2))|0\rangle = \langle 0|T(\phi(x_1)\phi(x_2))|0\rangle_c$$

$$= x_1 \text{——} x_2 + x_1 \text{——} x_2 . \tag{9.59}$$

Similarly, for the 4-point function, we have

$$\langle 0|T(\phi(x_1)\phi(x_2)\phi(x_3)\phi(x_4))|0\rangle_c$$
$$= \langle 0|T(\phi(x_1)\phi(x_2)\phi(x_3)\phi(x_4))|0\rangle$$
$$\quad - \langle 0|T(\phi(x_1)\phi(x_2))|0\rangle \langle 0|T(\phi(x_3)\phi(x_4))|0\rangle$$
$$\quad - \langle 0|T(\phi(x_1)\phi(x_3))|0\rangle \langle 0|T(\phi(x_2)\phi(x_4))|0\rangle$$
$$\quad - \langle 0|T(\phi(x_1)\phi(x_4))|0\rangle \langle 0|T(\phi(x_2)\phi(x_3))|0\rangle$$

$$= \tag{9.60}$$

Thus, we see that $W[J]$ indeed generates connected Green's functions (given by connected Feynman diagrams). Given this, we can write down the diagrammatic expansion of the connected 2-point Green's function up to order λ^2 in this theory simply as

$$x_1 \text{—⊘—} x_2$$

$$= x_1 \text{——} x_2 + x_1 \text{——} x_2 + x_1 \text{——} x_2$$

$$+ x_1 \text{—⊖—} x_2 + x_1 \text{——} x_2. \tag{9.61}$$

The organization of the perturbation series is now quite straightforward. Note that while $W[J]$ generates connected diagrams, it contains diagrams that are reducible to two connected diagrams upon cutting a single internal line (propagator). Thus, in the 2-point function represented in Eq. (9.61), the third graph is reducible upon cutting the internal propagator connecting the two bubbles. Such diagrams are called 1P (one particle) reducible. It is clear that the 1P irreducible diagrams are in some sense more fundamental since we can construct all the connected diagrams from them. We will take up the study of the 1PI (one particle irreducible) diagrams in the next chapter.

9.5 References

A. Das, *Lectures on Quantum Field Theory*, World Scientific Publishing.

K. Huang, *Quarks, Leptons and Gauge Fields*, World Scientific Publishing.

J. Jona-Lasinio, Nuovo Cimento **34**, 1790 (1964).

J. Schwinger, Proc. Natl. Sci. USA **37**, 452 (1951); *ibid* **37**, 455 (1951).

CHAPTER 10

Effective action

10.1 The classical field

As we saw in the last chapter, the generating functional for Green's functions in a scalar field theory is given by

$$Z[J] = e^{\frac{i}{\hbar}W[J]} = N \int \mathcal{D}\phi \, e^{\frac{i}{\hbar}S[\phi,J]} \,, \tag{10.1}$$

where $W[J]$ generates connected Green's functions. We note that the one point function in the presence of an external source is given by

$$\frac{\delta W[J]}{\delta J(x)} = \frac{(-i\hbar)}{Z[J]} \frac{\delta Z[J]}{\delta J(x)} = \langle 0|\phi(x)|0\rangle^J \,. \tag{10.2}$$

For the ϕ^4-theory, we have seen in Eq. (9.38) that the vacuum expectation value of the field operator vanishes in the absence of external sources (namely, when $J = 0$). In general, however, let us note that we can write

$$\langle 0|\phi(x)|0\rangle = \langle 0|e^{\frac{i}{\hbar}P\cdot x}\phi(0)e^{-\frac{i}{\hbar}P\cdot x}|0\rangle \,, \tag{10.3}$$

where we have identified $e^{-\frac{i}{\hbar}P\cdot x}$ with the operator for space-time translations (P_μ corresponds to the generator of infinitesimal translations in quantum field theory). Assuming that the vacuum state in our Hilbert space is unique and that it is Poincaré invariant in the absence of sources, namely, that it satisfies

$$e^{-\frac{i}{\hbar}P\cdot a}|0\rangle = |0\rangle \,,$$

$$\text{or,} \quad P_\mu|0\rangle = 0 \,, \tag{10.4}$$

we obtain from Eqs. (10.3) and (10.4) that

$$\langle 0|\phi(x)|0\rangle = \langle 0|\phi(0)|0\rangle = \text{constant}. \tag{10.5}$$

Thus, from symmetry arguments alone, we conclude that the one point function can, in general, be a constant independent of space-time coordinates when sources are not present. For the ϕ^4-theory, this constant coincides with zero, namely, we have

$$\langle 0|\phi(x)|0\rangle = 0. \tag{10.6}$$

The value of the one point function is quite important in the study of symmetries. As we will see later, a nonvanishing value of this quantity signals the spontaneous breakdown of some symmetry in the theory.

In the presence of an external source, however, the vacuum expectation value of the field operator becomes a functional of the source and need not be a constant. Let us denote this by

$$\frac{\delta W[J]}{\delta J(x)} = \frac{(-i\hbar)}{Z[J]} \frac{\delta Z[J]}{\delta J(x)} = \langle 0|\phi(x)|0\rangle^J = \phi_{\text{c}}(x), \tag{10.7}$$

and it is clear from Eq. (10.7) that it is indeed a functional of the external source. The field $\phi_{\text{c}}(x)$ which is only a classical variable is known as the classical field. To understand the meaning of the classical field as well as the reason for its name, let us analyze the generating functional in Eq. (10.1)

$$Z[J] = N \int \mathcal{D}\phi \, e^{\frac{i}{\hbar} S[\phi, J]}, \tag{10.8}$$

in some detail. Since $Z[J]$ is independent of $\phi(x)$, an arbitrary, infinitesimal change in $\phi(x)$ ($\phi(x) \rightarrow \phi(x) + \delta\phi(x)$) in the integrand on the right hand side will leave the generating functional invariant. Namely, under such a change,

$$\delta Z[J] = N \int \mathcal{D}\phi \, \frac{i}{\hbar} \delta S[\phi, J] e^{\frac{i}{\hbar} S[\phi, J]}$$

$$= \frac{iN}{\hbar} \int \mathcal{D}\phi \left(\int d^4x \, \delta\phi(x) \frac{\delta S[\phi, J]}{\delta\phi(x)} \right) e^{\frac{i}{\hbar} S[\phi, J]}$$

$$= 0. \tag{10.9}$$

Here we are assuming that the functional integration measure in Eq. (10.8) does not change under the redefinition of the field variable. (If it does change, then we will have an additional term coming from the change in the measure (Jacobian). Such a term plays an important role in the study of anomalies.) Since the relation in Eq. (10.9) must be true for any arbitrary infinitesimal variation $\delta\phi(x)$ of the field variable, it now follows that we must have

$$N \int \mathcal{D}\phi \, \frac{\delta S[\phi, J]}{\delta\phi(x)} \, e^{\frac{i}{\hbar} S[\phi, J]} = 0. \tag{10.10}$$

This merely expresses what we already know from the study of quantum mechanics. Namely, that the Euler-Lagrange equations of a theory hold only as an expectation value equation (Ehrenfest's theorem) in a quantum theory, or more explicitly,

$$\left\langle 0 \left| \frac{\delta S[\phi, J]}{\delta\phi(x)} \right| 0 \right\rangle^J = 0. \tag{10.11}$$

The Euler-Lagrange equation (the classical equation) has the generic form (see, for example, Eq. (9.8) or (9.17))

$$-\frac{\delta S[\phi, J]}{\delta\phi(x)} = F(\phi(x)) - J(x) = 0, \tag{10.12}$$

where the quantity $F(\phi(x))$ depends on the specific dynamics of the system and for the ϕ^4-theory, we note from Eq. (9.15) that it has the particular form

$$F(\phi(x)) \equiv -\frac{\delta S[\phi]}{\delta\phi(x)} = (\partial_\mu \partial^\mu + m^2)\phi(x) + \frac{\lambda}{3!} \phi^3(x). \tag{10.13}$$

Using Eq. (10.12) in Eq. (10.10), then, we obtain

$$-N \int \mathcal{D}\phi \, \frac{\delta S[\phi, J]}{\delta \phi(x)} \, e^{\frac{i}{\hbar} S[\phi, J]}$$

$$= N \int \mathcal{D}\phi \, (F(\phi(x)) - J(x)) \, e^{\frac{i}{\hbar} S[\phi, J]} = 0 \,. \tag{10.14}$$

Let us recall that

$$-i\hbar \frac{\delta Z[J]}{\delta J(x)} = N \int \mathcal{D}\phi \, \frac{\delta S[\phi, J]}{\delta J(x)} \, e^{\frac{i}{\hbar} S[\phi, J]}$$

$$= N \int \mathcal{D}\phi \, \phi(x) \, e^{\frac{i}{\hbar} S[\phi, J]} \,, \tag{10.15}$$

which allows us to make the identification

$$\phi(x) \to -i\hbar \frac{\delta}{\delta J(x)} \,, \tag{10.16}$$

in the sense that a factor of $\phi(x)$ inside the path integral is equivalent to $(-i\hbar \frac{\delta}{\delta J(x)})$ acting on $Z[J]$. Using this, we can write Eq. (10.14) also as

$$\left[F\left(-i\hbar \frac{\delta}{\delta J(x)}\right) - J(x) \right] Z[J] = 0,$$

$$\text{or,} \quad \left[F\left(-i\hbar \frac{\delta}{\delta J(x)}\right) - J(x) \right] e^{\frac{i}{\hbar} W[J]} = 0,$$

$$\text{or,} \quad e^{-\frac{i}{\hbar} W[J]} \left[F\left(-i\hbar \frac{\delta}{\delta J(x)}\right) - J(x) \right] e^{\frac{i}{\hbar} W[J]} = 0,$$

$$\text{or,} \quad \left[F\left(\frac{\delta W[J]}{\delta J(x)} - i\hbar \frac{\delta}{\delta J(x)}\right) - J(x) \right] 1 = 0,$$

$$\text{or,} \quad F\left(\phi_c(x) - i\hbar \frac{\delta}{\delta J(x)}\right) 1 - J(x) = 0 \,. \tag{10.17}$$

Here we have made it explicit that any free operator $(\frac{\delta}{\delta J(x)})$ on the left hand side acts on 1.

This is the full dynamical equation of the theory at the quantum level. It is quite different from the classical Euler-Lagrange equation in Eq. (10.12). But let us note that in the limit $\hbar \to 0$, the complete equation in Eq. (10.17) reduces to the form

$$F(\phi_c(x)) - J(x) = 0, \tag{10.18}$$

which is the familiar classical Euler-Lagrange equation in Eq. (10.12). It is for this reason that $\phi_c(x)$ is called the classical field. To get a better feeling for the quantum equation in Eq. (10.17) as well as Eq. (10.18) in the $\hbar \to 0$ limit, let us consider specifically the ϕ^4-theory. In this case, as we have noted earlier in Eq. (10.13), the classical Euler-Lagrange equation is given by

$$F(\phi(x)) - J(x) = (\partial_\mu \partial^\mu + m^2)\phi(x) + \frac{\lambda}{3!}\phi^3(x) - J(x) = 0. \tag{10.19}$$

Consequently, the quantum equation takes the form

$$F\left(\phi_c(x) - i\hbar \frac{\delta}{\delta J(x)}\right) - J(x) = 0,$$

$$\text{or,} \quad (\partial_\mu \partial^\mu + m^2)\left(\phi_c(x) - i\hbar \frac{\delta}{\delta J(x)}\right)$$

$$+ \frac{\lambda}{3!}\left(\phi_c(x) - i\hbar \frac{\delta}{\delta J(x)}\right)^3 - J(x) = 0,$$

$$\text{or,} \quad (\partial_\mu \partial^\mu + m^2)\phi_c(x) + \frac{\lambda}{3!}\phi_c^3(x) - J(x)$$

$$- \frac{i\lambda\hbar}{2!}\phi_c(x)\frac{\delta\phi_c(x)}{\delta J(x)} - \frac{\lambda\hbar^2}{3!}\frac{\delta^2\phi_c(x)}{\delta J(x)\delta J(x)} = 0. \tag{10.20}$$

In terms of $W[J]$, this can also be written as

$$(\partial_\mu \partial^\mu + m^2)\frac{\delta W[J]}{\delta J(x)} + \frac{\lambda}{3!}\left(\frac{\delta W[J]}{\delta J(x)}\right)^3 - J(x)$$

$$- \frac{i\lambda\hbar}{2!}\frac{\delta W[J]}{\delta J(x)}\frac{\delta^2 W[J]}{\delta J^2(x)} - \frac{\lambda\hbar^2}{3!}\frac{\delta^3 W[J]}{\delta J^3(x)} = 0. \tag{10.21}$$

We can think of Eq. (10.21) as the master equation governing the full dynamics of the quantum theory. By taking higher functional derivatives of (10.21), we can determine the dynamical equations satisfied by various connected Green's functions of the theory. These equations are known as the Schwinger-Dyson equations and the Bethe-Salpeter equations are special cases of these equations. We should note here that in a quantum field theory whenever there are products of operators at the same space-time point, the corresponding Green's functions become ill defined and have to be regularized (defined) in some manner. The manifestation of this problem is quite clear in Eq. (10.21). The quantum dynamical equation involves second and third functional derivatives of $W[J]$ at the same space-time point which are not at all well defined. One needs to develop a systematic regularization procedure to handle these difficulties. The discussion of these topics lies outside the scope of these lectures.

On the other hand, let us note that in the limit $\hbar \to 0$, all such ill defined terms vanish and we have from Eq. (10.20)

$$(\partial_\mu \partial^\mu + m^2)\phi_c(x) + \frac{\lambda}{3!}\phi_c^3(x) - J(x) = 0 \,, \tag{10.22}$$

which is the classical Euler-Lagrange equation. Furthermore, let us note that we can solve this equation iteratively through the use of the propagator (Green's function) defined in Eqs. (9.23) and (9.24) as follows. First, we note that the solution for $\phi_c(x)$ can be written as an integral equation

$$\phi_c(x) = \int d^4x_1 \, G_F(x - x_1)\left(-J(x_1) + \frac{\lambda}{3!}\phi_c^3(x_1)\right)$$

$$= -\int d^4x_1 G_F(x - x_1)J(x_1) + \frac{\lambda}{3!}\int d^4x_1 \, G_F(x - x_1)\phi_c^3(x_1) \,, \tag{10.23}$$

which can be solved iteratively. The iterative solution has the form

$$\phi_c(x) \simeq -\int d^4x_1 \, G_F(x - x_1)J(x_1) + \frac{\lambda}{3!}\int d^4x_1 G_F(x - x_1)$$

$$\times \left(\int d^4x_2 G_F(x_1 - x_2)\left(-J(x_2) + \frac{\lambda}{3!}\phi_c^3(x_2)\right)\right)^3$$

$$= -\int d^4x_1 G_F(x-x_1)J(x_1) - \frac{\lambda}{3!}\int d^4x_1 d^4x_2 d^4x_3 d^4x_4 G_F(x-x_1)$$

$$\times G_F(x_1-x_2)G_F(x_1-x_3)G_F(x_1-x_4)J(x_2)J(x_3)J(x_4)+\cdots$$

$$(10.24)$$

We can diagrammatically represent this iterative solution if we introduce a vertex describing the interaction of the field with the external source as

$$\xrightarrow{\hspace{2cm}}_{x}\!\! \times \;=\; \frac{i}{\hbar}\left.\frac{\delta S[\phi, J]}{\delta \phi(x)}\right|_{\phi=0} = \frac{i}{\hbar}J(x)\,. \qquad (10.25)$$

In this case, the iterative solution for the classical field can be written diagrammatically as

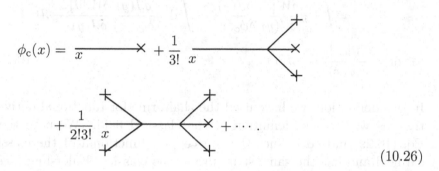

$$(10.26)$$

In other words, the classical field in the limit $\hbar \to 0$ generates all the tree diagrams or the Born diagrams of the theory. It is, therefore, also known as the Born functional or the generating functional for tree diagrams. The combinatoric factors cancel out when taking functional derivatives of ϕ_c with respect to sources yielding the tree level n-point amplitudes.

10.2 Effective action

The classical field, as we have seen in Eq. (10.7), is defined as

$$\frac{\delta W[J]}{\delta J(x)} = \phi_c(x)\,. \qquad (10.27)$$

This relation indicates that the variables $J(x)$ and $\phi_c(x)$ are in some sense conjugate variables. As we have argued earlier, the classical field is a functional of the source $J(x)$. However, we can also invert the defining relation for the classical field and solve for the source $J(x)$ as a functional of $\phi_c(x)$ at least perturbatively. In fact, let us define a new functional through a Legendre transformation as

$$\Gamma[\phi_c] = W[J] - \int d^4x \, J(x)\phi_c(x). \tag{10.28}$$

It is clear, then, that

$$\frac{\delta\Gamma[\phi_c]}{\delta\phi_c(x)} = \frac{\delta W[J]}{\delta\phi_c(x)} - \int d^4y \frac{\delta J(y)}{\delta\phi_c(x)}\phi_c(y) - J(x)$$

$$= \int d^4y \frac{\delta W[J]}{\delta J(y)}\frac{\delta J(y)}{\delta\phi_c(x)} - \int d^4y \frac{\delta J(y)}{\delta\phi_c(x)}\frac{\delta W[J]}{\delta J(y)} - J(x),$$

$$\text{or,} \quad \frac{\delta\Gamma[\phi_c]}{\delta\phi_c(x)} = -J(x). \tag{10.29}$$

In this derivation, we have used the chain rule for functional derivatives as well as the definition of the classical field. We note that Eq. (10.29) indeed defines the source as a functional of the classical field and has the same structure as the classical Euler-Lagrange equation for a system in the presence of an external source. Let us recall from Eqs. (10.12) and (10.13) that the tree level Euler-Lagrange equation is given by

$$\frac{\delta S[\phi]}{\delta\phi(x)} = -J(x). \tag{10.30}$$

It is for this reason that $\Gamma[\phi_c]$ is also known as the effective action functional for the theory (which includes all the quantum corrections). Note that as we have discussed earlier in Eq. (10.5), when

$$J \to 0, \quad \phi_c(x) \to \text{constant}. \tag{10.31}$$

In the framework of the effective action, we see from Eq. (10.29) that the value of this constant is determined from the equation

$$\frac{\delta\Gamma[\phi_c]}{\delta\phi_c(x)}\bigg|_{\phi_c(x)=\text{const.}} = 0.\tag{10.32}$$

This is an extremum equation which is much easier to analyze in order to determine whether a symmetry is spontaneously broken.

To understand the meaning of this new functional $\Gamma[\phi_c]$, let us note that if we treat $\phi_c(x)$ as our independent variable, then we can write

$$\frac{\delta}{\delta J(x)} = \int d^4y\, \frac{\delta\phi_c(y)}{\delta J(x)}\frac{\delta}{\delta\phi_c(y)}$$

$$= \int d^4y\, \frac{\delta^2 W[J]}{\delta J(x)\delta J(y)}\frac{\delta}{\delta\phi_c(y)}.\tag{10.33}$$

Using this in Eq. (10.29) we obtain

$$\frac{\delta}{\delta J(y)}\left(\frac{\delta\Gamma[\phi_c]}{\delta\phi_c(x)}\right) = -\delta^4(x-y),$$

or, $$\int d^4z\, \frac{\delta^2 W[J]}{\delta J(y)\delta J(z)}\frac{\delta^2\Gamma[\phi_c]}{\delta\phi_c(z)\delta\phi_c(x)} = -\delta^4(x-y).\tag{10.34}$$

Introducing the compact notation

$$W^{(n)} = \frac{\delta^n W[J]}{\delta J(x_1)\cdots\delta J(x_n)},$$

$$\Gamma^{(n)} = \frac{\delta^n\Gamma[\phi_c]}{\delta\phi_c(x_1)\cdots\delta\phi_c(x_n)},\tag{10.35}$$

we can write Eq. (10.34) also in the compact form (we can view this as an operator relation where the appropriate coordinate dependencies arise by taking matrix elements in the coordinate basis)

$$W^{(2)}\Gamma^{(2)} = -1.\tag{10.36}$$

We recall from Eqs. (9.33) and (9.55) that the full Feynman Green's function of the interacting theory is defined to be (we reserve the notation G_F for the free Feynman Green's function for simplicity)

$$W^{(2)}\Big|_{J=0} = -G. \tag{10.37}$$

Furthermore, recalling that when $J = 0$, $\phi_c(x) = \phi_c = $ constant, we have from Eqs. (10.36) and (10.37)

$$W^{(2)}\Big|_{J=0} \Gamma^{(2)}\Big|_{\phi_c} = -1,$$

$$\text{or,} \quad G\Gamma^{(2)}\Big|_{\phi_c} = 1. \tag{10.38}$$

In other words, $\Gamma^{(2)}\big|_{\phi_c}$ corresponds to the inverse of the propagator at every order of the perturbation theory. Thus, writing

$$\Gamma^{(2)}\Big|_{\phi_c} = \Gamma_0^{(2)} - \Sigma = G_F^{-1} - \Sigma, \tag{10.39}$$

where $\Gamma_0^{(2)}$ corresponds to the tree level (free) 2-point (vertex) function and Σ, known as the self-energy, denotes the quantum corrections in $\Gamma^{(2)}\big|_{\phi_c}$ (to any order), we have from Eq. (10.38)

$$G(G_F^{-1} - \Sigma) = 1,$$

$$\text{or,} \quad G = \frac{1}{G_F^{-1} - \Sigma}$$

$$= \frac{1}{G_F^{-1}} + \frac{1}{G_F^{-1}}\Sigma\frac{1}{G_F^{-1}} + \frac{1}{G_F^{-1}}\Sigma\frac{1}{G_F^{-1}}\Sigma\frac{1}{G_F^{-1}} + \cdots$$

$$= G_F + G_F\Sigma G_F + G_F\Sigma G_F\Sigma G_F + \cdots. \tag{10.40}$$

(Equation (10.40) describes a form of what is known as the Lippmann-Schwinger equation.) Introducing the diagrammatic representation

$$-\frac{i}{\hbar}\Sigma = \quad \text{—O—} \quad , \tag{10.41}$$

we have the diagrammatic relation for the complete propagator of the theory as (recall that the propagator is given by $i\hbar$ times the Green's function)

$$(10.42)$$

It is clear from this relation that Σ is nothing other than the 1P irreducible (1PI) 2-point vertex function. It is also known as the proper self energy for the given theory.

Given the relation in Eq. (10.34)

$$\int \mathrm{d}^4 z \, \frac{\delta^2 W[J]}{\delta J(y)\delta J(z)} \, \frac{\delta^2 \Gamma[\phi_c]}{\delta\phi_c(z)\delta\phi_c(x)} = -\delta^4(x-y) \,, \tag{10.43}$$

we can differentiate this with respect to $\frac{\delta}{\delta J(w)}$ to obtain (with the use of Eq. (10.33))

$$\int \mathrm{d}^4 z \frac{\delta^3 W[J]}{\delta J(w)\delta J(y)\delta J(z)} \, \frac{\delta^2 \Gamma[\phi_c]}{\delta\phi_c(z)\delta\phi_c(x)} \tag{10.44}$$

$$= -\int \mathrm{d}^4 z \mathrm{d}^4 \sigma \frac{\delta^2 W[J]}{\delta J(y)\delta J(z)} \, \frac{\delta^3 \Gamma[\phi_c]}{\delta\phi_c(z)\delta\phi_c(x)\delta\phi_c(\sigma)} \, \frac{\delta^2 W[J]}{\delta J(\sigma)\delta J(w)} \,.$$

Recalling that

$$W^{(2)}\Gamma^{(2)} = -1 \,, \tag{10.45}$$

we can also rewrite this equation as

$$\frac{\delta^3 W[J]}{\delta J(x)\delta J(y)\delta J(z)}$$

$$= \int \mathrm{d}^4 x' \mathrm{d}^4 y' \mathrm{d}^4 z' \frac{\delta^2 W[J]}{\delta J(x)\delta J(x')} \, \frac{\delta^2 W[J]}{\delta J(y)\delta J(y')} \, \frac{\delta^2 W[J]}{\delta J(z)\delta J(z')}$$

$$\times \frac{\delta^3 \Gamma[\phi_c]}{\delta\phi_c(x')\delta\phi_c(y')\delta\phi_c(z')} \,. \tag{10.46}$$

In compact notation, we can write this equation also as

$$W^{(3)} = W^{(2)}W^{(2)}W^{(2)}\Gamma^{(3)}. \tag{10.47}$$

Furthermore, introducing the diagrammatic representations (recall the definition of connected Green's functions from (9.57)),

$$= (-i\hbar)\, W^{(2)}(x,y)\Big|_{J=0},$$

$$= (-i\hbar)^2\, W^{(3)}(x,y,z)\Big|_{J=0},$$

$$= \frac{i}{\hbar}\, \Gamma^{(3)}(x,y,z)\Big|_{\phi_c}, \tag{10.48}$$

we note that Eq. (10.46) or (10.47) can also be diagrammatically represented as (when $J = 0$ and $\phi_c(x) = \phi_c = $ constant)

$$\tag{10.49}$$

This shows that $\Gamma^{(3)}\big|_{\phi_c}$ gives the proper 3-point vertex function (in other words, it is 1PI).

A similar calculation for the 4-point connected Green's function leads to the diagrammatic relation

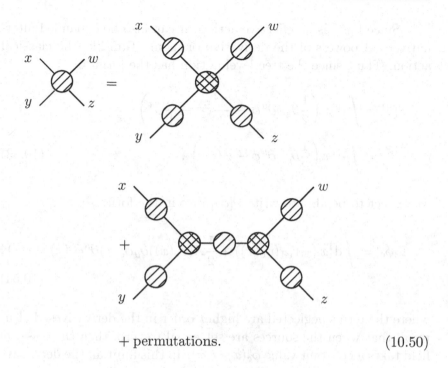

+ permutations.   (10.50)

Thus, we see that we can expand the effective action functional as

$$\Gamma[\phi_c] = \sum_{n=1}^{\infty} \int d^4x_1 \cdots d^4x_n \frac{1}{n!} \left. \Gamma^{(n)}(x_1, \cdots, x_n) \right|_{\phi_c} \phi_c(x_1) \cdots \phi_c(x_n),$$

(10.51)

where (in the above expansion we are expanding around $\phi_c = 0$ which we assume to be the solution of Eq. (10.32))

$$\left. \Gamma^{(n)}(x_1, \cdots, x_n) \right|_{\phi_c} = \left. \Gamma^{(n)}(x_1, \cdots, x_n) \right|_{\phi_c=0},$$

(10.52)

is the proper (1PI) n-point vertex function of the theory. It is for this reason that $\Gamma[\phi_c]$ is known as the 1PI generating functional. Let us note here that the 1PI vertex functions with suitable external wave functions lead to the scattering matrix elements of the theory.

Since $\Gamma[\phi_c]$ is an effective action, it can also be expanded alternatively in powers of the derivative or momentum like the classical action. Thus, since the tree level action has the form

$$S[\phi] = \int d^4x \left(\frac{1}{2} \partial_\mu \phi \partial^\mu \phi - \frac{m^2}{2} \phi^2 - \frac{\lambda}{4!} \phi^4 \right)$$

$$= \int d^4x \left(\frac{1}{2} \partial_\mu \phi \partial^\mu \phi - V(\phi) \right), \tag{10.53}$$

we expect to be able to write $\Gamma[\phi_c]$ also in the form

$$\Gamma[\phi_c] = \int d^4x \left(-V_{\text{eff}}(\phi_c(x)) + \frac{1}{2} A(\phi_c(x)) \partial_\mu \phi_c(x) \partial^\mu \phi_c(x) + \cdots \right), \tag{10.54}$$

where the terms neglected are higher order in the derivatives. Let us recall that when the sources are set equal to zero, then the classical field takes a constant value $\phi_c(x) = \phi_c$. In this limit all the derivative terms in the expansion in Eq. (10.54) vanish, leading to

$$\Gamma[\phi_c] = - \int d^4x \, V_{\text{eff}}(\phi_c) = -V_{\text{eff}}(\phi_c) \int d^4x. \tag{10.55}$$

In other words, in this limit, the effective action simply picks out the effective potential including quantum corrections to all orders. The quantity $\int d^4x$ which represents the space-time volume is conventionally also written as

$$\int d^4x = (2\pi)^4 \delta^4(0). \tag{10.56}$$

Let us also note that the constant value of $\phi_c = \langle \phi \rangle$ when the sources are turned off is obtained from the extremum equation in Eq. (10.32)

$$\left. \frac{\delta \Gamma[\phi_c]}{\delta \phi_c(x)} \right|_{\phi_c(x) = \phi_c} = 0. \tag{10.57}$$

In terms of the effective potential, this condition becomes equivalent to

$$\frac{\partial V_{\text{eff}}(\phi_{\text{c}})}{\partial \phi_{\text{c}}}\bigg|_{\phi_{\text{c}}=\langle\phi\rangle} = 0\,, \tag{10.58}$$

which is the familiar extremization condition from the study of classical mechanics. Note here that since ϕ_{c} is an ordinary variable (namely, it does not depend on space-time coordinates), only ordinary partial derivatives are involved in the above extremum condition. In this sense, this equation is much easier to analyze than a functional equation. We also note that the renormalized values of the masses and the coupling constants (including all quantum corrections) can be obtained (in this theory) from the effective potential simply as

$$\frac{\partial^2 V_{\text{eff}}}{\partial \phi_{\text{c}}^2}\bigg|_{\phi_{\text{c}}=\langle\phi\rangle} = m_R^2\,,$$

$$\frac{\partial^4 V_{\text{eff}}}{\partial \phi_{\text{c}}^4}\bigg|_{\phi_{\text{c}}=\langle\phi\rangle} = \lambda_R\,. \tag{10.59}$$

The study of the effective potential is, therefore, quite important. It is particularly useful in analyzing when the quantum corrections can change the qualitative tree level behavior or the classical behavior of a theory.

10.3 Loop expansion

We have already described the Feynman rules in the coordinate space. However, for most practical calculations it is quite useful to work in the momentum space. Given the Feynman rules in the coordinate space, these rules can be readily generalized to the momentum space. (This simply involves taking Fourier transforms.) For the ϕ^4 theory, for example, the momentum space Feynman rules take the form

$$\xrightarrow{\quad p \quad} = i\hbar G_F(p) = \lim_{\epsilon \to 0} \frac{i\hbar}{p^2 - m^2 + i\epsilon},$$

$$\begin{array}{c} p_1 \\ \\ p_2 \end{array} \!\!\!\!\times\!\!\!\! \begin{array}{c} p_4 \\ \\ p_3 \end{array} = -\frac{i\lambda}{\hbar}\, \delta^4(p_1 + p_2 + p_3 + p_4)\,, \qquad\qquad (10.60)$$

where all momenta at the vertex are assumed to be incoming. In evaluating a Feynman diagram, we should integrate over each of the intermediate momenta, namely, the momenta of the internal propagators (with the normalization factor $\frac{1}{(2\pi)^4}$). Thus, for example, let us evaluate the 1PI 2-point vertex function at order λ

$$I = \quad \begin{array}{c} k \\ \\ p_1 \xrightarrow{\hspace{2cm}} p_2 \end{array} . \qquad\qquad (10.61)$$

According to our rules, we obtain

$$I = \frac{1}{2}\left(-\frac{i\lambda}{\hbar}\right)\int \frac{\mathrm{d}^4 k}{(2\pi)^4} \frac{i\hbar}{k^2 - m^2}\, \delta^4(p_1 - p_2 + k - k)$$

$$= \frac{\lambda}{2}\, \delta^4(p_1 - p_2) \int \frac{\mathrm{d}^4 k}{(2\pi)^4} \frac{1}{k^2 - m^2}\,. \qquad\qquad (10.62)$$

Note that the factor of $\frac{1}{2}$ in front of the integral in Eq. (10.62) simply corresponds to the symmetry factor of the diagram which we discussed earlier. In writing the propagator, we have not explicitly included the $i\epsilon$ term although it should always be kept in mind in evaluating the integral. We will discuss the actual evaluation of the integrals later. For the present, let us simply note that the calculations indeed take a simpler form in the momentum space.

Let us next recognize from the form of the exponent in the path integral that the quantity which determines the dynamics of the system is

$$\frac{1}{\hbar} \mathcal{L}(\phi, \partial_\mu \phi) = \frac{1}{\hbar} \left(\frac{1}{2} \partial_\mu \phi \partial^\mu \phi - \frac{m^2}{2} \phi^2 - \frac{\lambda}{4!} \phi^4 \right) . \tag{10.63}$$

The Planck's constant which measures the quantum nature of an amplitude comes as a multiplicative factor in the exponent. As we have seen in Eqs. (9.49) and (10.60), the consequence of this is that each vertex has a factor of $\frac{1}{\hbar}$ associated with it. On the other hand, the propagator which is the inverse of the operator in the quadratic part of the Lagrangian (two point function), comes multiplied with a factor of $\hbar$. Thus, suppose we are considering a proper vertex diagram (1PI diagram) with V vertices and I internal lines or propagators, then the total number of $\hbar$ factors associated with such a diagram is given by

$$P = I - V . \tag{10.64}$$

In other words, such a diagram will behave like $\sim \hbar^P$.

Let us also calculate the number of independent momentum integrations associated with such a diagram. First, let us note that in a proper vertex diagram, there are no external propagators or legs. Second, the momenta associated with each of the internal lines must be integrated. Since there are I internal lines, there will, therefore, be I momentum integrations. Of course, not all such momenta will be independent since at each vertex there is a momentum conserving δ-function. Each such δ-function will reduce the number of momentum integration by one. Since there are V vertices, there will be as many momentum conserving δ-functions. However, we will need to have an overall momentum conserving δ-function for the amplitude (vertex). Hence, the δ-functions will effectively reduce the number of momentum integrations by $V - 1$. Therefore, the number of independent internal momentum integrations associated with the diagram will be given by

$$L = I - (V - 1) = I - V + 1 = P + 1 . \tag{10.65}$$

But the number of independent momentum integrations precisely defines the number of loops in a diagram and from the above relation we note that the number of loops associated with a diagram is related to the power of $\hbar$ associated with a diagram. In fact, the number of loops exceeds the power of $\hbar$ by one. (An amplitude with L loops goes as $\sim \hbar^{L-1}$.) Therefore, expanding an amplitude in powers of $\hbar$ is also equivalent to an expansion in the number of loops.

The loop expansion provides a valid perturbative expansion simply because $\hbar$ is a small quantity. (We have already introduced the concept of such an expansion in powers of $\hbar$ in discussing the WKB approximation.) This expansion is quite useful and is very different from expanding in powers of the coupling constant. This follows mainly from the fact that the expansion parameter $\hbar$ multiplies the entire Lagrangian density (action). Consequently, such an expansion is insensitive to how we divide the Lagrangian density into a free part and an interaction part. The loop expansion is, therefore, unaffected by any such separation. This is particularly useful if the theory exhibits spontaneous symmetry breakdown in which case, as we will see later, the vacuum expectation value $\phi_c = \langle \phi \rangle$ becomes dependent on the coupling constants of the theory. Shifting the fields around such a value complicates perturbation in powers of the coupling constants. However, as we have argued, the loop expansion is unaffected by such a shift.

10.4 Effective potential at one loop

Let us next calculate the effective potential for the ϕ^4 theory at one loop. In this case, the classical action is given by

$$S[\phi] = \int \mathrm{d}^4x \left(\frac{1}{2} \partial_\mu \phi \partial^\mu \phi - \frac{m^2}{2} \phi^2 - \frac{\lambda}{4!} \phi^4 \right). \tag{10.66}$$

As we have seen earlier in Eq. (10.30), in this case, the classical dynamical equations are given by

$$-\frac{\delta S[\phi]}{\delta \phi(x)} = (\partial_\mu \partial^\mu + m^2)\phi(x) + \frac{\lambda}{3!} \phi^3(x) = J(x). \tag{10.67}$$

Furthermore, we note from Eq. (10.20) that the classical field, $\phi_c(x)$, satisfies the equation

$$(\partial_\mu \partial^\mu + m^2)\phi_c(x) + \frac{\lambda}{3!}\phi_c^3(x)$$

$$- \frac{i\lambda\hbar}{2!}\phi_c(x)\frac{\delta\phi_c(x)}{\delta J(x)} - \frac{\lambda\hbar^2}{3!}\frac{\delta^2\phi_c(x)}{\delta J^2(x)} = J(x). \tag{10.68}$$

If we now use the relations (see Eqs. (10.7) and (10.29))

$$\frac{\delta\Gamma[\phi_c]}{\delta\phi_c(x)} = -J(x),$$

$$\frac{\delta\phi_c(x)}{\delta J(y)} = \frac{\delta^2 W[J]}{\delta J(x)\delta J(y)} = -G(x-y, \phi_c), \tag{10.69}$$

and remember that we are interested only in one loop effects, then keeping terms up to linear power in $\hbar$, we obtain from Eq. (10.68)

$$\frac{\delta\Gamma[\phi_c]}{\delta\phi_c(x)} = -(\partial_\mu \partial^\mu + m^2)\phi_c(x) - \frac{\lambda}{3!}\phi_c^3(x)$$

$$+ \frac{i\lambda\hbar}{2!}\phi_c(x)\frac{\delta^2 W[J]}{\delta J^2(x)} + O(\hbar^2)$$

$$= \frac{\delta S[\phi_c]}{\delta\phi_c(x)} - \frac{i\lambda\hbar}{2}\phi_c(x)G(x-x, \phi_c) + O(\hbar^2), \tag{10.70}$$

which can be written as

$$\frac{\delta\left(\Gamma[\phi_c] - S[\phi_c]\right)}{\delta\phi_c(x)} = -\frac{i\lambda\hbar}{2}\phi_c(x)G(0, \phi_c) + O(\hbar^2). \tag{10.71}$$

If we expand the effective action in a power series in $\hbar$ as

$$\Gamma[\phi_c] = S[\phi_c] + \hbar S_1[\phi_c] + O(\hbar^2), \tag{10.72}$$

then, Eq. (10.71) leads to

$$\frac{\delta\left(\hbar S_1[\phi_c]\right)}{\delta\phi_c(x)} + O(\hbar^2) = -\frac{i\lambda\hbar}{2}\phi_c(x)G(0, \phi_c) + O(\hbar^2). \tag{10.73}$$

Therefore, to linear order in $\hbar$, we can consistently write

$$\frac{\delta S_1[\phi_c]}{\delta \phi_c(x)} = -\frac{i\lambda}{2}\phi_c(x)G(0,\phi_c). \tag{10.74}$$

From the structure of the action (see Eq. (10.54))

$$S_1[\phi_c] = \int d^4x \left(-V_1(\phi_c(x)) + \cdots\right), \tag{10.75}$$

where V_1 denotes the one loop potential and the dots represent terms with derivatives, we obtain

$$\left.\frac{\delta S_1[\phi_c]}{\delta \phi_c(x)}\right|_{\phi_c(x)=\phi_c} = -\frac{\partial V_1(\phi_c)}{\partial \phi_c}. \tag{10.76}$$

Thus, if we restrict to $\phi_c(x) = \phi_c = $ constant, then Eq. (10.74) takes the form

$$\frac{\partial V_1(\phi_c)}{\partial \phi_c} = \frac{i\lambda}{2}\,\phi_c G(0,\phi_c). \tag{10.77}$$

Let us note that although the Green's function $G(x-y,\phi_c)$ can itself have a power series expansion in $\hbar$, consistency requires that we only use the lowest order expression for the Green's function in the above equation. Furthermore, noting from Eq. (10.34) that

$$\int d^4z\, \frac{\delta^2\Gamma[\phi_c]}{\delta\phi_c(x)\delta\phi_c(z)}\frac{\delta^2 W[J]}{\delta J(z)\delta J(y)} = -\delta^4(x-y), \tag{10.78}$$

this relation leads to the lowest order of the Green's function that we are interested in as

$$\int d^4z\, \left.\frac{\delta^2 S[\phi_c]}{\delta\phi_c(x)\delta\phi_c(z)}\right|_{\phi_c} (-G(z-y,\phi_c)) = -\delta^4(x-y),$$

$$\text{or,} \int d^4z((\partial_\mu\partial^\mu + m^2 + \frac{\lambda}{2}\phi_c^2)\delta^4(x-z))G(z-y,\phi_c) = -\delta^4(x-y),$$

$$\text{or,} \ (\partial_\mu\partial^\mu + m^2 + \frac{\lambda}{2}\phi_c^2)G(x-y,\phi_c) = -\delta^4(x-y),$$

$$\text{or,} \ (\partial_\mu\partial^\mu + m_{\text{eff}}^2)G(x-y,\phi_c) = -\delta^4(x-y). \tag{10.79}$$

Here we have used the form of the action from Eq. (9.15) as well as Eq. (10.37) and have defined

$$m_{\text{eff}}^2 = m^2 + \frac{\lambda}{2}\phi_c^2 . \tag{10.80}$$

The Green's function can now be trivially determined and as we have seen before in Eq. (9.23) it has the form (the "$i\epsilon$" term is understood)

$$G(x - y, \phi_c) = \int \frac{d^4k}{(2\pi)^4} \frac{1}{k^2 - m_{\text{eff}}^2} e^{-ik\cdot(x-y)} . \tag{10.81}$$

Substituting this back into Eq. (10.77), we obtain

$$\frac{\partial V_1(\phi_c)}{\partial \phi_c} = \frac{i\lambda}{2}\phi_c G(0, \phi_c) = \frac{i\lambda}{2}\phi_c \int \frac{d^4k}{(2\pi)^4} \frac{1}{k^2 - m_{\text{eff}}^2},$$

$$\text{or,} \quad V_1(\phi_c) = \frac{i\lambda}{2} \int_0^{\phi_c} d\phi_c\, \phi_c \int \frac{d^4k}{(2\pi)^4} \frac{1}{k^2 - m^2 - \frac{\lambda}{2}\phi_c^2} . \tag{10.82}$$

Here, we are assuming that $V_1(\phi_c = 0) = 0$. If that is not true, then we have to add a constant term to the above expression. The one loop correction to the potential can now be determined by going over to Euclidean space and doing the integral. First, we note that if we interchange the orders of integration in Eq. (10.82), we obtain

$$V_1(\phi_c) = \frac{i\lambda}{2} \int \frac{d^4k}{(2\pi)^4} \int_0^{\phi_c} d\phi_c \frac{\phi_c}{k^2 - m^2 - \frac{\lambda}{2}\phi_c^2}$$

$$= -\frac{i}{2} \int \frac{d^4k}{(2\pi)^4} \int_0^{\phi_c} \frac{d\left(\frac{\lambda}{2}\phi_c^2\right)}{\frac{\lambda}{2}\phi_c^2 + m^2 - k^2}$$

$$= -\frac{i}{2} \int \frac{d^4k}{(2\pi)^4} \ln\left(\frac{\lambda}{2}\phi_c^2 + m^2 - k^2\right)\Big|_0^{\phi_c}$$

$$= -\frac{i}{2} \int \frac{d^4k}{(2\pi)^4} \ln\left(\frac{\frac{\lambda}{2}\phi_c^2 + m^2 - k^2}{m^2 - k^2}\right) . \tag{10.83}$$

Now, rotating to Euclidean space (see Section 4.1), we obtain

$$V_1(\phi_c) = -\frac{i}{2} \int \frac{id^4 k_E}{(2\pi)^4} \ln \left(\frac{\frac{\lambda}{2}\phi_c^2 + m^2 + k_E^2}{m^2 + k_E^2} \right)$$

$$= \frac{1}{2} \int \frac{d^3\Omega}{(2\pi)^4} k_E^3 dk_E \ln \left(\frac{\frac{\lambda}{2}\phi_c^2 + m^2 + k_E^2}{m^2 + k_E^2} \right). \qquad (10.84)$$

Since the integrand does not depend on the angular variables, the angular integration can be done trivially and has the value

$$\int d^3\Omega = 2\pi^2, \qquad (10.85)$$

so that we have

$$V_1(\phi_c) = \frac{1}{2} \frac{1}{(2\pi)^4} (2\pi^2) \int_0^\infty \frac{1}{2} (dk_E^2) k_E^2 \ln \left(\frac{k_E^2 + m_{\text{eff}}^2}{k_E^2 + m^2} \right)$$

$$= \frac{1}{32\pi^2} \int_0^\infty dk_E^2 k_E^2 \ln \left(\frac{k_E^2 + m_{\text{eff}}^2}{k_E^2 + m^2} \right). \qquad (10.86)$$

Defining

$$x = k_E^2, \qquad (10.87)$$

we note that the one loop potential takes the form

$$V_1(\phi_c) = \frac{1}{32\pi^2} \int_0^\infty dx\, x \left(\ln(x + m_{\text{eff}}^2) - \ln(x + m^2) \right). \qquad (10.88)$$

Clearly, the integral is divergent and, therefore, we have to cut off the integral at some high momentum scale to obtain

$$V_1(\phi_c) = \frac{1}{32\pi^2} \int_0^{\Lambda^2} dx\, x \left(\ln(x + m_{\text{eff}}^2) - \ln(x + m^2) \right)$$

$$\simeq \frac{1}{32\pi^2} \left\{ \left[\frac{\Lambda^4}{2} \left(\ln \Lambda^2 - \frac{1}{2} \right) + m_{\text{eff}}^2 \Lambda^2 \right.\right.$$

$$\left. - \frac{m_{\text{eff}}^4}{2} \ln \Lambda^2 + \frac{m_{\text{eff}}^4}{2} \left(\ln m_{\text{eff}}^2 - \frac{1}{2} \right) \right]$$

$$- \left[\frac{\Lambda^4}{2} \left(\ln \Lambda^2 - \frac{1}{2} \right) + m^2 \Lambda^2 \right.$$

$$\left.\left. - \frac{m^4}{2} \ln \Lambda^2 + \frac{m^4}{2} \left(\ln m^2 - \frac{1}{2} \right) \right] \right\}$$

$$= \frac{1}{32\pi^2} \left[(m_{\text{eff}}^2 - m^2)\Lambda^2 - \frac{m_{\text{eff}}^4}{2} \ln \frac{\Lambda^2}{\mu^2} + \frac{m^4}{2} \ln \frac{\Lambda^2}{\mu^2} \right.$$

$$\left. + \frac{m_{\text{eff}}^4}{2} \left(\ln \frac{m_{\text{eff}}^2}{\mu^2} - \frac{1}{2} \right) - \frac{m^4}{2} \left(\ln \frac{m^2}{\mu^2} - \frac{1}{2} \right) \right]$$

$$= \frac{1}{32\pi^2} \left[\frac{\lambda}{2} \phi_c^2 \Lambda^2 - \frac{\lambda}{4} \phi_c^2 \left(2m^2 + \frac{\lambda}{2} \phi_c^2 \right) \ln \frac{\Lambda^2}{\mu^2} \right.$$

$$+ \frac{1}{2} \left(m^2 + \frac{\lambda}{2} \phi_c^2 \right)^2 \left(\ln \frac{(m^2 + \frac{\lambda}{2}\phi_c^2)}{\mu^2} - \frac{1}{2} \right)$$

$$\left. - \frac{m^4}{2} \left(\ln \frac{m^2}{\mu^2} - \frac{1}{2} \right) \right]. \tag{10.89}$$

Here we have introduced an arbitrary mass scale μ to write the expression in a meaningful manner. Note that the one loop potential, as it stands, diverges in the limit $\Lambda \to \infty$ which is the physical limit for the true value of the integral. This brings out one of the essential features of quantum field theory. Namely, point-like interactions necessarily induce divergences simply because the Heisenberg uncertainty principle, in this case, allows for an infinite uncertainty in the momentum being exchanged. This necessitates a systematic procedure for eliminating divergences in such theories. This is known as the renormalization theory which we will not go into. Let us simply note here that up to one loop, then, we can write the effective potential of the ϕ^4 theory to be

$$V_{\text{eff}}^{(1)} = V + V_1 . \tag{10.90}$$

10.5 References

A. Das, *Lectures on Quantum Field Theory*, World Scientific Publishing (2008).

J. Goldstone, A. Salam and S. Weinberg, Phys. Rev. **127**, 965 (1962).

J. Jona-Lasinio, Nuovo Cimento **34**, 1790 (1964).

Y. Nambu, Phys. Lett. **26B**, 626 (1966).

J. Zinn-Justin, *Quantum Field Theory and Critical Phenomena*, Oxford University Press.

CHAPTER 11

Invariances and their consequences

11.1 Symmetries of the action

Let us continue with the ϕ^4 theory and note that, in this case, we have

$$S[\phi] = \int d^4x \, \mathcal{L}(\phi, \partial_\mu \phi) \,, \tag{11.1}$$

where the Lagrangian density has the form

$$\mathcal{L}(\phi, \partial_\mu \phi) = \frac{1}{2} \partial_\mu \phi \partial^\mu \phi - \frac{m^2}{2} \phi^2 - \frac{\lambda}{4!} \phi^4, \qquad \lambda > 0. \tag{11.2}$$

We can write the action in Eq. (11.1) also in terms of the Lagrangian in the form

$$S[\phi] = \int dt \, L \,, \tag{11.3}$$

where

$$L = \int d^3x \, \mathcal{L}(\phi, \partial_\mu \phi) \,. \tag{11.4}$$

Given this theory where the basic variables are the fields $\phi(x)$, we can define the momentum (density) conjugate to the field variables in a straightforward manner as

$$\Pi(x) = \frac{\partial \mathcal{L}}{\partial \dot{\phi}(x)} = \dot{\phi}(x) \,. \tag{11.5}$$

This is the analogue of the relation between the momentum and velocity in classical mechanics, namely, $p = \dot{x}$ (for $m = 1$). In quantum field theory, this then is the starting point for quantization in the operator language. However, in the path integral formalism, we treat all variables classically. Therefore, let us analyze various concepts in the classical language. First, let us note that given the Lagrangian in Eq. (11.3), we can obtain the Hamiltonian through a Legendre transformation as

$$H = \int d^3x \, \Pi(x)\dot{\phi}(x) - L \,. \tag{11.6}$$

In the present case, we can write this out in detail as

$$H = \int d^3x \left(\Pi(x)\dot{\phi}(x) - \frac{1}{2}\dot{\phi}^2(x) + \frac{1}{2}\nabla\phi(x) \cdot \nabla\phi(x) \right.$$
$$\left. + \frac{m^2}{2}\phi^2(x) + \frac{\lambda}{4!}\phi^4(x) \right)$$
$$= \int d^3x \left(\frac{1}{2}\Pi^2(x) + \frac{1}{2}\nabla\phi \cdot \nabla\phi(x) + \frac{m^2}{2}\phi^2(x) + \frac{\lambda}{4!}\phi^4(x) \right). \tag{11.7}$$

Sometimes, this is also written as

$$H = \int d^3x \left(\frac{1}{2}\dot{\phi}^2(x) + \frac{1}{2}\nabla\phi(x) \cdot \nabla\phi(x) + \frac{m^2}{2}\phi^2(x) + \frac{\lambda}{4!}\phi^4(x) \right). \tag{11.8}$$

However, this is not strictly correct since the Hamiltonian is not a function of velocity.

Given a Lagrangian density, which depends only on $\phi(x)$ and $\partial_\mu\phi(x)$, the Euler-Lagrange equation is obtained to be (this is simply the generalization of Eq. (1.31) to the case of a field theory)

$$-\frac{\delta S[\phi]}{\delta\phi(x)} = \partial_\mu \frac{\partial\mathcal{L}}{\partial\partial_\mu\phi(x)} - \frac{\partial\mathcal{L}}{\partial\phi(x)} = 0 \,, \tag{11.9}$$

which gives the dynamics of the system. Given the dynamical equations, we can ask how unique is the Lagrangian density for the system. The answer, not surprisingly, turns out to be that the Lagrangian density is unique only up to a total (four) divergence. Namely, both

$$\mathcal{L}(\phi, \partial_\mu \phi), \quad \text{and} \quad \mathcal{L}(\phi, \partial_\mu \phi) + \partial_\mu K^\mu(\phi, \partial_\lambda \phi), \tag{11.10}$$

give the same Euler-Lagrange equation. We can, of course, check this directly. But a more intuitive way to understand this is to note that with the usual assumptions about the asymptotic fall off of the field variables, we have

$$S_K = \int d^4x \, \mathcal{L}_K = \int d^4x \, \partial_\mu K^\mu(\phi, \partial_\lambda \phi) = 0. \tag{11.11}$$

In other words, a total divergence in the Lagrangian density does not change the action. Consequently, the variation of S_K cannot contribute to the dynamical equations. (We note here that even when the asymptotic fall off of the fields is not fast enough, this statement remains true.) We can, of course, check this explicitly for specific examples. For example, choosing

$$\mathcal{L}_K = \partial_\mu K^\mu = \partial_\mu(\phi \partial^\mu \phi) = \partial_\mu \phi \partial^\mu \phi + \phi \partial_\mu \partial^\mu \phi, \tag{11.12}$$

the Euler-Lagrange equation gives

$$\frac{\delta S_K[\phi]}{\delta \phi(x)} = \partial_\mu \partial_\nu \frac{\partial \mathcal{L}_K}{\partial \partial_\mu \partial_\nu \phi} - \partial_\mu \frac{\partial \mathcal{L}_K}{\partial \partial_\mu \phi} + \frac{\partial \mathcal{L}_K}{\partial \phi}$$

$$= \partial_\mu \partial^\mu \phi - 2\partial_\mu \partial^\mu \phi + \partial_\mu \partial^\mu \phi \equiv 0. \tag{11.13}$$

With this analysis, therefore, it is clear that a given system of dynamical equations will remain invariant under a set of infinitesimal transformations of the field variables of the form

$$\phi \to \phi + \delta\phi, \tag{11.14}$$

if and only if the corresponding Lagrangian density changes, at the most, by a total divergence under the same transformations. Namely, if

$$\mathcal{L} \to \mathcal{L} + \partial_\mu K^\mu \,, \tag{11.15}$$

under a field transformation, then the transformation defines an invariance of the dynamical equations. Note that in the special case when $K^\mu = 0$, the Lagrangian density itself is invariant under the set of field transformations in Eq. (11.14) and, therefore, also defines a symmetry of the system. However, this defines a very special class of symmetries. In general, if under

$$\phi \to \phi + \delta\phi \,,$$
$$S[\phi] \to S[\phi + \delta\phi] = S[\phi] \,, \tag{11.16}$$

then we say that the field transformations define an invariance or a symmetry of the action and lead to a symmetry of the dynamical equations.

Continuous transformations, by definition, depend on a parameter of transformation continuously. This parameter can be a space-time independent parameter or it can depend on the space-time coordinates. In the first case, the transformations would change the field variables by the same amount at every space-time point. On the other hand, in the second case, the change in the field variables will be different at different space-time points depending on the value of the parameter at that point. Accordingly, the two kinds of transformations are known respectively as global and local transformations. The basic symmetries in gauge theories are local symmetries.

11.2 Nöther's theorem

In simple terms Nöther's theorem says that, for every continuous global symmetry of a system, there exists a current density which is conserved. More specifically, it says that for a system described by a Lagrangian density $\mathcal{L}(\phi, \partial_\mu \phi)$, if the infinitesimal global transformations

$$\phi(x) \to \phi(x) + \delta_\epsilon \phi(x) \,, \tag{11.17}$$

where ϵ is the infinitesimal constant parameter of transformation (the tensor indices of the fields as well as the parameters have been suppressed), define a symmetry of the system, in the sense that under these transformations

$$\mathcal{L} \to \mathcal{L} + \delta_\epsilon \mathcal{L} = \mathcal{L} + \partial_\mu K^\mu(\phi, \partial_\lambda \phi, \delta_\epsilon \phi) \,, \tag{11.18}$$

then,

$$J_\epsilon^\mu(x) = \frac{\partial \mathcal{L}}{\partial \partial_\mu \phi(x)} \delta_\epsilon \phi(x) - K^\mu(x) \,, \tag{11.19}$$

defines a current density which is conserved.

To see that $J_\epsilon^\mu(x)$ is indeed conserved, let us note that

$$\begin{aligned}
\partial_\mu J_\epsilon^\mu(x) &= \partial_\mu \left(\frac{\partial \mathcal{L}}{\partial \partial_\mu \phi(x)} \delta_\epsilon \phi(x) \right) - \partial_\mu K^\mu(x) \\
&= \left(\partial_\mu \frac{\partial \mathcal{L}}{\partial \partial_\mu \phi(x)} \right) \delta_\epsilon \phi(x) + \frac{\partial \mathcal{L}}{\partial \partial_\mu \phi(x)} \partial_\mu \delta_\epsilon \phi(x) - \partial_\mu K^\mu(x) \\
&= \frac{\partial \mathcal{L}}{\partial \phi(x)} \delta_\epsilon \phi(x) + \frac{\partial \mathcal{L}}{\partial \partial_\mu \phi(x)} \delta_\epsilon(\partial_\mu \phi(x)) - \partial_\mu K^\mu(x) \,.
\end{aligned} \tag{11.20}$$

Here, we have used the Euler-Lagrange equation, Eq. (11.9), as well as the fact that $\partial_\mu \delta_\epsilon \phi(x) = \delta_\epsilon \partial_\mu \phi(x)$ for a global transformation. We note next that the first two terms in Eq. (11.20) simply give the infinitesimal change in the Lagrangian density under the transformations. Therefore, we can also write using Eq. (11.18)

$$\partial_\mu J_\epsilon^\mu(x) = \delta_\epsilon \mathcal{L} - \partial_\mu K^\mu(x) = 0 \,. \tag{11.21}$$

This shows that the current density given in Eq. (11.19) is indeed conserved. The current density defined in Eq. (11.19) depends on the parameter of transformation in addition to the field variables. A more fundamental quantity is the current density without the parameter of transformation and let us denote this symbolically as

$$J_\epsilon^\mu = \epsilon J^\mu\,. \tag{11.22}$$

We have to remember that this is only a symbolic relation simply because the parameter ϵ may itself have a tensor structure depending on the transformation. In that case, the current density without the parameter will have a more complicated tensor structure as we will see shortly. (The conserved current density has a tensor structure one rank higher than the parameter of transformation.) It is worth pointing out here that the conserved Nöther current is unique up to the addition of a transverse vector. Namely, both

$$J^\mu(x), \quad \text{and} \quad J^\mu_{\text{improved}}(x) = J^\mu(x) + \tilde{K}_T^\mu, \quad \partial_\mu \tilde{K}_T^\mu = 0, \tag{11.23}$$

can be considered as the conserved current associated with the symmetry. (This simply corresponds to the fact that K^μ has an arbitrariness up to adding a transverse part.) This ambiguity is profitably used in some theories, mainly involving gravitation, to define improved currents which have matrix elements which are better behaved. But, we will not pursue this further in these lectures.

As in classical electrodynamics, we know that given a conserved current density, we can define a charge which is a constant of motion as

$$Q = \int \mathrm{d}^3 x\, J^0(\mathbf{x}, t)\,. \tag{11.24}$$

The fact that this charge is a constant independent of time can be seen simply as

$$\frac{\mathrm{d}Q}{\mathrm{d}t} = \frac{\mathrm{d}}{\mathrm{d}t} \int \mathrm{d}^3 x\, J^0(\mathbf{x}, t) = \int \mathrm{d}^3 x\, \partial_0 J^0(\mathbf{x}, t)$$

$$= \int \mathrm{d}^3 x\, \left(\partial_0 J^0(\mathbf{x}, t) + \boldsymbol{\nabla} \cdot \mathbf{J}(\mathbf{x}, t)\right)\,. \tag{11.25}$$

Here, we have added a total divergence term whose integral vanishes under our assumptions on the asymptotic behavior of the field variables. Thus, we have

$$\frac{dQ}{dt} = \int d^3x \, \partial_\mu J^\mu = 0 \,, \tag{11.26}$$

which follows from the conservation of the current density discussed in Eqs. (11.21) and (11.22). This shows that the charge is a constant of motion.

Since Q is a constant of motion, this implies classically that the Poisson bracket of Q with H vanishes. Quantum mechanically, it is the commutator of the two operators which vanishes

$$[Q, H] = 0 \,. \tag{11.27}$$

But this is precisely the symmetry condition in quantum mechanics. Namely, we know from our studies in quantum mechanics that a transformation is a symmetry of the system if the generator of the infinitesimal symmetry transformation commutes with the Hamiltonian (which implies that they can have simultaneous eigenstates). Conversely, any operator which commutes with the Hamiltonian is the generator of an infinitesimal global symmetry transformation which leaves the system invariant. Thus, we recognize Q to be the generator of the infinitesimal symmetry transformations in the present case. This simply means that the infinitesimal change in any variable (operator) can be obtained from the commutator (we are setting $\hbar = 1$)

$$\delta_\epsilon \phi = -i[\epsilon Q, \phi] \,. \tag{11.28}$$

(Classically, we should use appropriate Poisson bracket relations.) It is now clear that the vanishing of the commutator between Q and H simply reflects the fact that the Hamiltonian is invariant under the symmetry transformations as we would expect.

In quantum field theory, the unitary operator implementing finite symmetry transformations can be written in terms of the generators of infinitesimal transformations as

$$U(\alpha) = e^{-\frac{i}{\hbar}\alpha Q}, \tag{11.29}$$

where α is the constant parameter of finite transformation. Under such a transformation a field operator is supposed to transform as

$$\phi(x) \to U(\alpha)\phi(x)U^{-1}(\alpha) = e^{-\frac{i}{\hbar}\alpha Q}\phi(x)e^{\frac{i}{\hbar}\alpha Q}. \tag{11.30}$$

And, furthermore, a true symmetry transformation of the physical system (theory) is supposed to leave the ground state of the system or the vacuum invariant (in addition to leaving the Hamiltonian invariant), namely,

$$U(\alpha)|0\rangle = e^{-i\alpha Q}|0\rangle = |0\rangle. \tag{11.31}$$

Equivalently, it follows from Eq. (11.31) that

$$Q|0\rangle = 0. \tag{11.32}$$

In other words, for a true symmetry, the conserved charge annihilates the vacuum (the vacuum carries zero charge quantum number). Therefore, in such a case, we note from Eq. (11.28) that for any field operator in the theory

$$\langle 0|\delta_\epsilon\phi(x)|0\rangle = -i\langle 0|[\epsilon Q, \phi(x)]|0\rangle = 0, \tag{11.33}$$

where we have used Eq. (11.32). As we will see shortly, if there is a spontaneous breakdown of a symmetry, then, the conserved charge Q does not annihilate the vacuum and that under the transformation the vacuum expectation value of the change of some operator in the theory becomes nonzero.

11.2.1 Example

As an example of Nöther's theorem, let us study global space-time translations as a symmetry of quantum field theories. Let us continue to use the ϕ^4 theory for this discussion.

Let us define the infinitesimal translations

$$x^\mu \to x^\mu + \epsilon^\mu,$$

or, $\quad \delta_\epsilon x^\mu = \epsilon^\mu,$ $\hspace{6cm}$ (11.34)

as the global transformations where ϵ^μ is the infinitesimal constant parameter of transformation. In such a case (repeated indices are being summed),

$$\delta_\epsilon \phi(x) = \phi(x + \epsilon) - \phi(x) = \epsilon^\mu \partial_\mu \phi(x),$$
$$\delta_\epsilon \partial_\mu \phi(x) = \partial_\mu \phi(x + \epsilon) - \partial_\mu \phi(x)$$
$$= \partial_\mu(\delta_\epsilon \phi(x)) = \epsilon^\nu \partial_\mu \partial_\nu \phi(x). \hspace{2cm} (11.35)$$

Given this, we can obtain the infinitesimal change in the Lagrangian density in Eq. (11.2)

$$\mathcal{L} = \frac{1}{2}\partial_\mu \phi(x)\partial^\mu \phi(x) - \frac{m^2}{2}\phi^2(x) - \frac{\lambda}{4!}\phi^4(x), \hspace{1.5cm} (11.36)$$

in a straightforward manner. However, a much simpler way to evaluate the change is to note that the Lagrangian density is effectively (implicitly) a function of x, namely, $\mathcal{L} = \mathcal{L}(x)$. Thus,

$$\delta_\epsilon \mathcal{L} = \mathcal{L}(x + \epsilon) - \mathcal{L}(x) = \epsilon^\mu \partial_\mu \mathcal{L}(x) = \partial_\mu K^\mu, \hspace{1cm} (11.37)$$

where we have determined

$$K^\mu = \epsilon^\mu \mathcal{L}(x) = \epsilon^\mu \mathcal{L}(\phi, \partial_\mu \phi), \hspace{2.5cm} (11.38)$$

in the present case.

On the other hand, we see from Eq. (11.2) that for this theory

$$\frac{\partial \mathcal{L}}{\partial \partial_\mu \phi(x)} = \partial^\mu \phi(x). \hspace{4cm} (11.39)$$

As a result, we see from Eqs. (11.35), (11.38) and (11.19), that the Nöther current density defined in Eq. (11.19), in this case, follows to be

$$
\begin{aligned}
J_\epsilon^\mu(x) &= \frac{\partial \mathcal{L}}{\partial \partial_\mu \phi(x)} \, \delta_\epsilon \phi(x) - K^\mu \\
&= \partial^\mu \phi(x) \left(\epsilon^\nu \partial_\nu \phi(x) \right) - \epsilon^\mu \mathcal{L} \\
&= \epsilon^\nu \left(\partial^\mu \phi(x) \partial_\nu \phi(x) - \delta_\nu^\mu \mathcal{L} \right) \\
&= \epsilon_\nu \left(\partial^\mu \phi(x) \partial^\nu \phi(x) - \eta^{\mu\nu} \mathcal{L} \right) .
\end{aligned}
\tag{11.40}
$$

This is the conserved current density associated with translations and the current density without the parameter of transformation has the form (see Eq. (11.22))

$$
J_\epsilon^\mu(x) = \epsilon_\nu T^{\mu\nu} ,
\tag{11.41}
$$

where we see from Eq. (11.40) that

$$
T^{\mu\nu} = \partial^\mu \phi(x) \partial^\nu \phi(x) - \eta^{\mu\nu} \mathcal{L} .
\tag{11.42}
$$

There are several comments in order here. First, let us note that the fundamental conserved current density (in this case $T^{\mu\nu}$) is not necessarily a vector. Its tensorial character depends completely on the parameter of transformation (in the present case a vector). Second, we note from Eq. (11.42) that the conserved current density, in this case, is a symmetric second rank tensor density which is conserved, namely,

$$
T^{\mu\nu} = T^{\nu\mu}, \qquad \partial_\mu T^{\mu\nu} = 0.
\tag{11.43}
$$

This is known as the stress tensor of the theory.

Let us also note from Eqs. (11.24) and (11.42) that the conserved charge, in this case, is a four vector and has the form

$$
P^\mu = \int \mathrm{d}^3 x \, T^{0\mu} .
\tag{11.44}
$$

To understand the meaning of the charges in Eq. (11.44) (there are, in fact, four of them), let us write them out explicitly. We note from Eqs. (11.42) and (11.44) that

$$P^0 = \int d^3x\, T^{00}$$

$$= \int d^3x\, ((\dot\phi(x))^2 - \mathcal{L})$$

$$= \int d^3x \left((\dot\phi(x))^2 - \frac{1}{2}\dot\phi^2(x) + \frac{1}{2}\boldsymbol{\nabla}\phi\cdot\boldsymbol{\nabla}\phi \right.$$
$$\left. + \frac{m^2}{2}\phi^2(x) + \frac{\lambda}{4!}\phi^4(x) \right)$$

$$= \int d^3x \left(\frac{1}{2}\dot\phi^2(x) + \frac{1}{2}\boldsymbol{\nabla}\phi\cdot\boldsymbol{\nabla}\phi + \frac{m^2}{2}\phi^2(x) + \frac{\lambda}{4!}\phi^4(x) \right)$$

$$= H,$$

$$P^i = \int d^3x\, T^{0i} = \int d^3x\, \dot\phi\partial^i\phi,$$

$$\text{or,} \quad \mathbf{P} = -\int d^3x\, \dot\phi(x)\boldsymbol{\nabla}\phi(x). \tag{11.45}$$

We recognize the first conserved quantity (namely, P^0) as the Hamiltonian of the system obtained in Eq. (11.8) and from relativistic invariance we conclude that $\mathbf{P}$ must represent the total momentum of the system. Thus, we recover the familiar result that infinitesimal space-time translations are generated by the energy-momentum operators of the theory. Furthermore, if translations are true symmetries of the quantum field theory, the conserved charge operators, P^μ, must annihilate the vacuum

$$P^\mu|0\rangle = 0,$$

which we have talked about earlier. (In quantum field theory, the charge operators are normal ordered so that the annihilation operators stand to the right of the creation operators.)

11.3 Complex scalar field

So far we have discussed the ϕ^4 theory where the basic field variable is real. As we have mentioned before, such a theory can describe spin zero mesons which are charge neutral. Let us next consider a scalar field theory where the basic field variable is complex. Namely, in this case,

$$\phi^*(x) \neq \phi(x)\,. \tag{11.46}$$

One way to study such a theory is to expand the complex field in terms of two real fields as

$$\phi(x) = \frac{1}{\sqrt{2}}(\phi_1(x) + i\phi_2(x))\,. \tag{11.47}$$

However, let us continue with the complex field $\phi(x)$ as the basic variable. The real Lagrangian density describing quartic interactions can be generalized from Eq. (11.2) and has the form

$$\mathcal{L}(\phi, \phi^*) = \partial_\mu \phi^* \partial^\mu \phi - m^2 \phi^* \phi - \frac{\lambda}{4}(\phi^* \phi)^2\,, \tag{11.48}$$

with $\lambda > 0$. We can treat ϕ and ϕ^* as independent dynamical variables. Correspondingly, the two Euler-Lagrange equations following from Eq. (11.48) are given by

$$-\frac{\delta S[\phi, \phi^*]}{\delta \phi^*(x)} = \partial_\mu \frac{\partial \mathcal{L}}{\partial \partial_\mu \phi^*} - \frac{\partial \mathcal{L}}{\partial \phi^*} = 0,$$

$$\text{or,} \quad (\partial_\mu \partial^\mu + m^2)\phi + \frac{\lambda}{2}(\phi^* \phi)\phi = 0\,, \tag{11.49}$$

and

$$-\frac{\delta S[\phi, \phi^*]}{\delta \phi(x)} = \partial_\mu \frac{\partial \mathcal{L}}{\partial \partial_\mu \phi} - \frac{\partial \mathcal{L}}{\partial \phi} = 0,$$

$$\text{or,} \quad (\partial_\mu \partial^\mu + m^2)\phi^* + \frac{\lambda}{2}(\phi^* \phi)\phi^* = 0\,. \tag{11.50}$$

Thus, the two dynamical equations in Eqs. (11.49) and (11.50) correspond to two coupled scalar field equations. We should have expected this since having a complex field doubles the number of degrees of freedom.

Let us next note that if we make a finite phase transformation of the form

$$\phi(x) \rightarrow e^{-i\alpha}\phi(x),$$
$$\phi^*(x) \rightarrow e^{i\alpha}\phi^*(x), \tag{11.51}$$

where α is a real, constant (global) parameter of transformation or equivalently, an infinitesimal transformation of the form (ϵ is infinitesimal)

$$\delta_\epsilon \phi(x) = -i\epsilon\phi(x),$$
$$\delta_\epsilon \phi^*(x) = i\epsilon\phi^*(x), \tag{11.52}$$

then, under such a transformation

$$\phi^*\phi \rightarrow e^{i\alpha}\phi^* e^{-i\alpha}\phi = \phi^*\phi. \tag{11.53}$$

Equivalently, under the infinitesimal transformations of Eq. (11.52), we note that

$$\delta_\epsilon(\phi^*\phi) = (\delta_\epsilon\phi^*)\phi + \phi^*(\delta_\epsilon\phi) = i\epsilon\phi^*\phi - i\epsilon\phi^*\phi = 0. \tag{11.54}$$

Namely, under the transformation in Eq. (11.51) or (11.52), the product $\phi^*\phi$ remains unchanged. Similarly, we note that under the transformation of Eq. (11.51)

$$\partial_\mu\phi^*\partial^\mu\phi \rightarrow \partial_\mu(e^{i\alpha}\phi^*)\partial^\mu(e^{-i\alpha}\phi) = \partial_\mu\phi^*\partial^\mu\phi. \tag{11.55}$$

Alternatively, from the form of the infinitesimal transformations in Eq. (11.52), we obtain

$$\begin{aligned}
\delta_\epsilon(\partial_\mu\phi^*\partial^\mu\phi) &= \big(\delta_\epsilon(\partial_\mu\phi^*)\big)\partial^\mu\phi + \partial_\mu\phi^*\big(\delta_\epsilon(\partial^\mu\phi)\big) \\
&= \partial_\mu(\delta_\epsilon\phi^*)\partial^\mu\phi + \partial_\mu\phi^*(\partial^\mu(\delta_\epsilon\phi)) \\
&= i\epsilon\partial_\mu\phi^*\partial^\mu\phi - i\epsilon\partial_\mu\phi^*\partial^\mu\phi = 0.
\end{aligned} \tag{11.56}$$

(It is important to recognize that the invariance in Eqs. (11.55) and (11.56) results because the parameter of transformation is assumed to be independent of space-time coordinates.)

In this case, therefore, we see that the constant phase transformations define a symmetry of the theory, in the sense that,

$$\mathcal{L} = \partial_\mu \phi^* \partial^\mu \phi - m^2 \phi^* \phi - \frac{\lambda}{4}(\phi^* \phi)^2 \to \mathcal{L} \,. \tag{11.57}$$

Equivalently, under the infinitesimal transformations

$$\delta_\epsilon \mathcal{L} = 0 \,. \tag{11.58}$$

Such a symmetry transformation, where the space-time coordinates do not change, is called an internal symmetry. For such an invariance, we note that

$$K^\mu = 0 \,. \tag{11.59}$$

Therefore, the conserved current density constructed through the Nöther procedure has the form (see Eq. (11.19))

$$
\begin{aligned}
J_\epsilon^\mu &= \frac{\partial \mathcal{L}}{\partial \partial_\mu \phi^*} \delta_\epsilon \phi^* + \frac{\partial \mathcal{L}}{\partial \partial_\mu \phi} \delta_\epsilon \phi \\
&= \partial^\mu \phi (i\epsilon \phi^*) + \partial^\mu \phi^* (-i\epsilon \phi) \\
&= i\epsilon \left(\phi^* \partial^\mu \phi - (\partial^\mu \phi^*)\phi \right) \\
&= \epsilon (i\phi^* \overset{\leftrightarrow}{\partial^\mu} \phi) = \epsilon \, J^\mu,
\end{aligned}
\tag{11.60}
$$

where we have defined

$$J^\mu = i\phi^* \overset{\leftrightarrow}{\partial^\mu} \phi \equiv i\left(\phi^* \partial^\mu \phi - (\partial^\mu \phi^*)\,\phi \right). \tag{11.61}$$

The conserved current density J^μ, in this case, is a four vector very much like the electromagnetic current density (the parameter of transformation is a scalar). Therefore, it can be identified with the electromagnetic current density associated with this system. Consequently, this theory can describe charged spin zero mesons. The conserved charge, for the present case, can be written as

$$
Q = \int d^3x \, J^0 = \int d^3x \, i(\phi^* \dot{\phi} - \dot{\phi}^* \phi)
$$

$$
= i \int d^3x \, (\phi^* \dot{\phi} - \dot{\phi}^* \phi) \,. \tag{11.62}
$$

In quantum field theory, Q would represent the electric charge operator. As we have mentioned earlier, if the phase transformations in Eq. (11.51) or (11.52) define a true symmetry of the system, then the charge operator in Eq. (11.62) (normal ordered) would annihilate the vacuum. In other words, in such a case, we would have

$$
Q|0\rangle = 0 \,, \tag{11.63}
$$

corresponding to the fact that the vacuum is charge neutral. (Vacuum is an eigenstate of Q with zero eigenvalue.)

11.4 Ward identities

Symmetries are quite important in the study of physical theories for various reasons. First of all, they lead to conserved quantities and conserved quantum numbers. But more importantly, in quantum field theories they give rise to relations between various Green's functions and, therefore, between the transition amplitudes. Thus, as an example, let us consider the generating functional for the complex scalar field in the presence of sources

$$
Z[J, J^*] = e^{iW[J,J^*]} = N \int \mathcal{D}\phi \mathcal{D}\phi^* \, e^{iS[\phi,\phi^*,J,J^*]} \,. \tag{11.64}
$$

Let us note here that we have set $\hbar = 1$ for simplicity and that we have defined

$$
S[\phi, \phi^*, J, J^*] = S[\phi, \phi^*] + \int d^4x (J^*\phi + J\phi^*) \,, \tag{11.65}
$$

with $S[\phi, \phi^*]$ representing the dynamical action for the system. Here, we note that J^* is the source for the field ϕ whereas J corresponds to the source for ϕ^*. Note also that even though the action $S[\phi, \phi^*]$ is invariant under the global phase transformations of Eq. (11.52), the complete action $S[\phi, \phi^*, J, J^*]$ is not symmetric unless we simultaneously change J and J^* also. In fact, let us note that infinitesimally,

$$
\begin{aligned}
\delta_\epsilon S[\phi, \phi^*, J, J^*] &= \delta_\epsilon S[\phi, \phi^*] + \delta_\epsilon \left(\int d^4x \, (J^*\phi + J\phi^*) \right) \\
&= \int d^4x \, (J^* \delta_\epsilon \phi + J \delta_\epsilon \phi^*) \\
&= -i\epsilon \int d^4x \, (J^*\phi - J\phi^*).
\end{aligned}
\tag{11.66}
$$

Since the generating functional does not depend on the field variables (that is, the fields are all integrated out), making a field redefinition in the integrand of the path integral should not change the generating functional. In particular, if the redefinition corresponds to the infinitesimal phase transformations defined in Eq. (11.52), then we will have

$$
\begin{aligned}
\delta_\epsilon Z[J, J^*] = 0 &= N \int \mathcal{D}\phi \mathcal{D}\phi^* \, i\delta_\epsilon S \, e^{iS[\phi, \phi^*, J, J^*]} \\
&= N \int \mathcal{D}\phi \mathcal{D}\phi^* \left(\epsilon \int d^4x \, (J^*\phi - J\phi^*) \right) e^{iS[\phi, \phi^*, J, J^*]},
\end{aligned}
\tag{11.67}
$$

where we have used Eq. (11.66). In general, one should also worry about the change coming from the Jacobian under a field redefinition. In the present case, the Jacobian for the infinitesimal field redefinition can be easily checked to be unity.

Let us recall that by definition,

$$
\begin{aligned}
\frac{\delta Z[J, J^*]}{\delta J(x)} &= iN \int \mathcal{D}\phi \mathcal{D}\phi^* \, \phi^*(x) \, e^{iS[\phi, \phi^*, J, J^*]}, \\
\frac{\delta Z[J, J^*]}{\delta J^*(x)} &= iN \int \mathcal{D}\phi \mathcal{D}\phi^* \, \phi(x) \, e^{iS[\phi, \phi^*, J, J^*]}.
\end{aligned}
\tag{11.68}
$$

Using this then, Eq. (11.67) becomes

$$\epsilon \int d^4x \left(J^*(x) \left(-i\frac{\delta Z}{\delta J^*(x)} \right) - J(x) \left(-i\frac{\delta Z}{\delta J(x)} \right) \right) = 0. \quad (11.69)$$

This must hold for any arbitrary value of the parameter ϵ and, therefore, we conclude that

$$\int d^4x \left(J^*(x)\frac{\delta}{\delta J^*(x)} - J(x)\frac{\delta}{\delta J(x)} \right) Z[J, J^*] = 0,$$

or, $$\int d^4x \left(J^*(x)\frac{\delta}{\delta J^*(x)} - J(x)\frac{\delta}{\delta J(x)} \right) e^{iW[J,J^*]} = 0,$$

or, $$\int d^4x \left(J^*(x)\frac{\delta W}{\delta J^*(x)} - J(x)\frac{\delta W}{\delta J(x)} \right) = 0. \quad (11.70)$$

This is the master equation for defining symmetry relations between connected Green's functions. By taking higher functional derivatives of Eq. (11.70), we can obtain relations between various connected n-point Green's functions as a result of the symmetry in the problem. In this case, the symmetry relations are quite simple (simply because the symmetry transformations in Eq. (11.52) are simple), but in the case of more complicated symmetries such as gauge symmetries, such relations are extremely useful and go under the name of Ward-Takahashi identities of the theory (also known as Slavnov-Taylor identities particularly in the case of non-Abelian gauge symmetries). It is interesting to note that we could also have derived the Ward identities from a combined set of transformations of the form

$$\delta_\epsilon \phi = -i\epsilon\phi, \qquad \delta_\epsilon \phi^* = i\epsilon\phi^*,$$
$$\delta_\epsilon J = -i\epsilon J, \qquad \delta_\epsilon J^* = i\epsilon J^*. \quad (11.71)$$

In such a case, it is easy to see that the complete action in Eq. (11.65) is invariant. Namely,

$$\delta_\epsilon S[\phi, \phi^*, J, J^*] = 0. \quad (11.72)$$

Therefore, from

$$Z[J, J^*] = e^{iW[J,J^*]} = N \int \mathcal{D}\phi \mathcal{D}\phi^* \, e^{iS[\phi,\phi^*,J,J^*]} \,, \tag{11.73}$$

we obtain

$$\delta_\epsilon Z[J, J^*] = N \int \mathcal{D}\phi \mathcal{D}\phi^* \, (i\delta_\epsilon S) \, e^{iS[\phi,\phi^*,J,J^*]} = 0,$$

or, $i\delta_\epsilon W[J, J^*] e^{iW[J,J^*]} = 0,$

or, $\delta_\epsilon W[J, J^*] = 0,$

or, $\int \mathrm{d}^4 x \left(\dfrac{\delta W}{\delta J^*(x)} \delta_\epsilon J^*(x) + \dfrac{\delta W}{\delta J(x)} \delta_\epsilon J(x) \right) = 0,$

or, $i\epsilon \int \mathrm{d}^4 x \left(J^*(x) \dfrac{\delta W}{\delta J^*(x)} - J(x) \dfrac{\delta W}{\delta J(x)} \right) = 0,$

or, $\int \mathrm{d}^4 x \left(J^*(x) \dfrac{\delta W}{\delta J^*(x)} - J(x) \dfrac{\delta W}{\delta J(x)} \right) = 0 \,. \tag{11.74}$

This is, of course, the same relation as in Eq. (11.70).

Let us note that in the case of a complex scalar field we will have a complex classical field defined by (see Eq. (10.7))

$$\phi_c(x) = \frac{\delta W}{\delta J^*(x)} = \langle 0|\phi(x)|0\rangle^{J,J^*},$$

$$\phi_c^*(x) = \frac{\delta W}{\delta J(x)} = \langle 0|\phi^*(x)|0\rangle^{J,J^*} \,. \tag{11.75}$$

From the transformation properties of the fields $\phi(x)$ and $\phi^*(x)$ in Eq. (11.52), we can immediately determine the transformation properties of the vacuum expectation values in Eq. (11.75). Namely, we obtain (this corresponds to asking, by how much would $\phi_c(x)$ change if we change $\phi(x)$ according to Eq. (11.52))

$$\delta_\epsilon \phi_c(x) = \langle 0|\delta_\epsilon \phi(x)|0\rangle^{J,J^*} = -i\epsilon \langle 0|\phi(x)|0\rangle^{J,J^*} = -i\epsilon \phi_c(x),$$

$$\delta_\epsilon \phi_c^*(x) = \langle 0|\delta_\epsilon \phi^*(x)|0\rangle^{J,J^*} = i\epsilon \langle 0|\phi^*(x)|0\rangle^{J,J^*} = i\epsilon \phi_c^*(x), \tag{11.76}$$

where we have assumed the invariance of the vacuum state under such a transformation. From the transformation properties of the classical fields in Eq. (11.76), we can now work out the Ward-Takahashi identities for the 1PI vertex functions. In the present case, we note that the effective action is given by

$$\Gamma[\phi_c, \phi_c^*] = W[J, J^*] - \int d^4x \left(J^*(x)\phi_c(x) + J(x)\phi_c^*(x) \right), \quad (11.77)$$

from which it follows that

$$\delta_\epsilon \Gamma[\phi_c, \phi_c^*] = -\int d^4x \left(J^*(x)\delta_\epsilon\phi_c(x) + J(x)\delta_\epsilon\phi_c^*(x) \right)$$

$$= i\epsilon \int d^4x \left(J^*(x)\phi_c(x) - J(x)\phi_c^*(x) \right)$$

$$= i\epsilon \int d^4x \left(J^*(x)\frac{\delta W}{\delta J^*(x)} - J(x)\frac{\delta W}{\delta J(x)} \right)$$

$$= 0. \qquad (11.78)$$

Here in the last step, we have used the relation in Eq. (11.70). This shows that the infinitesimal transformations define a symmetry of the effective action as well. On the other hand, we note that using Eq. (11.76) we can write

$$\delta_\epsilon \Gamma[\phi_c, \phi_c^*] = \int d^4x \left(\frac{\delta\Gamma}{\delta\phi_c(x)}\delta_\epsilon\phi_c(x) + \frac{\delta\Gamma}{\delta\phi_c^*(x)}\delta_\epsilon\phi_c^*(x) \right)$$

$$= -i\epsilon \int d^4x \left(\frac{\delta\Gamma}{\delta\phi_c(x)}\phi_c(x) - \frac{\delta\Gamma}{\delta\phi_c^*(x)}\phi_c^*(x) \right). \quad (11.79)$$

Therefore, following Eq. (11.78), we can set this to zero and noting that the parameter ϵ is arbitrary, we obtain

$$\int d^4x \left(\frac{\delta\Gamma}{\delta\phi_c(x)}\phi_c(x) - \frac{\delta\Gamma}{\delta\phi_c^*(x)}\phi_c^*(x) \right) = 0. \qquad (11.80)$$

This is the master equation and by taking higher order functional derivatives we can derive relations between various 1PI vertex functions, as a consequence of the symmetry in the theory. We note here that the relation in Eq. (11.80) could have been derived directly from the master equation in Eq. (11.70) using various definitions as follows

$$\int d^4x \left(J^*(x) \frac{\delta W}{\delta J^*(x)} - J(x) \frac{\delta W}{\delta J(x)} \right) = 0,$$

or, $$\int d^4x \left(-\frac{\delta \Gamma}{\delta \phi_c(x)} \phi_c(x) - \left(-\frac{\delta \Gamma}{\delta \phi_c^*(x)} \right) \phi_c^*(x) \right) = 0,$$

or, $$\int d^4x \left(\frac{\delta \Gamma}{\delta \phi_c(x)} \phi_c(x) - \frac{\delta \Gamma}{\delta \phi_c^*(x)} \phi_c^*(x) \right) = 0. \tag{11.81}$$

11.5 Spontaneous symmetry breaking

Let us next consider the complex scalar field theory defined by the Lagrangian density (if the fields are treated as operators, ϕ^* should be replaced by $\phi^\dagger$)

$$\mathcal{L} = \partial_\mu \phi^* \partial^\mu \phi + m^2 \phi^* \phi - \frac{\lambda}{4}(\phi^*\phi)^2, \qquad \lambda > 0. \tag{11.82}$$

This is the same Lagrangian density as in Eq. (11.48) except for the sign in the mass term which is opposite. It is clear that this Lagrangian density is also invariant under the global phase transformations in Eq. (11.51) or (11.52) since each individual term is. Therefore, the phase transformations define a symmetry of this theory as well and according to Nöther's theorem, there exists a conserved charge which is the same as given in Eq. (11.61).

For constant field configurations, if we look at the potential of this theory, namely,

$$V(\phi, \phi^*) = -m^2 \phi^* \phi + \frac{\lambda}{4}(\phi^*\phi)^2, \tag{11.83}$$

we note that the extrema of the potential occur at

$$\frac{\partial V}{\partial \phi^*} = \left(-m^2 + \frac{\lambda}{2}\phi^*\phi\right)\phi = 0\,,$$

$$\frac{\partial V}{\partial \phi} = \left(-m^2 + \frac{\lambda}{2}\phi^*\phi\right)\phi^* = 0\,. \tag{11.84}$$

There are two solutions of these extremum conditions which are easily obtained to be

$$\phi_c = \phi_c^* = 0\,,$$

$$\phi_c^*\phi_c = \frac{2m^2}{\lambda}\,. \tag{11.85}$$

However, it is quite easy to see that

$$\left.\frac{\partial^2 V}{\partial \phi^* \partial \phi}\right|_{\phi=\phi^*=0} = -m^2\,, \tag{11.86}$$

whereas

$$\left.\frac{\partial^2 V}{\partial \phi^* \partial \phi}\right|_{\phi^*\phi=\frac{2m^2}{\lambda}} = \left.\frac{\lambda}{2}\phi^*\phi\right|_{\phi^*\phi=\frac{2m^2}{\lambda}} = m^2\,. \tag{11.87}$$

Consequently, the extremum at $\phi_c = \phi_c^* = 0$ is really a local maximum of the potential whereas the true minimum occurs at

$$\phi_c^*\phi_c = \frac{2m^2}{\lambda}\,. \tag{11.88}$$

Note that since for constant field configurations the derivative terms vanish, this also defines the field configuration for the true minimum of energy or the true ground state of this theory.

To better understand what is involved here, let us express the complex field in terms of two real scalar fields. Namely, let us write

$$\phi = \frac{1}{\sqrt{2}}\left(\sigma + i\rho\right)\,, \tag{11.89}$$

where we assume that σ and ρ are real (Hermitian) scalar fields. In terms of these variables, then, the minimum of the potential occurs at

$$\phi_c^* \phi_c = \frac{1}{2}(\sigma_c^2 + \rho_c^2) = \frac{2m^2}{\lambda},$$

$$\text{or,} \quad \sigma_c^2 + \rho_c^2 = \frac{4m^2}{\lambda}. \tag{11.90}$$

It is clear that, in this case, there is an infinite number of degenerate minima lying on a circle in the σ-ρ plane. For simplicity, let us choose $\rho_c = 0$. Then, the minimum of the potential is determined from

$$\sigma_c^2 = \frac{4m^2}{\lambda}$$

$$\text{or,} \quad \sigma_c = \pm\frac{2m}{\sqrt{\lambda}}. \tag{11.91}$$

Let us, in fact choose the minimum to be at (the choice of a minimum breaks the symmetry of the theory)

$$\sigma_c = \frac{2m}{\sqrt{\lambda}}, \qquad \rho_c = 0. \tag{11.92}$$

In this case, therefore, we see that one of the fields develops a vacuum expectation value, namely,

$$\sigma_c = \langle 0|\sigma(x)|0\rangle = \frac{2m}{\sqrt{\lambda}},$$

$$\rho_c = \langle 0|\rho(x)|0\rangle = 0. \tag{11.93}$$

To understand further the meaning of this, let us plot the potential in Eq. (11.83) as a function of σ and ρ for constant values of the fields which is shown in Fig. 11.1. We note that

$$V(\sigma, \rho) = -\frac{m^2}{2}(\sigma^2 + \rho^2) + \frac{\lambda}{16}(\sigma^2 + \rho^2)^2, \qquad \lambda > 0. \tag{11.94}$$

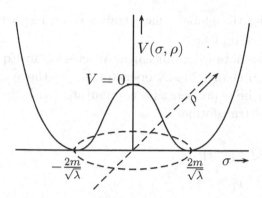

Figure 11.1: Mexican hat potential exhibiting a nontrivial minimum.

Thus, the potential, in the present case, is very much like the instanton potential in Eq. (7.54), but the minima are infinitely degenerate. Popularly, such a potential is known as the Mexican hat potential. Let us also note that since

$$\phi = \frac{1}{\sqrt{2}}(\sigma + i\rho),$$
(11.95)

we can deduce from the transformation rule in Eq. (11.52)

$$\delta\phi = -i\epsilon\phi,$$
(11.96)

that

$$\frac{1}{\sqrt{2}}(\delta\sigma + i\delta\rho) = -i\epsilon \frac{1}{\sqrt{2}}(\sigma + i\rho),$$

$$\text{or,} \quad \delta\sigma + i\delta\rho = \epsilon(\rho - i\sigma).$$
(11.97)

From this we conclude that under the global phase transformations the real scalar fields transform as

$$\delta\sigma = \epsilon\rho, \quad \delta\rho = -\epsilon\sigma.$$
(11.98)

In other words, the global phase transformations correspond to a rotation in the σ-ρ plane.

Let us also note from our earlier discussion in Eq. (11.28) that the infinitesimal change in any operator, under the symmetry transformation, can be expressed as a commutator with the charge associated with the transformation as

$$\delta\sigma = -i\epsilon\,[Q,\sigma] = \epsilon\rho\,,$$

$$\delta\rho = -i\epsilon\,[Q,\rho] = -\epsilon\sigma\,. \tag{11.99}$$

Therefore, since with our choice in Eq. (11.93)

$$\langle 0|\sigma(x)|0\rangle = \sigma_{\mathrm{c}} = \frac{2m}{\sqrt{\lambda}}\,, \tag{11.100}$$

we conclude using Eq. (11.98) that

$$\langle 0|\delta\rho|0\rangle = -\epsilon\langle 0|\sigma|0\rangle = -\epsilon\,\frac{2m}{\sqrt{\lambda}}\,,$$

$$\text{or,}\quad -i\epsilon\langle 0|[Q,\rho]|0\rangle = -\epsilon\,\frac{2m}{\sqrt{\lambda}}\,. \tag{11.101}$$

It is clear, therefore, that in the present case we must have

$$Q|0\rangle \neq 0\,, \tag{11.102}$$

in order that the relation in Eq. (11.101) is consistent. In such a case, we say that the symmetry of the Hamiltonian (or the theory) is spontaneously broken. (As we mentioned earlier, the choice of a minimum and, therefore, a vacuum state breaks the symmetry of the theory.)

Since Q does not annihilate the vacuum of the present theory, let us assume that the charge takes the vacuum to a state

$$Q|0\rangle = |\chi\rangle\,. \tag{11.103}$$

We know from Eq. (11.27) that the symmetry of the Hamiltonian implies that

$$[Q, H] = 0. \tag{11.104}$$

Assuming that the vacuum state has zero energy (i.e. $H|0\rangle = 0$), we then obtain using Eq. (11.103)

$$[Q, H]|0\rangle = 0,$$

or, $(QH - HQ)|0\rangle = 0,$

or, $HQ|0\rangle = 0,$

or, $H|\chi\rangle = 0. \tag{11.105}$

In other words, the state $|\chi\rangle$ defined in Eq. (11.103) would appear to be degenerate with the vacuum in energy (even when the vacuum state does not have zero energy). We may, therefore, conclude that this is another vacuum state in the Hilbert space (namely, it would suggest that the vacuum state of the theory is not unique). The problem with this interpretation is that this state is not normalizable. This can be easily seen from Eq. (11.103) and (11.24) as follows. (Q is seen from Eq. (11.62) to be Hermitian and we set $\hbar = 1$ for simplicity.)

$$\langle \chi | \chi \rangle = \langle 0|QQ|0\rangle$$

$$= \left\langle 0 \left| \int d^3x \, J^0(\mathbf{x}, t) Q \right| 0 \right\rangle$$

$$= \int d^3x \, \langle 0|e^{iP \cdot x} J^0(0) e^{-iP \cdot x} Q|0\rangle. \tag{11.106}$$

We have already seen that the Hamiltonian commutes with Q expressing the fact that the charge is independent of time. Since Q does not depend on spatial coordinates either, it follows that the momentum operator also commutes with Q. In fact, in general, we can write

$$[P_\mu, Q] = 0. \tag{11.107}$$

There are many ways of obtaining this result besides the argument given above. The most intuitive way is to note that P_μ generates space-time translations whereas Q generates a phase transformation in the internal Hilbert space. Both these transformations are independent of each other and, therefore, their order should not matter which is equivalent to saying that the generators must commute. A consequence of their commutativity is that we have

$$e^{-iP\cdot x}Q = Qe^{-iP\cdot x}. \tag{11.108}$$

Using this in Eq. (11.106), then, we obtain

$$\langle\chi|\chi\rangle = \int d^3x \, \langle 0|e^{iP\cdot x}J^0(0)Qe^{-iP\cdot x}|0\rangle$$

$$= \int d^3x \, \langle 0|J^0(0)Q|0\rangle$$

$$= \langle 0|J^0(0)Q|0\rangle \int d^3x \to \infty, \tag{11.109}$$

where we have used the property of the ground state, namely,

$$P_\mu|0\rangle = 0. \tag{11.110}$$

In other words, the state $|\chi\rangle$ is not normalizable and hence cannot be thought of as another vacuum in the Hilbert space. This analysis also shows that the finite transformation operator

$$U(\alpha) = e^{-i\alpha Q}, \tag{11.111}$$

does not act unitarily on the Hilbert space. (It takes states out of the Hilbert space.) In fact, it is straightforward to show that the charge Q may not exist when there is spontaneous breakdown of symmetry. Let us note, however, that even though Q may not exist, commutators such as

$$[Q, \phi(x)], \tag{11.112}$$

are well defined in such a theory and as a result finite field transformations such as

$$U(\alpha)\phi(x)U^{-1}(\alpha) = e^{-i\alpha Q}\phi(x)e^{i\alpha Q}, \tag{11.113}$$

are also well defined. Another way of saying this is to note that while the operator $U(\alpha)$ defines unitary transformations for the field variables (operators), it does not act unitarily on the Hilbert space. This is another manifestation of spontaneous symmetry breaking.

To analyze further the consequences of spontaneous symmetry breaking, let us note that even classically, if the potential has a nontrivial minimum, then a stable perturbation would require us to expand the theory around the stable minimum. Correspondingly, let us define

$$\sigma \to \langle \sigma \rangle + \sigma = \frac{2m}{\sqrt{\lambda}} + \sigma,$$

$$\rho \to \langle \rho \rangle + \rho = \rho. \tag{11.114}$$

In this case, the Lagrangian density of the theory in Eq. (11.82) becomes

$$
\begin{aligned}
\mathcal{L} &= \partial_\mu \phi^* \partial^\mu \phi + m^2 \phi^* \phi - \frac{\lambda}{4}(\phi^* \phi)^2 \\
&= \frac{1}{2}\partial_\mu \sigma \partial^\mu \sigma + \frac{1}{2}\partial_\mu \rho \partial^\mu \rho + \frac{m^2}{2}(\sigma^2 + \rho^2) - \frac{\lambda}{16}(\sigma^2 + \rho^2)^2 \\
&\to \frac{1}{2}\partial_\mu \left(\sigma + \frac{2m}{\sqrt{\lambda}}\right)\partial^\mu \left(\sigma + \frac{2m}{\sqrt{\lambda}}\right) + \frac{1}{2}\partial_\mu \rho \partial^\mu \rho \\
&\quad + \frac{m^2}{2}\left(\left(\sigma + \frac{2m}{\sqrt{\lambda}}\right)^2 + \rho^2\right) - \frac{\lambda}{16}\left(\left(\sigma + \frac{2m}{\sqrt{\lambda}}\right)^2 + \rho^2\right)^2 \\
&= \frac{1}{2}\partial_\mu \sigma \partial^\mu \sigma + \frac{1}{2}\partial_\mu \rho \partial^\mu \rho + \frac{m^2}{2}\left(\sigma^2 + \rho^2 + \frac{4m}{\sqrt{\lambda}}\sigma + \frac{4m^2}{\lambda}\right) \\
&\quad - \frac{\lambda}{16}\left(\sigma^2 + \rho^2 + \frac{4m}{\sqrt{\lambda}}\sigma + \frac{4m^2}{\lambda}\right)^2
\end{aligned}
$$

$$= \frac{1}{2}\partial_\mu\sigma\partial^\mu\sigma + \frac{1}{2}\partial_\mu\rho\partial^\mu\rho + \sigma^2\left(\frac{m^2}{2} - m^2 - \frac{m^2}{2}\right)$$

$$+ \rho^2\left(\frac{m^2}{2} - \frac{m^2}{2}\right) + \sigma\left(\frac{2m^3}{\sqrt{\lambda}} - \frac{2m^3}{\sqrt{\lambda}}\right)$$

$$+ \left(\frac{2m^4}{\lambda} - \frac{m^4}{\lambda}\right) - \frac{m\sqrt{\lambda}}{2}\sigma(\sigma^2 + \rho^2) - \frac{\lambda}{16}(\sigma^2 + \rho^2)^2$$

$$= \frac{1}{2}\partial_\mu\sigma\partial^\mu\sigma + \frac{1}{2}\partial_\mu\rho\partial^\mu\rho - m^2\sigma^2 + \frac{m^4}{\lambda}$$

$$- \frac{m\sqrt{\lambda}}{2}\sigma(\sigma^2 + \rho^2) - \frac{\lambda}{16}(\sigma^2 + \rho^2)^2. \tag{11.115}$$

We now see the interesting fact that while the field σ is massive with the right sign for the mass term, the field ρ has become massless. This is a general feature of spontaneous symmetry breaking, namely, whenever a continuous symmetry is spontaneously broken in a manifestly Lorentz invariant theory with a positive metric, there necessarily arise massless fields (particles). These are known as the Goldstone fields (Nambu-Goldstone fields) or Goldstone modes (particles). In the present case, we note that ρ corresponds to the Goldstone field and let us recall our earlier result, namely,

$$\langle 0|\delta\rho|0\rangle = -\epsilon\frac{2m}{\sqrt{\lambda}}. \tag{11.116}$$

This is also a general feature of theories with spontaneously broken symmetries. Namely, in such theories, the change in the Goldstone field under the symmetry transformation acquires a nonzero vacuum expectation value. In terms of the plot of the potential in Fig. 11.1, it is easier to understand the Goldstone mode intuitively. The minimum of the potential occurs along a valley and the Goldstone mode simply reflects the motion along the valley of the potential.

In particle physics, one does not know of elementary spin zero particles which are massless. The closest that comes to being massless is the pi-meson ($\approx$ 140MeV). The Goldstone particles were, therefore, not well received by the particle physics community in the beginning. However, in the presence of gauge fields like the photon field, the Goldstone modes get absorbed into the longitudinal components of the gauge bosons effectively making them (gauge bosons) massive. This is known as the Higgs mechanism and is widely used in the physical models of fundamental interactions.

A massless field or a massless particle, of course, has associated with it an infinite characteristic length (Compton length). The most familiar massless field is the photon field and we know that as a consequence of the photon being massless, the Coulomb force has an infinite range. In fact, we recognize that the two point function in such a theory has an infinite correlation signifying that two charged particles feel the presence of each other even at infinite separation. Similarly, we conclude that when Goldstone modes are present, certain correlation lengths will become infinite.

11.6 Goldstone theorem

In a manifestly Lorentz invariant quantum field theory with a positive metric for the Hilbert space, the Goldstone theorem states that if there is spontaneous breakdown of a continuous symmetry, then there must exist massless particles (Nambu-Goldstone particles) in the theory.

To see a general proof of this theorem, let us assume that we have a theory of n-scalar fields described by the Lagrangian density

$$\mathcal{L} = \mathcal{L}(\phi_i, \partial_\mu \phi_i), \qquad i = 1, 2, \cdots, n. \qquad (11.117)$$

Furthermore, let us assume that the global transformations

$$\delta_\epsilon \phi_i = T_{ij}(\epsilon)\phi_j, \qquad (11.118)$$

where we assume summation over repeated indices and where the global parameter of transformation, ϵ, may itself have an index, define a symmetry of the theory described by the Lagrangian density in Eq. (11.117). In this case, we can define the generating functional with appropriate sources as ($\hbar = 1$)

$$Z[J_i] = e^{iW[J_i]} = N \int \mathcal{D}\phi_i \, e^{iS[\phi_i, J_i]}. \qquad (11.119)$$

Furthermore, the classical fields are defined to be

$$\phi_{ic}(x) = \frac{\delta W}{\delta J_i(x)} = \langle 0|\phi_i(x)|0\rangle^{J_k} \,. \tag{11.120}$$

The case of spontaneous symmetry breaking, of course, corresponds to having nontrivial ϕ_{ic}'s when the sources are turned off. Namely, even if for one of the values of i,

$$\phi_{ic} = \phi_{ic}(x)|_{J_k=0} = \left.\frac{\delta W}{\delta J_i(x)}\right|_{J_k=0} \neq 0 \,, \tag{11.121}$$

then, we will have spontaneous breakdown of the symmetry.

Let us note from Eqs. (11.76) and (11.118) that the classical fields would transform under the symmetry transformations as

$$\delta_\epsilon \phi_{ic}(x) = T_{ij}(\epsilon)\phi_{jc}(x) \,. \tag{11.122}$$

We also know from Eq. (10.29) that the 1PI vertex functional satisfies the defining relation

$$\frac{\delta \Gamma[\phi_{ic}]}{\delta \phi_{ic}(x)} = -J_i(x) \,. \tag{11.123}$$

When the sources are turned off, this defines the extremum equation whose solutions, ϕ_{ic}, will have at least one nonzero value if the symmetry is spontaneously broken. Given the above relation, we can also obtain

$$\begin{aligned}
-\delta_\epsilon J_i(x) &= \int \mathrm{d}^4 y \, \frac{\delta^2 \Gamma[\phi_{ic}]}{\delta \phi_{ic}(x)\delta \phi_{jc}(y)} \, \delta_\epsilon \phi_{jc}(y) \\
&= \int \mathrm{d}^4 y \, \frac{\delta^2 \Gamma[\phi_{ic}]}{\delta \phi_{ic}(x)\delta \phi_{jc}(y)} \, T_{jk}(\epsilon)\phi_{kc}(y) \,.
\end{aligned} \tag{11.124}$$

When we switch off the sources, consistency of Eq. (11.124) will lead to (in this case $\phi_{ic}(x) = \phi_{ic} = $ constant)

$$\int d^4y \left. \frac{\delta^2 \Gamma}{\delta\phi_{ic}(x)\delta\phi_{jc}(y)} \right|_{\phi_{ic}} T_{jk}(\epsilon)\phi_{kc} = 0,$$

or, $\int d^4y \left(G^{-1}(x-y)\right)_{ij} T_{jk}(\epsilon)\phi_{kc} = 0,$

or, $\left(G^{-1}(p_\mu = 0)\right)_{ij} T_{jk}(\epsilon)\phi_{kc} = 0.$ $\qquad$ (11.125)

We note that this system of equations will have a nontrivial solution (namely, at least one ϕ_{ic} would be nontrivial, leading to spontaneous breaking of the symmetry) only if

$$\det \left(G_F^{-1}(p_\mu = 0)\right)_{ij} = 0.$$ $\qquad$ (11.126)

We recognize that $(G^{-1}(p_\mu = 0))_{ij}$ defines the complete (to all orders) mass matrix of the theory. Therefore, there must exist at least one massless particle in the theory for Eq. (11.125) to have a nontrivial solution. This proves the Goldstone theorem.

11.7 References

A. Das, *Lectures on Quantum Field Theory*, World Scientific (2008).

J. Goldstone, Nuovo Cimento, **19**, 154 (1961).

G. S. Guralnik *et al.*, in *Advances in Particle Physics*, Ed. R. L. Cool and R. E. Marshak, John-Wiley Publishing.

E. L. Hill, Rev. Mod. Phys., **23**, 253 (1957).

C. Itzykson and J.-B. Zuber, *Quantum Field Theory*, McGraw-Hill Publishing.

Y. Nambu and G. Jona-Lasinio, Phys. Rev., **122**, 345 (1961).

CHAPTER 12

Gauge theories

Gauge theories are very fundamental in our present understanding of physical forces and, in this chapter, we will study how such theories are described in the path integral formalism. Gauge theories are defined to be theories with a local symmetry and the symmetry is based on some relevant symmetry group. As we have seen in the last chapter, in general, symmetries have important consequences and in the case of a local symmetry, the consequences are even more powerful. However, along with beautiful structures, local invariances also bring in difficulties. For theories with simple local symmetries such as Maxwell's theory, these difficulties can be handled with ease, but problems become quite severe when the relevant local symmetries are more complex such as a non-Abelian symmetry. In this case, the path integral description needs to be analyzed carefully. (Similar difficulties also manifest in the canonical quantization of such theories.) In this chapter, we will start with Maxwell's theory where the difficulties can be simply handled and then we will go into a detailed discussion of non-Abelian gauge theories within the context of the path integral formalism.

12.1 Maxwell theory

Let us recall that the Lagrangian density for Maxwell's theory is given by

$$\mathcal{L} = -\frac{1}{4}F_{\mu\nu}F^{\mu\nu}\,, \tag{12.1}$$

where the field strength tensor, $F_{\mu\nu}$, describing the electric and the magnetic fields is defined in terms of the vector potential A_μ as

$$F_{\mu\nu} = \partial_\mu A_\nu - \partial_\nu A_\mu = -F_{\nu\mu} \,. \tag{12.2}$$

As we know, Maxwell's theory is invariant under gauge transformations which are local transformations. Explicitly, under a local change of the vector potential by a gradient,

$$A_\mu(x) \rightarrow A_\mu(x) + \partial_\mu\alpha(x) \,, \tag{12.3}$$

the field strength tensor remains invariant

$$
\begin{aligned}
F_{\mu\nu} &= \partial_\mu A_\nu - \partial_\nu A_\mu \\
&\rightarrow \partial_\mu \left(A_\nu + \partial_\nu\alpha(x) \right) - \partial_\nu \left(A_\mu + \partial_\mu\alpha(x) \right) \\
&= \partial_\mu A_\nu - \partial_\nu A_\mu = F_{\mu\nu} \,.
\end{aligned}
\tag{12.4}
$$

Namely, the physically observable electric and magnetic fields are not sensitive to a redefinition of the vector potential by a gauge transformation. Since Maxwell's equations can be described in terms of the field strength tensors, the gauge transformation of Eq. (12.3) defines a symmetry (invariance) of the Maxwell theory. These are simple local gauge transformations belonging to the Abelian $U(1)$ group and correspondingly, Maxwell's theory is known as an Abelian gauge theory. For completeness, we note here that the gauge transformation of Eq. (12.3) can be written in terms of a local $U(1)$ phase transformation as

$$
\begin{aligned}
A_\mu(x) &\rightarrow A_\mu(x) + i(\partial_\mu e^{-i\alpha(x)})e^{i\alpha(x)} \\
&= U(x)A_\mu(x)U^{-1}(x) + i(\partial_\mu U(x))U^{-1}(x) \,,
\end{aligned}
\tag{12.5}
$$

where $U(x) = e^{-i\alpha(x)} \in U(1)$. (Here we have set the coupling constant to unity.)

The gauge invariance, of course, restricts the structure of the theory (Lagrangian density) and leads to difficulties that can be seen as follows. Let us note that the canonical momenta conjugate to the field variables A_μ can be calculated from the Lagrangian density Eq. (12.1) and take the forms

$$\Pi^\mu = \frac{\partial \mathcal{L}}{\partial \dot{A}_\mu} = -F^{0\mu} , \tag{12.6}$$

which implies the constraint

$$\Pi^0 = -F^{00} = 0 . \tag{12.7}$$

This is, in fact, a general feature of gauge theories, namely, the local invariance leads to constraints. As a result, the canonical quantization of such theories is nontrivial. However, we note that since Maxwell's theory as well as physical quantities such as the field strength tensor are invariant under the gauge transformation in Eq. (12.3), we can choose a gauge (a particular form of the vector potential defined by the gauge choice) to work with. If we choose the gauge $\nabla \cdot \mathbf{A} = 0$ (Coulomb gauge), then the equations of motion

$$\partial_\mu F^{\mu\nu} = 0 , \tag{12.8}$$

lead to (for $\nu = 0$)

$$\partial_\mu F^{\mu 0} = \partial_i F^{i0} = 0,$$

$$\text{or,} \quad \partial_i \left(\partial^i A^0 - \partial^0 A^i \right) = -\nabla^2 A^0 = 0,$$

$$\text{or,} \quad A^0 = A_0 = 0 . \tag{12.9}$$

On the other hand, if sources (charges and currents) are present, the gauge

$$\nabla \cdot \mathbf{A} = 0 , \tag{12.10}$$

would lead to (we assume $\mathcal{L}_I = J^\mu A_\mu$)

$$\nabla^2 A^0 = J^0,$$

$$\text{or,} \quad A^0 = \frac{1}{\nabla^2} J^0 . \tag{12.11}$$

In either case, the true dynamics of the theory is, therefore, contained in the transverse physical degrees of freedom and the longitudinal (as well as the temporal) degrees of freedom can be expressed in terms of these. The canonical quantization can now be carried out, but we lose manifest Lorentz invariance in the process. Let us emphasize here that the final result for the calculation of any amplitude in the canonical formalism remains manifestly Lorentz invariant. However, there is no manifest Lorentz invariance in the intermediate steps.

We can ask what would happen if we were to treat the theory in the path integral formalism as opposed to the canonical formalism. Here we note that we can write the Lagrangian density for the Maxwell theory also as

$$\mathcal{L} = -\frac{1}{4}F_{\mu\nu}F^{\mu\nu} = \frac{1}{2}A_\mu P^{\mu\nu}A_\nu + \text{ total derivatives}, \qquad (12.12)$$

where

$$P^{\mu\nu} = \eta^{\mu\nu}\Box - \partial^\mu\partial^\nu,$$

$$P^{\mu\nu}P_\nu^{\ \lambda} = (\eta^{\mu\nu}\Box - \partial^\mu\partial^\nu)\left(\delta_\nu^\lambda\Box - \partial_\nu\partial^\lambda\right)$$

$$= \eta^{\mu\lambda}\Box^2 - \partial^\mu\partial^\lambda\Box - \partial^\mu\partial^\lambda\Box + \partial^\mu\partial^\lambda\Box$$

$$= \Box\left(\eta^{\mu\lambda}\Box - \partial^\mu\partial^\lambda\right)$$

$$= \Box P^{\mu\lambda}. \qquad (12.13)$$

With a suitable normalization $P^{\mu\nu}$ can be thought of as a projection operator. ($\overline{P}^{\mu\nu} = \frac{1}{\Box}P^{\mu\nu}$ is the normalized projection operator.) In fact, we note that

$$\partial_\mu P^{\mu\nu} = \partial_\mu\left(\eta^{\mu\nu}\Box - \partial^\mu\partial^\nu\right)$$

$$= (\partial^\nu\Box - \Box\partial^\nu) = 0 = \partial_\nu P^{\mu\nu}. \qquad (12.14)$$

Therefore, this is the transverse projection operator which projects out the components of any vector transverse (perpendicular) to the gradient operator ∂^μ. As a result, the inverse of $P^{\mu\nu}$ does not exist and the Green's function and, therefore, the Feynman propagator of the theory cannot be defined. This implies that if we were to apply the path integral formalism naively, neither the generating functional will exist nor can we carry out perturbation theory.

We note here that whenever the determinant of the matrix of highest derivatives in the Lagrangian density vanishes, the system is singular and contains constraints among the field variables. In such a case, without any further input, the Cauchy initial value problem cannot be uniquely solved, simply because the Green's function does not exist. As a consequence, we see that the naive canonical quantization has unpleasant features in the case of Maxwell's theory since the fields are constrained and the momentum corresponding to A_0 vanishes. In a physical gauge such as the Coulomb gauge, we can solve for the constraints and quantize only the true dynamical degrees of freedom. However, in this process, we give up manifest Lorentz invariance since we single out the transverse degrees of freedom.

We can, of course, take an alternative approach. Namely, since we realize that the difficulties in quantization arise because of the singular nature of the Lagrangian density, we could modify the theory to make it nonsingular. Let us consider, for example, the Lagrangian density (this formulation of the theory is due to Fermi and this gauge is known as the Feynman-Fermi gauge)

$$\mathcal{L} = -\frac{1}{4} F_{\mu\nu} F^{\mu\nu} - \frac{1}{2} (\partial_\mu A^\mu)^2 + J^\mu A_\mu \,, \tag{12.15}$$

where J^μ represents a conserved current (the sources in Maxwell's theory)

$$\partial_\mu J^\mu = 0 \,. \tag{12.16}$$

Here we have generalized Maxwell's theory to include a conserved current. But more than that we have also added a term $-\frac{1}{2} (\partial_\mu A^\mu)^2$ to Maxwell's Lagrangian density. This term breaks gauge invariance and, consequently, leads to a nonsingular theory. But clearly this would appear to be different from Maxwell's theory. Therefore, at this point there is no justification for adding this new term to the Lagrangian density. But to understand the issue better, let us look at the equation of motion following from the action in the present case

$$\partial_\mu \frac{\partial \mathcal{L}}{\partial \partial_\mu A_\nu} - \frac{\partial \mathcal{L}}{\partial A_\nu} = 0,$$

or, $-\partial_\mu F^{\mu\nu} - \partial^\nu (\partial \cdot A) - J^\nu = 0,$

or, $\partial_\mu F^{\mu\nu} + \partial^\nu (\partial \cdot A) = -J^\nu .$ (12.17)

Without the second term on the left hand side in Eq. (12.17), this is just Maxwell's equations in the presence of conserved sources. If we now write out the left hand side, Eq. (12.17) takes the form

$$\partial_\mu \left(\partial^\mu A^\nu - \partial^\nu A^\mu \right) + \partial^\nu (\partial \cdot A) = -J^\nu ,$$

or, $\Box A^\nu = -J^\nu ,$

or, $\Box \partial \cdot A = -\partial \cdot J = 0 .$ (12.18)

An alternative way to see this is to note that if we take the divergence of the equations of motion in (12.17), we have

$$\partial_\nu \partial_\mu F^{\mu\nu} + \Box (\partial \cdot A) = -\partial_\nu J^\nu ,$$

or, $\Box (\partial \cdot A) = 0 ,$ (12.19)

where we have used the antisymmetry of the field strength tensor as well as the conservation of J^μ.

Thus we see that although the presence of the term $-\frac{1}{2}\chi^2$ in the Lagrangian density where

$$\chi = \partial \cdot A ,$$ (12.20)

seems to modify the theory, χ is in reality a free field and, therefore, the presence of this additional term in the Lagrangian density would not change the physics of Maxwell's theory. Furthermore, we realize that if we restrict classically

$$\chi = 0, \quad \text{at} \quad t = 0$$

and

$$\frac{\partial \chi}{\partial t} = 0, \quad \text{at} \quad t = 0,$$ (12.21)

then, Eq. (12.19) would determine that $\chi = 0$ at all times and we get back our familiar Maxwell's theory. Thus, classically we can think of Maxwell's theory as described by the modified Lagrangian density of Eq. (12.15) with the supplementary condition

$$\partial \cdot A = 0.$$ (12.22)

In the quantum theory, however, the field variables, A_μ, are operators and Eq. (12.22) is hard to impose on the theory as an operator relation. One can think of imposing such a condition on the vector space of states to select out the physical Hilbert space, namely,

$$\partial_\mu A^\mu |\text{phys}\rangle = 0.$$ (12.23)

However, this, too, turns out to be too stringent a condition. This not only demands that certain kinds of photons are not present in the physical state, but it also requires that those photons cannot be emitted (created) either. Gupta and Bleuler weakened the supplementary condition on the physical states to have the form

$$\partial_\mu A^{\mu (+)}(x)|\text{phys}\rangle = 0,$$ (12.24)

where $A^{\mu (+)}(x)$ is the positive frequency part of the Maxwell's field and contains only the destruction operator. (This is commonly known as the Gupta-Bleuler quantization (condition).) We remark here that since (see Eq. (12.19))

$$\Box \partial_\mu A^\mu(x) = 0,$$ (12.25)

$\partial_\mu A^\mu(x)$ is like a free scalar field. Therefore, it can be decomposed into positive and negative frequency parts uniquely in a relativistically invariant manner and this decomposition is preserved under time evolution. Furthermore, we note that the theory has four degrees of freedom resulting from the four components of A_μ and since the components can be time-like or space-like, the resulting vector space of states has the problem of a negative metric. On the other hand, the Gupta-Bleuler condition (12.24) selects out three kinds of photon states as being physical, two transverse photon states with a positive norm and a linear combination of the time-like and the longitudinal photon states with zero norm. The state with zero norm, however, is orthogonal to every other state (including itself) and, therefore, decouples from the theory and the physical Hilbert space effectively consists of the physical transverse photon states. We can think of the Gupta-Bleuler supplementary condition as imposing the Lorenz condition

$$\partial_\mu A^\mu(x) = 0$$

on the physical quantum states since it implies

$$\langle \psi | \partial_\mu A^\mu | \psi \rangle = 0 \,, \tag{12.26}$$

where $|\psi\rangle$ represents a physical state.

Thus, we see that the physical subspace of the theory selected by the supplementary condition contains states with positive semi-definite norm (negative norm states are eliminated by the supplementary condition or the physical state condition and, consequently, there is no problem with a probabilistic interpretation). Since the zero norm states are orthogonal to all the states including themselves, if we further mod out the states by the zero norm states, we have the true physical subspace of the theory where the norm of states is positive definite, namely,

$$\overline{V}_{\text{phys}} = \frac{V_{\text{phys}}}{V_0} \,, \tag{12.27}$$

where V_0 represents the set of states with zero norm.

Since the modified theory in Eq. (12.15) (which is equivalent to Maxwell's theory) is nonsingular, it can be described in the path integral formalism in the standard manner. In this simple gauge theory, therefore, there is an easy solution to the problem associated with the local invariance of the theory. As we will see in the next section, the difficulties become more severe in a non-Abelian gauge theory.

12.2 Non-Abelian gauge theory

As we have mentioned earlier, non-Abelian gauge theories are based on nontrivial symmetry groups where the generators of the group satisfy a non-commutative Lie algebra. Let us take a brief digression into the structure of $SU(n)$. If we assume that $T^a, a = 1, 2, \cdots, n^2 - 1$ (dim $SU(n)$) denote the generators of the group, then they satisfy the Lie algebra of the form (we take the generators to be Hermitian and repeated indices are summed)

$$\left[T^a, T^b\right] = if^{abc}T^c,\tag{12.28}$$

where f^{abc} denote the completely antisymmetric, real structure constants of the algebra. The Jacobi identity for the algebra is given by

$$\left[\left[T^a, T^b\right], T^c\right] + \left[\left[T^c, T^a\right], T^b\right] + \left[\left[T^b, T^c\right], T^a\right] = 0,\tag{12.29}$$

and imposes a restriction on the structure constants of the form

$$if^{abp}\left[T^p, T^c\right] + if^{cap}\left[T^p, T^b\right] + if^{bcp}\left[T^p, T^a\right] = 0,$$

$$\text{or,}\quad f^{abp}f^{pcq}T^q + f^{cap}f^{pbq}T^q + f^{bcp}f^{paq}T^q = 0,$$

$$\text{or,}\quad f^{abp}f^{pcq} + f^{cap}f^{pbq} + f^{bcp}f^{paq} = 0.\tag{12.30}$$

We can find the forms of the generators in various representations of the group much like in the case of angular momentum. However, a particular representation that is very important as well as useful in the study of gauge theories is given by

$$(T^a)_{bc} = -if^{abc}.\tag{12.31}$$

This is consistent with the hermiticity requirement for the generators

$$\left(T^{a\dagger}\right)_{bc} = ((T^a)_{cb})^* = \left(-if^{acb}\right)^* = if^{acb} = -if^{abc}$$

$$= (T^a)_{bc}.\tag{12.32}$$

Furthermore, we can easily check that this representation naturally satisfies the Lie algebra Eq. (12.28),

$$
\begin{aligned}
\left[T^a, T^b\right]_{cq} &= \left(T^a T^b - T^b T^a\right)_{cq} \\
&= (T^a)_{cp}\left(T^b\right)_{pq} - \left(T^b\right)_{cp}(T^a)_{pq} \\
&= \left(-if^{acp}\right)\left(-if^{bpq}\right) - \left(-if^{bcp}\right)\left(-if^{apq}\right) \\
&= -f^{acp} f^{bpq} + f^{bcp} f^{apq} \\
&= -f^{cap} f^{pbq} - f^{bcp} f^{paq} \\
&= f^{abp} f^{pcq} \qquad \text{(by Jacobi identity)} \\
&= if^{abp}\left(-if^{pcq}\right) \\
&= if^{abp}\left(T^p\right)_{cq} ,
\end{aligned}
\tag{12.33}
$$

so that the identification

$$
\left(T^a_{(\text{adj})}\right)_{bc} = -if^{abc},
\tag{12.34}
$$

indeed defines a representation of the Lie algebra known as the adjoint representation.

For any representation of a simple Lie group we can write

$$
\text{Tr}\, T^a T^b = C_2 \delta^{ab},
\tag{12.35}
$$

where C_2 is a normalization constant which determines the values of the structure constants. It depends on the representation but not on the indices a and b. To prove this let us note that we can always diagonalize the tensor $\text{Tr}(T^a T^b)$ such that (this is a symmetric real matrix)

$$
\text{Tr}\left(T^a T^b\right) = \begin{cases} 0 & \text{if} \quad a \neq b, \\ K_a & \text{if} \quad a = b. \end{cases}
\tag{12.36}
$$

Let us next note that the quantity

$$h^{abc} = \text{Tr}\left(\left[T^a, T^b\right]T^c\right)$$
$$= \text{Tr}\left(T^a T^b T^c\right) - \text{Tr}\left(T^b T^a T^c\right),   \tag{12.37}$$

is completely antisymmetric in all its indices. Furthermore, using the commutation relations in the definition in Eq. (12.37), we have

$$h^{abc} = \text{Tr}\left([T^a, T^b]T^c\right) = \text{Tr}\left(if^{abp}T^p T^c\right)$$
$$= if^{abp}\,\text{Tr}\,(T^p T^c)$$
$$= if^{abp}K_p\delta^{pc}$$
$$= iK_c f^{abc},   \quad \text{(no sum on } c\text{)}.   \tag{12.38}$$

On the other hand, we note that

$$h^{acb} = \text{Tr}\left(\left[T^a, T^c\right]T^b\right)$$
$$= if^{acp}\,\text{Tr}\left(T^p T^b\right)$$
$$= if^{acp}K_p\delta^{pb}$$
$$= iK_b f^{acb} = -iK_b f^{abc},   \quad \text{(no sum on } b\text{)}.   \tag{12.39}$$

However, since h^{abc} is completely antisymmetric

$$h^{acb} = -h^{abc}.   \tag{12.40}$$

Comparing the two results in Eqs. (12.38) and (12.39) we see that

$$K_b = K_c = K.   \tag{12.41}$$

(Alternatively, we note that

$$\text{Tr } T^a T^b = K_a \delta^{ab},$$

$$\text{Tr } T^b T^a = K_b \delta^{ab}.$$

$$(12.42)$$

By cyclicity of the trace, the two must equal and hence we have

$$K_a = K_b,$$

$$(12.43)$$

which proves Eq. (12.41).) In other words, we see that we can write

$$\text{Tr } \left(T^a T^b \right) = C_2 \delta^{ab},$$

$$(12.44)$$

where the constant C_2 depends only on the representation. (It is chosen to be $\frac{1}{2}$ for the fundamental representation to which the fermions belong in $SU(n)$.) We note that if we write

$$\text{Tr } T^a T^b = T(R) \, \delta^{ab},$$

$$(12.45)$$

then $T(R)$ is known as the index of the representation R. Similarly, we have

$$(T^a T^a)_{mn} = C(R) \, \delta_{mn},$$

$$(12.46)$$

where $C(R)$ is known as the Casimir of the representation R. The two are clearly related as

$$T(R) \dim G = C(R) \dim R,$$

$$(12.47)$$

where $\dim G$, $\dim R$ denote respectively the dimensionalities of the group G and the representation R.

To construct a gauge theory with $SU(n)$ symmetry, we note that the gauge potentials must belong to some representation of the symmetry group and, therefore, can be written as matrices. In fact, the gauge potentials can be expanded in terms of the generators of the Lie algebra (in some representation) as

$$A_\mu(x) = T^a A_\mu^a(x) \,, \tag{12.48}$$

where summation over repeated indices is understood. The gauge transformation in Eq. (12.5) can now be generalized to (we are setting the coupling constant to unity)

$$A_\mu(x) \to U(x)A_\mu(x)U^{-1}(x) + i(\partial_\mu U(x))U^{-1}(x) \,, \tag{12.49}$$

where, in the present case,

$$U(x) = e^{-i\alpha(x)} = e^{-i\alpha^a(x)T^a} \,, \tag{12.50}$$

belongs to the non-Abelian group $SU(n)$. Furthermore, since U and A_μ are now matrices, they do not commute in general and the actual gauge transformation in Eq. (12.49) is much more complicated than (12.3). In fact, for the simple case of an infinitesimal gauge transformation, Eq. (12.49) takes the form (here the parameter of transformation $\alpha(x) = \epsilon(x)$ is assumed to be infinitesimal)

$$\begin{aligned}
A_\mu(x) &\to A_\mu(x) + i\,[A_\mu(x), \epsilon(x)] + \partial_\mu \epsilon(x) \\
&= A_\mu(x) + D_\mu \epsilon(x) \,,
\end{aligned} \tag{12.51}$$

where we have defined (coupling constant has been set to unity)

$$D_\mu \epsilon(x) = \partial_\mu \epsilon(x) + i\,[A_\mu(x), \epsilon(x)] \,, \tag{12.52}$$

also known as the covariant derivative.

To construct the Lagrangian density for the gauge field in a gauge invariant manner, we note that under a gauge transformation given by Eq. (12.49), the tensor representing the Abelian field strength (see Eq. (12.2)) would transform as

$$f_{\mu\nu} = \partial_\mu A_\nu - \partial_\nu A_\mu$$
$$\rightarrow \partial_\mu \left[U A_\nu U^{-1} + i \left(\partial_\nu U \right) U^{-1} \right]$$
$$- \partial_\nu \left[U A_\mu U^{-1} + i \left(\partial_\mu U \right) U^{-1} \right]$$
$$= i \left[\left(\partial_\nu U \right) \left(\partial_\mu U^{-1} \right) - \left(\partial_\mu U \right) \left(\partial_\nu U^{-1} \right) \right]$$
$$+ \left(\partial_\mu U \right) A_\nu U^{-1} + U A_\nu \left(\partial_\mu U^{-1} \right)$$
$$- \left(\partial_\nu U \right) A_\mu U^{-1} - U A_\mu \left(\partial_\nu U^{-1} \right)$$
$$+ U \left(\partial_\mu A_\nu - \partial_\nu A_\mu \right) U^{-1} . \tag{12.53}$$

Namely, unlike QED, in the present case, $f_{\mu\nu} = \partial_\mu A_\nu - \partial_\nu A_\mu$ is not invariant under a gauge transformation. Let us also note that under a gauge transformation

$$i \left[A_\mu, A_\nu \right] = i \left(A_\mu A_\nu - A_\nu A_\mu \right)$$
$$\rightarrow i \left[\left(U A_\mu U^{-1} + i \left(\partial_\mu U \right) U^{-1} \right), \left(U A_\nu U^{-1} + i (\partial_\nu U) U^{-1} \right) \right]$$
$$= i U \left[A_\mu, A_\nu \right] U^{-1} - i \left(\left(\partial_\nu U \right) \left(\partial_\mu U^{-1} \right) - \left(\partial_\mu U \right) \left(\partial_\nu U^{-1} \right) \right)$$
$$- \left(\partial_\mu U \right) A_\nu U^{-1} - U A_\nu \left(\partial_\mu U^{-1} \right)$$
$$+ \left(\partial_\nu U \right) A_\mu U^{-1} + U A_\mu \left(\partial_\nu U^{-1} \right) . \tag{12.54}$$

Thus, it is clear that, in the present case, if we define the field strength tensor as (we are setting the coupling constant to unity)

$$F_{\mu\nu} = \partial_\mu A_\nu - \partial_\nu A_\mu + i \left[A_\mu, A_\nu \right] , \tag{12.55}$$

then, under a finite gauge transformation, $F_{\mu\nu}$ will transform covariantly, namely,

$$F_{\mu\nu} \rightarrow U \left(\partial_\mu A_\nu - \partial_\nu A_\mu + i \left[A_\mu, A_\nu \right] \right) U^{-1} = U F_{\mu\nu} U^{-1} . \tag{12.56}$$

It is now easy to construct the Lagrangian density for the gauge field as

$$\mathcal{L}_{\text{gauge}} = -\frac{1}{2}\text{Tr}\; F_{\mu\nu}F^{\mu\nu} = -\frac{1}{4}F^a_{\mu\nu}F^{\mu\nu\,a}\,, \tag{12.57}$$

where a particular normalization for the trace is assumed in the last relation (namely, T^a's are in the fundamental representation). This Lagrangian density is easily seen, from the cyclicity of trace, to be invariant under the gauge transformation Eq. (12.49) (see, for example, Eq. (12.56)). In components, the field strength tensor takes the form, (we have set the coupling constant to unity for simplicity)

$$F^a_{\mu\nu} = \partial_\mu A^a_\nu - \partial_\nu A^a_\mu - f^{abc}A^b_\mu A^c_\nu = -F^a_{\nu\mu}\,. \tag{12.58}$$

In components, the infinitesimal form of the transformation for the gauge fields in Eq. (12.51) take the forms

$$\begin{aligned}
\delta A^a_\mu &= A'^a_\mu - A^a_\mu \\
&= \left(\partial_\mu \epsilon^a - f^{abc}A^b_\mu\epsilon^c\right) \\
&= \left(\partial_\mu \epsilon^a + f^{bac}A^b_\mu\epsilon^c\right) \\
&= \left(\partial_\mu \epsilon^a + i\left(-if^{bac}\right)A^b_\mu\epsilon^c\right) \\
&= \left(\partial_\mu \epsilon^a + i\left(T^b_{(\text{adj})}\right)_{ac}A^b_\mu\epsilon^c\right)\,,
\end{aligned} \tag{12.59}$$

so that we can write

$$\delta A^a_\mu = \left(D^{(\text{adj})}_\mu \epsilon\right)^a = \partial_\mu \epsilon^a - f^{abc}A^b_\mu\epsilon^c\,, \tag{12.60}$$

which shows that the gauge field, A_μ, transforms according to the adjoint representation of the group. The infinitesimal transformation of the field strength tensor can be obtained from Eq. (12.56) and leads to

$$\delta F_{\mu\nu}^a = -f^{abc} F_{\mu\nu}^b \epsilon^c$$

$$= -i \left(-i f^{cab}\right) \epsilon^c F_{\mu\nu}^b$$

$$= -i \left(T_{(\text{adj})}^c\right)_{ab} \epsilon^c F_{\mu\nu}^b$$

$$= -i\epsilon^c \left(T_{(\text{adj})}^c\right)_{ab} F_{\mu\nu}^b. \qquad (12.61)$$

Thus, the field strength tensor $F_{\mu\nu}$ as well as the gauge field A_μ transform according to the adjoint representation of the group. (It does not matter what representationt the matter fields belong to, the gauge field must transform in the adjoint representation.) We note here that unlike the photon field, here the gauge field has self interaction. Physically we understand this in the following way. In the present case, the gauge field carries the charge of the non-Abelian symmetry group (they have a nontrivial symmetry index) in contrast to the photon field which is charge neutral. Since gauge fields couple to any particle carrying charge of the symmetry group, in the case of non-Abelian symmetry they must possess self interactions.

Let us next briefly examine the difficulties that arise in trying to canonically quantize the theory along the lines of Maxwell theory. The Euler-Lagrange equations following from Eq. (12.57) take the forms

$$\partial_\nu \frac{\partial \mathcal{L}}{\partial \partial_\nu A_\mu^a} - \frac{\partial \mathcal{L}}{\partial A_\mu^a} = 0$$

or, $\quad (D_\nu F^{\mu\nu})^a = 0, \qquad (12.62)$

where the covariant derivative is defined to be in the adjoint representation of the group and (coupling constant is unity)

$$F_{\mu\nu}^a = \partial_\mu A_\nu^a - \partial_\nu A_\mu^a - f^{abc} A_\mu^b A_\nu^c. \qquad (12.63)$$

For $\mu = 0$, we see that the Euler-Lagrange equations lead to a constraint of the form

$$D_i F^{0i\,a} = 0 \,, \tag{12.64}$$

which is a reflection of the gauge invariance of the theory. In fact, it can be explicitly checked the coefficient matrix of highest derivatives is the transverse projection operator just like in the Maxwell theory.

Let us define the momenta canonically conjugate to the field variables A_μ^a

$$\Pi^{\mu\,a}(x) = \frac{\partial \mathcal{L}}{\partial \dot{A}_\mu^a(x)} = -F^{0\mu\,a}(x) \,. \tag{12.65}$$

Noting that the field strength $F_{\mu\nu}^a$ is antisymmetric in the indices μ, ν, we have

$$\Pi^{0\,a}(x) = -F^{00\,a} = 0 \,, \tag{12.66}$$

and

$$\Pi^{i\,a}(x) = -F^{0i\,a} = E_i^a(x) \,, \tag{12.67}$$

where we can think of $E_i^a(x)$ as the non-Abelian electric field strength.

We note that the momentum conjugate to A_0^a does not exist much like in the Maxwell theory. This implies that A_0^a is like a c-number quantity which commutes with every other operator in the theory. Thus we can choose a physical gauge condition to suitably set it equal to zero, namely,

$$A_0^a(x) = 0 \,. \tag{12.68}$$

The analysis is parallel to what we have discussed in the case of Maxwell theory, although it is much more complicated. Barring technical issues such as the Gribov ambiguity (related to the existence of large gauge transformations), the theory can also be defined in terms of only physical transverse gauge fields in this case. Therefore, canonical quantization leads to a lack of manifest Lorentz invariance.

We can try to quantize a non-Abelian gauge theory covariantly, very much along the lines of the Abelian theory, namely, by modifying the theory. Thus, for example, let us look at the theory

$$\mathcal{L} = -\frac{1}{4}F_{\mu\nu}^a F^{\mu\nu\,a} - \frac{1}{2}(\partial_\mu A^{\mu a})^2 \,. \tag{12.69}$$

The additional term in this theory clearly breaks gauge invariance and, consequently, makes the theory nonsingular, much like in the Maxwell theory. However, in the present case, there are serious differences from Maxwell's theory. For example, let us note that in the Abelian theory,

$$\Box(\partial \cdot A) = 0\,, \tag{12.70}$$

and hence classically we can impose the condition

$$\partial \cdot A = 0\,, \tag{12.71}$$

which then leads to the Gupta-Bleuler condition on the physical states

$$\partial \cdot A^{(+)}|\text{phys}\rangle = 0\,. \tag{12.72}$$

In the case of the non-Abelian theory, however, even classically the equations of motion are given by

$$(D_\mu F^{\mu\nu})^a + \partial^\nu(\partial_\mu A^{\mu a}) = 0,$$
$$\text{or,} \quad \partial_\mu F^{\mu\nu\,a} - f^{abc}A_\mu^b F^{\mu\nu\,c} + \partial^\nu(\partial_\mu A^{\mu a}) = 0\,. \tag{12.73}$$

Contracting with ∂_ν, we obtain

$$\Box(\partial_\mu A^{\mu a}) = f^{abc}\partial_\nu\left(A_\mu^b F^{\mu\nu\,c}\right) \neq 0\,. \tag{12.74}$$

Thus, in contrast to the Abelian theory, we note that $(\partial_\mu A^{\mu a})$ does not behave like a free field and, consequently, the additional term has truly modified the theory. Furthermore, since $(\partial_\mu A^{\mu a})$ is not a free field, it cannot be uniquely decomposed into a positive and a negative frequency part (in a time invariant manner), nor can we think of a supplementary condition such as

$$\partial_\mu A^{\mu a(+)}|\text{phys}\rangle = 0\,, \tag{12.75}$$

in a physically meaningful manner, since it is not invariant under time evolution. (Namely, the physical subspace would keep changing with time which is not desirable.) Correspondingly the analog of the Gupta-Bleuler condition for non-Abelian gauge theories does not appear to exist. Therefore, we need to analyze the question of modifying the theory in a more systematic and detailed manner. We would see next how we can derive intuition on this important question from the path integral quantization.

12.3 Path integral for gauge theories

To understand better how gauge theories can be handled in the path integral formalism, let us go back to the Maxwell theory described by the Lagrangian density

$$\mathcal{L}^{(J)} = -\frac{1}{4}F_{\mu\nu}F^{\mu\nu} + J^\mu A_\mu\,, \tag{12.76}$$

where J^μ represents a conserved current (source). The generating functional in the path integral formalism is given by (in Eqs. (12.12) and (12.13), we have denoted $O^{\mu\nu}$ by $P^{\mu\nu}$)

$$Z[J_\mu] = e^{iW[J_\mu]} = N\int DA_\mu e^{iS^{(J)}[A_\mu]}$$

$$= N\int DA_\mu \, e^{i\left[\frac{1}{2}(A_\mu, O^{\mu\nu}A_\nu)+(J^\mu, A_\mu)\right]}\,,$$

where N is a normalization constant and

$$O^{\mu\nu}(x-y) = \left(\Box\eta^{\mu\nu} - \partial^\mu\partial^\nu\right)\delta^4(x-y)\,,$$

$$(J^\mu, A_\mu) = \int d^4x \, J^\mu(x)A_\mu(x),$$

$$(A_\mu, O^{\mu\nu}A_\nu) = \int d^4x d^4y \, A_\mu(x)O^{\mu\nu}(x-y)A_\nu(y)\,. \tag{12.77}$$

Here we have introduced a compact notation to describe integrations as described above. The functional integral is a Gaussian integral which we have worked out in earlier chapters and leads to

$$Z\left[J_\mu\right] = e^{iW[J_\mu]} = N \left(\det\left(-O^{\mu\nu}\right)\right)^{-\frac{1}{2}} e^{-\frac{i}{2}\left(J^\mu, O_{\mu\nu}^{-1} J^\nu\right)}. \qquad (12.78)$$

However, as we have seen before (see Eq. (12.13)), the operator $O^{\mu\nu}$ is a projection operator for transverse photons. The longitudinal vectors k_μ (or $\partial_\mu F$) are its eigenvectors with zero eigenvalue. Clearly therefore, the determinant of $O^{\mu\nu}$ vanishes. This implies that the generating functional does not exist. (The operator possesses zero modes and, consequently, the inverse of the matrix cannot be defined either.) Going over to the Euclidean space does not help either. In fact, as we can see even in the limit of vanishing sources, the generating functional does not exist.

The source of the difficulty is not hard to see. The Lagrangian density for Maxwell's theory is invariant under the gauge transformation

$$A_\mu \to A_\mu^{(\alpha)} = U A_\mu U^{-1} + i(\partial_\mu U)U^{-1}, \qquad (12.79)$$

where

$$U(\alpha) = e^{-i\alpha(x)}. \qquad (12.80)$$

For a fixed A_μ, all the $A_\mu^{(\alpha)}$'s that are obtained by making gauge transformations with all possible $\alpha(x)$ are said to lie on an "orbit" in the group space. The action S, on the other hand, is constant (invariant) on such orbits. Therefore, the generating functional, even in the absence of any sources, is proportional to the "volume" of the orbits denoted by

$$\int \prod_x d\alpha(x). \qquad (12.81)$$

(In the non-Abelian case, the measure should be replaced by the group invariant Haar measure $\prod_x dU(x)$.) This is an infinite factor (which is the reason why the Gaussian functional integral does not exist) and must be extracted out before doing any calculations. The method for extracting this factor out of the path integral is due to Faddeev and Popov and relies on the method of gauge fixing. We recognize that we should not integrate over all gauge field configurations because they are not really distinct. Rather we should integrate over each orbit only once.

The way this is carried out is by choosing a hypersurface which intersects each orbit only once, i.e., if

$$F\left(A_\mu\right) = 0,\tag{12.82}$$

defines the hypersurface which intersects the orbits once, then even if A_μ does not satisfy the condition, we can find a gauge transformed $A_\mu^{(\alpha)}$ which does, namely,

$$F\left(A_\mu^{(\alpha)}\right) = 0,\tag{12.83}$$

has a unique solution for $\alpha(x)$. This procedure is known as gauge fixing and the condition

$$F\left(A_\mu\right) = 0,\tag{12.84}$$

is known as the gauge condition. Physical quantities are, of course, gauge independent and do not depend on the choice of the hypersurface (gauge). Thus, for example,

$$
\begin{aligned}
F\left(A_\mu\right) = \partial_\mu A^\mu &= 0 \quad \text{is the Lorenz/Landau gauge,} \\
\boldsymbol{\nabla} \cdot \mathbf{A} &= 0 \quad \text{is the Coulomb gauge,} \\
A_0 &= 0 \quad \text{is the temporal gauge,} \\
A_3 &= 0 \quad \text{is the axial gauge,}
\end{aligned}\tag{12.85}
$$

and so on. We can already see the need for gauge fixing from the fact that because the action is gauge invariant so is the generating functional (if sources are transformed appropriately). Therefore, it would lead only to gauge invariant Green's functions. On the other hand, we know from ordinary perturbation theory that the Green's functions are, in general, gauge dependent although the S-matrix (the scattering matrix) elements are gauge independent. Thus one has to fix a gauge without which even the Cauchy initial value problem cannot be solved. (Only physical quantities need to be gauge independent.)

To extract out the infinite gauge volume factor, let us do the following trick due to Faddeev and Popov. Let us define

$$\Delta_{\text{FP}}\left[A_\mu\right] \int \prod_x \mathrm{d}\alpha(x)\, \delta\left(F\left(A_\mu^{(\alpha)}(x)\right)\right) = 1\,. \tag{12.86}$$

(The integration measure should be $\mathrm{d}U(x)$, the Haar measure, which is essential in the case of non-Abelian theories.) Note that the quantity $\Delta_{\text{FP}}\left[A_\mu\right]$ is gauge invariant which can be seen from the fact that

$$\Delta_{\text{FP}}^{-1}\left[A_\mu\right] = \int \prod_x \mathrm{d}\alpha(x)\, \delta\left(F\left(A_\mu^{(\alpha)}(x)\right)\right)\,. \tag{12.87}$$

Let us make a gauge transformation $A_\mu \to A_\mu^{(\alpha')}$. Then,

$$\begin{aligned}
\Delta_{\text{FP}}^{-1}\left[A_\mu^{(\alpha')}\right] &= \int \prod_x \mathrm{d}\alpha(x)\, \delta\left(F\left(A_\mu^{(\alpha+\alpha')}(x)\right)\right) \\
&= \int \prod_x \mathrm{d}\alpha(x)\, \delta\left(F\left(A_\mu^{(\alpha)}(x)\right)\right) \\
&= \Delta_{\text{FP}}^{-1}\left[A_\mu\right]\,.
\end{aligned} \tag{12.88}$$

This follows from the fact that the measure in the group space is invariant under a gauge transformation. That is

$$\int \mathrm{d}\alpha(x) = \int \mathrm{d}\left(\alpha(x) + \alpha'(x)\right)\,. \tag{12.89}$$

In the non-Abelian case, we should use the Haar measure which is gauge invariant, namely,

$$\int d(UU') = \int dU. \tag{12.90}$$

Remembering that $\Delta_{\text{FP}}[A_\mu]$ is gauge invariant we can now insert this identity factor into the generating functional to write

$$Z[J] = N \int \mathcal{D}A_\mu \left(\Delta_{\text{FP}}[A_\mu] \int \prod_x d\alpha(x) \delta\left(F\left(A_\mu^{(\alpha)} \right) \right) \right) e^{iS^{(J)}[A_\mu]}. \tag{12.91}$$

Furthermore, let us make an inverse gauge transformation

$$A_\mu \to A_\mu^{(-\alpha)}, \tag{12.92}$$

under which the generating functional takes the form

$$Z[J] = N \int \mathcal{D}A_\mu \Delta_{\text{FP}}[A_\mu] \int \prod_x d\alpha(x) \, \delta\left(F\left(A_\mu \right) \right) e^{iS^{(J)}[A_\mu]}$$

$$= N \left(\int \prod_x d\alpha(x) \right) \int \mathcal{D}A_\mu \Delta_{\text{FP}}[A_\mu] \, \delta\left(F\left(A_\mu \right) \right) e^{iS^{(J)}[A_\mu]}$$

$$= N \int \mathcal{D}A_\mu \Delta_{\text{FP}}[A_\mu] \, \delta\left(F\left(A_\mu \right) \right) e^{iS^{(J)}[A_\mu]}, \tag{12.93}$$

where the gauge volume has been factored and absorbed into the normalization constant N of the path integral.

This, therefore, gives the correct functional form for the generating functional. However, we still have to determine what $\Delta_{\text{FP}}[A_\mu]$ is. To do this let us note that

$$\Delta_{\text{FP}}^{-1}[A_\mu] = \int \prod_x d\alpha(x) \, \delta\left(F\left(A_\mu^{(\alpha)} \right) \right)$$

$$= \int \prod_x dF \, \delta\left(F\left(A_\mu^{(\alpha)} \right) \right) \left(\det \frac{\delta\alpha}{\delta F} \right)$$

$$= \det \left(\frac{\delta\alpha}{\delta F} \right)_{F\left(A_\mu^{(\alpha)} \right) = 0}. \tag{12.94}$$

We note that since $\Delta_{\mathrm{FP}}^{-1}[A_\mu]$ is gauge invariant we can make an inverse gauge transformation to make $F[A_\mu] = 0$ in the above derivation. On the other hand, for gauge fields which satisfy the condition

$$F(A_\mu) = 0, \tag{12.95}$$

we have

$$\alpha(x) = 0. \tag{12.96}$$

Thus, we determine

$$\Delta_{\mathrm{FP}}[A_\mu] = \det\left(\frac{\delta F\left(A_\mu^{(\alpha)}\right)}{\delta\alpha}\right)_{\alpha=0}. \tag{12.97}$$

The Faddeev-Popov determinant can, therefore, be thought of as the Jacobian that goes with a particular gauge choice. We see that the Faddeev-Popov determinant can be calculated simply by restricting to infinitesimal gauge transformations. (Here we are completely ignoring the problem of Gribov ambiguity associated with large gauge transformations.)

We can further generalize our derivation by noting that a general equation of the hypersurface has the form (physical results are not sensitive to the choice of the hypersurface)

$$F(A_\mu(x)) = f(x). \tag{12.98}$$

Here $f(x)$ is independent of A_μ. Then we can insert the identity

$$\Delta_{\mathrm{FP}}[A_\mu] \int \prod_x d\alpha(x)\, \delta\left(F\left(A_\mu^{(\alpha)}(x)\right) - f(x)\right) = 1, \tag{12.99}$$

into the functional integral. The Faddeev-Popov determinant is unchanged by this modification because $f(x)$ does not depend on $A_\mu(x)$. Thus the generating functional in this case is given by

$$Z[J] = N \int \mathcal{D}A_\mu \Delta_{\text{FP}}[A_\mu] \delta \left(F\left(A_\mu(x)\right) - f(x)\right) e^{iS^{(J)}[A_\mu]}.$$

$$(12.100)$$

Using the 't Hooft trick, we can now do the following. We note that physical quantities are independent of $f(x)$. Hence we can multiply the generating functional by a weight factor and integrate over all $f(x)$. Thus, the generating functional becomes

$$Z[J] = N \int \mathcal{D}A_\mu \Delta_{\text{FP}}[A_\mu] \int \mathcal{D}f \delta \left(F\left(A_\mu(x)\right) - f(x)\right)$$

$$\times \ e^{-\frac{i}{2\xi} \int \mathrm{d}^4 x (f(x))^2} e^{iS^{(J)}}$$

$$= N \int \mathcal{D}A_\mu \Delta_{\text{FP}}[A_\mu] e^{i\left[S^{(J)} - \frac{1}{2\xi} \int \mathrm{d}^4 x (F(A_\mu(x)))^2\right]}$$

$$= N \int \mathcal{D}A_\mu \Delta_{\text{FP}}[A_\mu] e^{i\left(S^{(J)} + S_{\text{GF}}\right)},$$

where

$$S_{\text{GF}} = \int \mathrm{d}^4 x \ \mathcal{L}_{\text{GF}} = -\frac{1}{2\xi} \int \mathrm{d}^4 x \left(F\left(A_\mu(x)\right)\right)^2 , \qquad (12.101)$$

and ξ is known as the gauge fixing parameter.

We furthermore, note that since

$$\Delta_{\text{FP}}[A_\mu] = \det \left(\frac{\delta F\left(A_\mu^{(\alpha)}\right)}{\delta \alpha}\right)_{\alpha=0} , \qquad (12.102)$$

we can write this as

$$\Delta_{\text{FP}}[A_\mu] = \det \left(\frac{\delta F\left(A_\mu^\alpha(x)\right)}{\delta \alpha(y)}\right)_{\alpha=0}$$

$$= \int \mathcal{D}\bar{c}\mathcal{D}c \ e^{-i\left(\bar{c},\left(\frac{\delta F}{\delta \alpha}\right)_{\alpha=0} c\right)}$$

$$= \int \mathcal{D}\bar{c}\mathcal{D}c \; e^{-i \int \mathrm{d}^4x \mathrm{d}^4y \, \bar{c}(x) \left(\frac{\delta F\left(A_\mu^\alpha(x)\right)}{\delta\alpha(y)} \right)_{\alpha=0} c(y)}$$

$$= \int \mathcal{D}\bar{c}\mathcal{D}c \; e^{iS_{\text{ghost}}}, \tag{12.103}$$

where

$$S_{\text{ghost}} = - \int \mathrm{d}^4x \mathrm{d}^4y \; \bar{c}(x) \left(\frac{\delta F\left(A_\mu^\alpha(x)\right)}{\delta\alpha(y)} \right)_{\alpha=0} c(y). \tag{12.104}$$

Here we have introduced two independent fictitious fields $c(x)$ and $\bar{c}(x)$ to write the determinant in the form of an action. We note here that, for Eq. (12.103) to hold, the ghost fields $c(x)$ and $\bar{c}(x)$ have to anti-commute (ghosts have the same Lorentz structure as the parameters of transformation, but opposite statistics), i.e.,

$$[c(x), c(y)]_+ = 0 \,,$$

$$[\bar{c}(x), \bar{c}(y)]_+ = 0 \,,$$

$$[c(x), \bar{c}(y)]_+ = 0 \,. \tag{12.105}$$

Thus although these behave as scalar objects under Lorentz transformations, they obey anti-commutation rules. These fields are known as Faddeev-Popov ghosts and, as is obvious from their anti-commutation relations, graphs involving these fictitious particles in closed loops must have an additional (-1) factor just like the fermions. Thus the generating functional now takes the form

$$Z[J] = e^{iW[J]} = N \int \mathcal{D}A_\mu \mathcal{D}\bar{c}\mathcal{D}c \; e^{iS_{\text{eff}}^{(J)}[A_\mu, c, \bar{c}]} \,, \tag{12.106}$$

where

$$S_{\text{eff}}^{(J)}[A_\mu, c, \bar{c}] = S^{(J)}[A_\mu] + S_{\text{GF}} + S_{\text{ghost}}$$

$$= \int \mathrm{d}^4x \; \mathcal{L}_{\text{eff}}^{(J)}[A_\mu, c, \bar{c}] \,. \tag{12.107}$$

Let us now look at a simple gauge fixing condition, for example, the covariant condition

$$F\left(A_\mu\right) = \partial_\mu A^\mu(x) = f(x).$$
(12.108)

Clearly the gauge fixing Lagrangian density has the form

$$\mathcal{L}_{\mathrm{GF}} = -\frac{1}{2\xi}\left(F\left(A_\mu(x)\right)\right)^2$$

$$= -\frac{1}{2\xi}\left(\partial_\mu A^\mu(x)\right)^2.$$
(12.109)

It is clear that this provides longitudinal components to the quadratic terms in fields and hence breaks gauge invariance. To obtain the corresponding ghost Lagrangian density for this gauge choice, we note that

$$F\left(A_\mu^{(\alpha)}\right) = \partial_\mu A^{\mu(\alpha)}(x) = \partial_\mu\left(A^\mu + \partial^\mu\alpha(x)\right),$$
(12.110)

so that

$$\left.\frac{\delta F\left(A_\mu^{(\alpha)}(x)\right)}{\delta\alpha(y)}\right|_{\alpha=0} = \Box_x\delta^4(x-y).$$
(12.111)

In this case, the ghost action is obtained to be

$$S_{\mathrm{ghost}} = -\int \mathrm{d}^4x\mathrm{d}^4y\,\bar{c}(x)\left.\frac{\delta F\left(A_\mu^\alpha(x)\right)}{\delta\alpha(y)}\right|_{\alpha=0} c(y)$$

$$= -\int \mathrm{d}^4x\mathrm{d}^4y\,\bar{c}(x)\left(\Box_x\delta^4(x-y)\right)c(y)$$

$$= \int \mathrm{d}^4x\,\partial_\mu\bar{c}(x)\partial^\mu c(x) = \int \mathrm{d}^4x\,\mathcal{L}_{\mathrm{ghost}},$$
(12.112)

where we have neglected total divergence terms.

Thus our effective Lagrangian density for this choice of gauge condition becomes

$$\mathcal{L}_{\text{eff}} = -\frac{1}{4}F_{\mu\nu}F^{\mu\nu} - \frac{1}{2\xi}\left(\partial_\mu A^\mu\right)^2 + \partial_\mu \bar{c}(x)\partial^\mu c(x) + J^\mu A_\mu. \quad (12.113)$$

We note here that the ghost fields are noninteracting in the case of the Abelian theory and, therefore, we may omit them and then for $\xi = 1$ our effective Lagrangian density coincides with Maxwell's theory in the Feynman-Fermi gauge (see Eq. (12.15)). In non-Abelian gauge theories, however, the ghost fields are interacting and have to be present. Furthermore, since these fields are really necessary for the unitarity of the S-matrix, we cannot neglect them even if they are noninteracting if we are doing calculations at finite temperature. It is also true that when Maxwell's theory is coupled to gravitational fields, the ghost fields automatically couple to the geometry also. Hence omitting the ghost Lagrangian density in such a case would lead to nonsensical results.

One way of looking at the ghost fields is as if they are there to subtract out the unphysical field degrees of freedom. For example, the A_μ field has four helicity components. On the other hand each of the ghost fields, being a scalar, has only one helicity component. Hence one can think of the effective Lagrangian density as having only two helicity states $(4 - 2 \times 1 = 2)$. This naive counting works pretty well as we will see later. (The ghost degrees of freedom subtract because they have the unphysical statistics.)

Let us now apply these ideas of path integral quantization to the case of non-Abelian gauge theories. Let us recall from our study in the Abelian theory that gauge invariance puts a very strong constraint on the structure of the Lagrangian density for the gauge field. In particular, the coefficient matrix of the quadratic terms in the Lagrangian density is singular and, therefore, non-invertible. As a result, if we take $\mathcal{L}_{\text{inv}}$ as describing the dynamics of the theory, then, we cannot define propagators and the entire philosophy of doing perturbative calculations with Feynman diagrams breaks down. In order to circumvent this difficulty, we add to the gauge invariant Lagrangian density a term which breaks gauge invariance and thereby allows us to define the propagator for the gauge field. Such a term is called a gauge fixing term and any term which makes the coefficient matrix of the quadratic terms nonsingular and maintains various global symmetries of the theory is allowed for this purpose. On the other hand, adding a gauge fixing Lagrangian density changes the theory, in general, and to compensate for that we have to add a Lagrangian density for the ghost fields following the prescription of Faddeev and Popov.

In the case of the non-Abelian gauge theory, as we have seen, the Lagrangian density for the gauge fields

$$\mathcal{L}_{\text{inv}} = -\frac{1}{4} F^a_{\mu\nu} F^{\mu\nu a}, \tag{12.114}$$

is invariant under the infinitesimal gauge transformation

$$A^a_\mu(x) \rightarrow A^{(\epsilon)a}_\mu(x) = A^a_\mu(x) + D_\mu \epsilon^a(x), \tag{12.115}$$

where $\epsilon^a(x)$ is the infinitesimal parameter of transformation and

$$D_\mu \epsilon^a(x) = \partial_\mu \epsilon^a(x) - f^{abc} A^b_\mu(x) \epsilon^c(x),$$
$$F^a_{\mu\nu}(x) = \partial_\mu A^a_\nu(x) - \partial_\nu A^a_\mu(x) - f^{abc} A^b_\mu A^c_\nu. \tag{12.116}$$

The standard covariant gauge fixing, in a non-Abelian gauge theory, consists of adding to the invariant Lagrangian density a term of the form

$$\mathcal{L}_{\text{GF}} = -\frac{1}{2\xi} \left(\partial_\mu A^{\mu a} \right)^2, \tag{12.117}$$

which would correspond to a gauge fixing condition of the form

$$F^a[A] = \partial_\mu A^{\mu a}(x) = f^a(x). \tag{12.118}$$

Here ξ represents an arbitrary constant parameter known as the gauge fixing parameter. Following the prescription of Faddeev and Popov (see Eq. (12.104)), we can write the ghost action corresponding to this gauge choice as

$$S_{\text{ghost}} = \int \mathrm{d}x \, \mathcal{L}_{\text{ghost}}$$

$$= - \int \mathrm{d}x \mathrm{d}y \, \bar{c}^a(x) \frac{\delta F^a[A^{(\alpha)}(x)]}{\delta \alpha^b(y)} \bigg|_{\alpha=0} c^b(y) \,. \qquad (12.119)$$

Here we have left the dimensionality of space-time arbitrary since our discussions apply in any dimension. We note that for the covariant gauge choice that we are assuming, we can write (we can think of $\alpha^a = \epsilon^a$ as infinitesimal since we are taking $\alpha^a = 0$ at the end)

$$F^a[A^{(\alpha)}(x)] = \partial^\mu A_\mu^{(\alpha)a} = \partial^\mu \left(A_\mu^a(x) + (D_\mu \alpha)^a(x) \right) \,. \qquad (12.120)$$

Consequently, the ghost Lagrangian density, for this choice of gauge fixing, follows to be

$$\mathcal{L}_{\text{ghost}} = - \int \mathrm{d}y \, \bar{c}^a(x) \left(\partial_x^\mu D_{\mu,x}^{ab} \delta(x-y) \right) c^b(y)$$

$$= - \left(D_\mu \left(\partial^\mu \bar{c} \right) \right)^a (x) c^a(x)$$

$$= \partial^\mu \bar{c}^a(x) \left(D_\mu c(x) \right)^a \,, \qquad (12.121)$$

where we have dropped a total derivative term in the last line of the above equation and have identified

$$D_{\mu,x}^{ab} = \delta^{ab} \partial_{\mu,x} - f^{acb} A_\mu^c(x) \,, \qquad (12.122)$$

which leads to (the ghosts are in the adjoint representation)

$$(D_\mu c)^a = D_{\mu,x}^{ab} c^b = \partial_\mu c^a - f^{acb} A_\mu^c c^b = \partial_\mu c^a - f^{abc} A_\mu^b c^c \,. \quad (12.123)$$

With all these modifications, the total Lagrangian density for a non-Abelian gauge theory has the form

$$\mathcal{L}_{\text{TOT}} = \mathcal{L}_{\text{inv}} + \mathcal{L}_{\text{GF}} + \mathcal{L}_{\text{ghost}} \,, \qquad (12.124)$$

which, in the covariant gauge that we are considering, is given by

$$\mathcal{L}_{\text{TOT}} = -\frac{1}{4} F^a_{\mu\nu} F^{\mu\nu a} - \frac{1}{2\xi} (\partial_\mu A^{\mu a})^2 + \partial^\mu \bar{c}^a (D_\mu c)^a . \qquad (12.125)$$

The path integral can now be defined with this $\mathcal{L}_{\text{TOT}}$ and will be well behaved.

12.4 BRST symmetry

As we have mentioned earlier, the gauge fixing and the ghost Lagrangian densities modify the original theory in a compensating manner which allows us to define the Feynman rules of the theory and carry out any perturbative calculation. In a deeper sense, the gauge fixing and the ghost Lagrangian densities, in the path integral formulation, merely correspond to a multiplicative factor of unity which does not change the physical content of the theory. However, without going into too much technical details, we can see that these additional terms in the Lagrangian density have no physical content in the following way.

The total Lagrangian density has been gauge fixed and, therefore, does not have the gauge invariance of the original theory. However, the total Lagrangian density, with gauge fixing and ghost terms, develops a global fermionic symmetry which, in some sense, remembers the gauge invariance of the original theory. It is easy to check that the total action is invariant under the global transformations

$$\delta A^a_\mu = \omega \left(D_\mu c \right)^a ,$$

$$\delta c^a = \frac{\omega}{2} f^{abc} c^b c^c,$$

$$\delta \bar{c}^a = -\frac{\omega}{\xi} \left(\partial_\mu A^{\mu a} \right) , \qquad (12.126)$$

where ω is the constant anti-commuting parameter of the global transformations ($\omega^2 = 0$). The invariance of the action can be seen by first noting that

$$\delta\left(D_\mu c^a\right) = \left(D_\mu \delta c\right)^a - f^{abc}\delta A_\mu^b c^c$$

$$= \frac{\omega}{2} D_\mu \left(f^{abc} c^b c^c\right) - \omega f^{abc} \left(D_\mu c^b\right) c^c = 0\,,$$

$$\delta\left(\frac{1}{2} f^{abc} c^b c^c\right) = f^{abc} \delta c^b c^c$$

$$= \frac{\omega}{2} f^{abc} f^{bpq} c^p c^q c^c$$

$$= \frac{\omega}{6} \left(f^{abc} f^{bpq} + f^{abp} f^{bqc} + f^{abq} f^{bcp}\right) c^p c^q c^c$$

$$= 0\,. \tag{12.127}$$

Here we have used the Jacobi identity for the symmetry algebra (see Eqs. (12.29) and (12.30)). Similarly, we obtain

$$\delta\left(\partial_\mu A^{\mu a}\right) = \omega \, \partial^\mu (D_\mu c)^a = 0\,, \tag{12.128}$$

when the ghost equations of motion are used. This shows that

$$\delta_2 \delta_1 \phi^a = 0\,, \tag{12.129}$$

for $\phi^a = A_\mu^a, c^a, \bar{c}^a$, where $\delta_{1,2}$ correspond to transformations with parameters $\omega_{1,2}$ respectively. We note that the nilpotency of the transformations holds off-shell only for the fields A_μ^a, c^a, while for $\bar{c}^a$ it is true only on-shell (when the ghost equations of motion are used).

The invariance of the action can now be easily checked. First, we note that the transformation for A_μ^a is really an infinitesimal gauge transformation with the parameter $\alpha^a = \omega c^a$ and, therefore, the invariant Lagrangian density is trivially invariant under these transformations, namely,

$$\delta \mathcal{L}_{\text{inv}} = 0\,. \tag{12.130}$$

Consequently, we need to worry only about the gauge fixing and the ghost Lagrangian densities which lead to

$$\delta \left(\mathcal{L}_{\text{GF}} + \mathcal{L}_{\text{ghost}} \right) = -\frac{1}{\xi} \left(\partial_\nu A^{\nu a} \right) \left(\partial^\mu \delta A_\mu^a \right) + \left(\partial^\mu \delta \bar{c}^a \right) \left(D_\mu c \right)^a$$

$$= -\frac{\omega}{\xi} \left(\partial_\nu A^{\nu a} \right) \partial^\mu (D_\mu c)^a - \frac{\omega}{\xi} \, \partial^\mu \left(\partial_\nu A^{\nu a} \right) (D_\mu c)^a$$

$$= -\partial^\mu \left(\frac{\omega}{\xi} \left(\partial_\nu A^{\nu a} \right) (D_\mu c)^a \right), \qquad (12.131)$$

so that the action is invariant. In this derivation, we have used the fact that $\delta \left(D_\mu c^a \right) = 0$ which we have seen earlier in Eq. (12.127). This shows that the total action is invariant under the global transformations with an anti-commuting constant parameter. This is known as the BRST (Becchi-Rouet-Stora-Tyutin) transformation for a gauge theory and arises in the presence the gauge fixing and the ghost Lagrangian densities. The present formulation of the BRST symmetry, however, is slightly unpleasant in the sense that the nilpotency of the anti-ghost field transformation holds only on-shell. This is a reflection of the fact that the theory is lacking some auxiliary fields and once the correct auxiliary fields are incorporated, the symmetry algebra will close off-shell.

We can write the gauge fixing Lagrangian density by introducing an auxiliary field which will also be quite useful for our later discussions. Let us rewrite

$$\mathcal{L}_{\text{GF}} = \partial^\mu F^a A_\mu^a + \frac{\xi}{2} F^a F^a, \qquad (12.132)$$

where F^a is an auxiliary field. It is clear from the form of $\mathcal{L}_{\text{GF}}$ that the equation of motion for the auxiliary field takes the form

$$\xi F^a = \partial^\mu A_\mu^a, \qquad (12.133)$$

and when we eliminate F^a using this equation, we recover the original gauge fixing Lagrangian density (if we ignore a total divergence term). Among other things, $\mathcal{L}_{\text{GF}}$ as written above, allows us to take such gauge choices as the Landau gauge which corresponds to simply taking the limit $\xi = 0$. The total Lagrangian density can now be written as

$$\mathcal{L}_{\text{TOT}} = \mathcal{L}_{\text{inv}} + \mathcal{L}_{\text{GF}} + \mathcal{L}_{\text{ghost}}$$

$$= -\frac{1}{4} F^a_{\mu\nu} F^{\mu\nu a} + \partial^\mu F^a A^a_\mu + \frac{\xi}{2} F^a F^a + \partial^\mu \bar{c}^a D_\mu c^a . \quad (12.134)$$

In this case, the BRST transformations take the form

$$\delta A^a_\mu = \omega \left(D_\mu c \right)^a ,$$

$$\delta c^a = \frac{\omega}{2} f^{abc} c^b c^c ,$$

$$\delta \bar{c}^a = -\omega F^a ,$$

$$\delta F^a = 0 , \quad\quad\quad\quad\quad\quad\quad\quad (12.135)$$

and it is straightforward to check that these transformations are all nilpotent off-shell, namely,

$$\delta_2 \delta_1 \phi^a = 0 , \quad\quad\quad\quad\quad\quad\quad\quad (12.136)$$

for $\phi^a = A^a_\mu, F^a, c^a, \bar{c}^a$. Therefore, F^a represents the missing auxiliary field that we had alluded to earlier.

We note that $\mathcal{L}_{\text{inv}}$ is invariant under the BRST transformation as we had argued earlier and the auxiliary field, F^a, does not transform at all which leads to

$$\delta \mathcal{L}_{\text{TOT}} = \delta \left(\mathcal{L}_{\text{GF}} + \mathcal{L}_{\text{ghost}} \right)$$

$$= \partial^\mu F^a \delta A^a_\mu + \partial^\mu \delta \bar{c}^a \left(D_\mu c \right)^a$$

$$= \omega \partial^\mu F^a \left(D_\mu c \right)^a - \omega \partial^\mu F^a \left(D_\mu c \right)^a$$

$$= 0 . \quad\quad\quad\quad\quad\quad\quad\quad (12.137)$$

Unlike the formulation without auxiliary fields, here we see that the total Lagrangian density is invariant under the BRST transformations (as opposed to changing by a total divergence).

These transformations, known as BRST transformations, which define a residual global symmetry of the full theory, in some sense, replace the original gauge invariance of the theory, and play a fundamental role in the study of non-Abelian gauge theories. There is also a second set of transformations involving the anti-ghost fields of the form

$$\bar{\delta} A_\mu^a = \bar{\omega} \, (D_\mu \bar{c})^a \,,$$

$$\bar{\delta} c^a = \bar{\omega} \left(F^a - f^{abc} c^b \bar{c}^c \right) \,,$$

$$\bar{\delta} \bar{c}^a = \frac{\bar{\omega}}{2} \, f^{abc} \bar{c}^b \bar{c}^c \,,$$

$$\bar{\delta} F^a = \bar{\omega} f^{abc} F^b \bar{c}^c \,, \tag{12.138}$$

which leave the total Lagrangian density invariant. These are known as the anti-BRST transformations. However, since these do not lead to any new information beyond what the BRST invariance provides, we will ignore this symmetry for the rest of our discussions. We note here that the BRST and the anti-BRST transformations are not quite symmetric in the ghost and the anti-ghost fields which is a reflection of the asymmetric nature of these fields in the ghost Lagrangian density. It is also worth noting here that these symmetries arise naturally in a superspace formulation of gauge theories.

In addition to these two anti-commuting symmetries, the total Lagrangian density is also invariant under the infinitesimal bosonic global symmetry transformations

$$\delta c^a = \epsilon c^a \,,$$

$$\delta \bar{c}^a = -\epsilon \bar{c}^a \,, \tag{12.139}$$

with all other fields remaining inert. Here ϵ represents a constant, bosonic infinitesimal parameter and the generator of the symmetry merely counts the ghost number of the fields. (The fact that these transformations are like scale transformations and not like phase transformations has to do with the particular hermiticity properties that the ghost and the anti-ghost fields must satisfy for consistent quantization of the theory.)

The vector space of the full theory, as is clear by now, contains many more states than the physical states alone. The physical

Hilbert space, therefore, needs to be properly selected for a discussion of physical questions. Furthermore, the physical space must be selected in such a way that it remains invariant under the time evolution of the system. In the covariant gauge in QED, for example, we have seen that the states in the physical space are selected as satisfying the Gupta-Bleuler condition

$$\partial^\mu A_\mu^{(+)}(x)|\text{phys}\rangle = 0 \,, \tag{12.140}$$

where the superscript, $(+)$, stands for the positive frequency part of the field. We recognize that even though this looks like one condition, in reality it is an infinite number of conditions since it has to hold for every value of the space-time coordinates. In QED, the Gupta-Bleuler condition works because $\partial^\mu A_\mu$ satisfies the free Klein-Gordon equation in the covariant gauge and hence the physical space so selected remains invariant under time evolution. The corresponding operator in a non-Abelian theory, on the other hand, does not satisfy a free equation as we have seen and hence is not suitable for identifying the physical subspace. On the other hand, the generators of the BRST symmetry, Q_{BRST}, and the ghost scaling symmetry, Q_{c}, are conserved and hence can be used to define a physical Hilbert space which would remain invariant under the time evolution of the system. (Q_{BRST} and Q_{c} are the charges constructed from the Nöther current for the respective transformations whose explicit forms can be obtained from the Nöther procedure.) Thus, we can identify the physical space of the gauge theory as satisfying

$$Q_{\text{BRST}}|\text{phys}\rangle = 0 \,,$$
$$Q_{\text{c}}|\text{phys}\rangle = 0 \,. \tag{12.141}$$

Note that even in the case of QED, these would appear to correspond to only two conditions and not an infinite number of conditions as we have seen is the case with the Gupta-Bleuler condition. It is, therefore, not clear *a priori* if these conditions are sufficient to reproduce the Gupta-Bleuler conditions in the case of QED.

To see that these conditions do indeed lead to the Gupta-Bleuler conditions in QED, let us note that the Nöther current associated with the BRST transformations has the form

$$J_{\text{BRST}}^{(\omega)\mu} = -F^{\mu\nu a}\delta A_\nu^a + \delta\overline{c}^a (D^\mu c)^a + (\partial^\mu \overline{c}^a)\delta c^a$$

$$= -\omega \left(F^{\mu\nu a}(D_\nu c)^a + F^a (D^\mu c)^a + \frac{1}{2} f^{abc} (\partial^\mu \overline{c}^a) c^b c^c \right),$$

$$\text{(12.142)}$$

where we have used the fact that the auxiliary field does not transform under the BRST transformations. From this, we can obtain the current without the parameter of transformations

$$J_{\text{BRST}}^\mu = F^{\mu\nu a}(D_\nu c)^a + F^a (D^\mu c)^a + \frac{1}{2} f^{abc} (\partial^\mu \overline{c}^a) c^b c^c, \quad \text{(12.143)}$$

and, in particular, for the Abelian theory where $f^{abc} = 0$, the BRST charge can be obtained to be (dx denotes integration over all space coordinates)

$$Q_{\text{BRST}} = \int d\mathbf{x} \left(F^{0\nu}\partial_\nu c + F\partial^0 c \right) = \int d\mathbf{x} \left(F^{0i}\partial_i c + F\dot{c} \right)$$

$$= \int d\mathbf{x} \left(\partial_i(F^{0i}c) - (\partial_i F^{0i})c + F\dot{c} \right)$$

$$= \int d\mathbf{x} \left(-(\partial^0 F)c + \int d\mathbf{x}\, F\dot{c} \right) = \int d\mathbf{x}\, F\overset{\leftrightarrow}{\partial^0} c,$$

$$\text{(12.144)}$$

where we have discarded the surface terms and used Maxwell's equations in the intermediate steps. If we use the field decomposition for the fields and normal order the BRST charge (so that the annihilation operators are to the right), this has the explicit form (dk denotes integration over all space components of the momentum)

$$Q_{\text{BRST}} = i \int d\mathbf{k} \left(c^{(-)}(-k)F^{(+)}(k) - F^{(-)}(-k)c^{(+)}(k) \right).$$

$$\text{(12.145)}$$

We recognize that the condition

$$Q_c|\text{phys}\rangle = 0\,, \tag{12.146}$$

implies that the physical states must have zero ghost number. This allows, in principle, states containing equal numbers of ghost and anti-ghost particles. Denoting the physical states of the theory as

$$|\text{phys}\rangle = |A_\mu\rangle \otimes |c, \bar{c}\rangle\,, \tag{12.147}$$

we note that if the physical states have to further satisfy the condition

$$Q_{\text{BRST}}|\text{phys}\rangle = i \int d\mathbf{k} \left(c^{(-)}(-k)F^{(+)}(k) - F^{(-)}(-k)c^{(+)}(k) \right)$$
$$\times \ |A_\mu\rangle \otimes |c, \bar{c}\rangle$$
$$= 0\,, \tag{12.148}$$

then this implies that

$$c^{(+)}(k)|c, \bar{c}\rangle = 0 = F^{(+)}(k)|A_\mu\rangle\,. \tag{12.149}$$

Namely, the physical states should have no ghost particles (remember zero ghost number),

$$|\text{phys}\rangle = |A_\mu\rangle \otimes |0, 0\rangle = |A_\mu\rangle\,, \tag{12.150}$$

and must satisfy

$$F^{(+)}(k)|\text{phys}\rangle = \left(k_\mu A^{\mu(+)}(k) \right)|\text{phys}\rangle = 0\,, \tag{12.151}$$

where we have used the equation of motion for the auxiliary field. This is precisely the Gupta-Bleuler condition in momentum space and this derivation shows how a single condition can give rise to an infinite number of conditions (in this case, for every momentum mode k). Thus, we feel confident that the physical state conditions are the right ones even for a non-Abelian theory.

We are now ready to show that the gauge fixing and the ghost Lagrangian densities lead to no physical consequences. We note that these extra terms in the Lagrangian density can, in fact, be written as a BRST variation (without the parameter of transformation), namely,

$$
\begin{aligned}
\mathcal{L}_{\mathrm{GF}} + \mathcal{L}_{\mathrm{ghost}} &= \delta \left(-\partial^{\mu} \overline{c}^{a} A_{\mu}^{a} - \frac{\xi}{2} \overline{c}^{a} F^{a} \right) \\
&= \left[Q_{\mathrm{BRST}}, \left(-\partial^{\mu} \overline{c}^{a} A_{\mu}^{a} - \frac{\xi}{2} \overline{c}^{a} F^{a} \right) \right]_{+}.
\end{aligned} \tag{12.152}
$$

Here we have used the fact that the BRST charge is the generator of the BRST transformations so that the transformations for fermionic operators are generated through the anti-commutator of the operators with the generator. It now follows from the physical condition that

$$
\begin{aligned}
&\langle \mathrm{phys}| \left(\mathcal{L}_{\mathrm{GF}} + \mathcal{L}_{\mathrm{ghost}} \right) |\mathrm{phys}'\rangle \\
&= -\langle \mathrm{phys}| \left[Q_{\mathrm{BRST}}, \left(\partial^{\mu} \overline{c}^{a} A_{\mu}^{a} + \frac{\xi}{2} \overline{c}^{a} F^{a} \right) \right]_{+} |\mathrm{phys}'\rangle \\
&= 0.
\end{aligned} \tag{12.153}
$$

This shows that the terms added to the original Lagrangian density have no contribution to the physical matrix elements of the theory. We can also show that all the physical matrix elements are independent of the choice of the gauge fixing parameter ξ in the following manner (the BRST variation denoted is without the parameter of transformation)

$$
\begin{aligned}
\frac{\partial}{\partial \xi} \langle 0|0\rangle^{J} &= \frac{\partial Z[J]}{\partial \xi} \\
&= \frac{i}{2} \langle 0| \int \mathrm{d}^{4}x \ F^{a} F^{a} |0\rangle^{J} = -\frac{i}{2} \int \mathrm{d}^{4}x \ \langle 0| \delta \left(\overline{c}^{a} F^{a} \right) |0\rangle^{J} \\
&= -\frac{i}{2} \int \mathrm{d}^{4}x \ \langle 0| [Q_{\mathrm{BRST}}, \overline{c}^{a} F^{a}]_{+} |0\rangle^{J} = 0, \tag{12.154}
\end{aligned}
$$

where we have used the fact that the vacuum belongs to the physical Hilbert space of the theory and as such is annihilated by the BRST charge.

12.5 Ward identities

The BRST invariance of the full theory leads to many relations between various scattering amplitudes of the theory. These are known as the Ward identities or the Slavnov-Taylor identities of the theory and are quite essential in establishing the renormalizability of gauge theories. These identities are best described within the context of path integrals which we will do next.

Let us consider an effective Lagrangian density which consists of $\mathcal{L}_{\text{TOT}}$ as well as source terms as follows

$$\mathcal{L}_{\text{eff}} = \mathcal{L}_{\text{TOT}} + J^{\mu a} A_\mu^a + J^a F^a + i \left(\overline{\eta}^a c^a - \overline{c}^a \eta^a \right)$$

$$+ K^{\mu a} (D_\mu c)^a + K^a \left(\frac{1}{2} f^{abc} c^b c^c \right) . \tag{12.155}$$

Here, we have not only introduced sources for all the field variables in the theory, but, in addition, we have also added sources for the composite variations under the BRST transformation (see Eq. (12.135)). The need for this will become clear shortly. Denoting all the fields and the sources generically by A and J respectively, we can write the generating functional for the theory as

$$Z[J] = e^{iW[J]} = N \int \mathcal{D}A \, e^{i \int dx \, \mathcal{L}_{\text{eff}}} . \tag{12.156}$$

The vacuum expectation values of operators, in the presence of sources, can now be written as

$$\langle A_\mu^a \rangle = A_\mu^{(c)a} = \frac{\delta W}{\delta J^{\mu a}},$$

$$\langle F^a \rangle = F^{(c)a} = \frac{\delta W}{\delta J^a},$$

$$\langle c^a \rangle = c^{(c)a} = -i \frac{\delta W}{\delta \overline{\eta}^a},$$

$$\langle \overline{c}^a \rangle = \overline{c}^{(c)a} = -i \frac{\partial W}{\delta \eta^a},$$

$$\langle (D_\mu c)^a \rangle = \frac{\delta W}{\delta K^{\mu a}},$$

$$\left\langle \frac{1}{2} f^{abc} c^b c^c \right\rangle = \frac{\delta W}{\delta K^a}. \qquad (12.157)$$

Here, we have assumed the convention of a left derivative for the anti-commuting fields. The fields $A^{(c)}$ are known as the classical fields and in what follows, we will ignore the superscript (c) for notational simplicity.

When the external sources are held fixed, the effective Lagrangian density is no longer invariant under the BRST transformations. In fact, recalling that $\mathcal{L}_{\text{TOT}}$ is BRST invariant and that the BRST transformations are nilpotent, we obtain the change in $\mathcal{L}_{\text{eff}}$ (remember that the parameter is anti-commuting) to be

$$\delta \mathcal{L}_{\text{eff}} = J^{\mu a} \delta A^a_\mu + J^a \delta F^a + i \left(\overline{\eta}^a \delta c^a - \delta \overline{c}^a \eta^a \right)$$

$$= \omega \left[J^{\mu a} \left(D_\mu c \right)^a + i \left(-\frac{1}{2} f^{abc} \overline{\eta}^a c^b c^c + F^a \eta^a \right) \right]. \qquad (12.158)$$

On the other hand, the generating functional is defined by integrating over all possible field configurations. Therefore, with a redefinition of the fields under a BRST transformation inside the path integral, the generating functional should be invariant. This immediately leads to

$$\delta Z[J] = 0 = N \int \mathcal{D}A \left(i \int \mathrm{d}^4 x \, \delta \mathcal{L}_{\text{eff}} \right) e^{i \int \mathrm{d}^4 x \, \mathcal{L}_{\text{eff}}}$$

$$= i\omega \int \mathrm{d}^4 x \left(J^{\mu a} \langle (D_\mu c)^a \rangle - i \overline{\eta}^a \left\langle \frac{1}{2} f^{abc} c^b c^c \right\rangle + i \eta^a \langle F^a \rangle \right). \qquad (12.159)$$

The measure can be easily checked to be invariant under such a fermionic transformation and we note that the above relation can also be written as (see Eq. (12.157))

$$\int \mathrm{d}^4 x \left(J^{\mu a}(x) \frac{\delta W}{\delta K^{\mu a}(x)} - i \overline{\eta}^a(x) \frac{\delta W}{\delta K^a(x)} + i \eta^a(x) \frac{\delta W}{\delta J^a(x)} \right) = 0. \qquad (12.160)$$

This is the Master equation from which one can derive all the identities relating the connected Green's functions of the theory. It is here that the usefulness of the sources for the composite operators becomes evident.

Most often, however, we are interested in the proper (1PI) vertices of the theory. These can be obtained by passing from the generating functional for the connected Green's functions, $W[J]$, to the generating functional for the proper vertices, $\Gamma[A]$, through a Legendre transformation. Defining a Legendre transformation involving (only) the field variables of the theory (the field variables are really the classical fields and we are dropping the superscript (c) for simplicity), we have

$$\Gamma[A, K] = W[J, K] - \int d^4x \left(J^{\mu a} A_\mu^a + J^a F^a + i \left(\overline{\eta}^a c^a - \overline{c}^a \eta^a \right) \right),$$

(12.161)

where K stands generically for the sources for the composite variations. From the definition of the generating functional for the proper vertices, it is clear that

$$\frac{\delta \Gamma}{\delta A_\mu^a} = -J^{\mu a},$$

$$\frac{\delta \Gamma}{\delta F^a} = -J^a,$$

$$\frac{\delta \Gamma}{\delta c^a} = i\overline{\eta}^a,$$

$$\frac{\delta \Gamma}{\delta \overline{c}^a} = i\eta^a,$$

$$\frac{\delta \Gamma}{\delta K_\mu^a} = \frac{\delta W}{\delta K_\mu^a},$$

$$\frac{\delta \Gamma}{\delta K^a} = \frac{\delta W}{\delta K^a}.$$

(12.162)

Using these definitions, we see that we can rewrite the Master equation in terms of the generating functional of the proper vertices as

$$\int d^4x \left(\frac{\delta\Gamma}{\delta A_\mu^a(x)} \frac{\delta\Gamma}{\delta K^{\mu a}(x)} + \frac{\delta\Gamma}{\delta c^a(x)} \frac{\delta\Gamma}{\delta K^a(x)} - F^a(x) \frac{\delta\Gamma}{\delta \bar{c}^a(x)} \right) = 0.$$

(12.163)

This is the Master equation from which we can derive all the relations, between various (1PI) proper vertices, resulting from the BRST invariance of the theory. This, in turn, is essential in proving the renormalizability of gauge theories. Thus, for example, let us note that we can write the Master identity in the momentum space as

$$\int d^4k \left(\frac{\delta\Gamma}{\delta A_\mu^a(-k)} \frac{\delta\Gamma}{\delta K^{\mu a}(k)} + \frac{\delta\Gamma}{\delta c^a(-k)} \frac{\delta\Gamma}{\delta K^a(k)} \right.$$
$$\left. - F^a(k) \frac{\delta\Gamma}{\delta \bar{c}^a(k)} \right) = 0.$$

(12.164)

Taking derivative of this with respect to $\frac{\delta^2}{\delta F^b(p)\delta c^c(-p)}$ and setting all field variables to zero gives

$$\frac{\delta^2\Gamma}{\delta F^b(p)\delta A_\mu^a(-p)} \frac{\delta^2\Gamma}{\delta c^c(-p)\delta K^{\mu a}(p)} - \frac{\delta^2\Gamma}{\delta c^c(-p)\delta \bar{c}^b(p)} = 0.$$

(12.165)

A simple analysis, then, shows that this relates the mixed two point function involving F-A_μ with the two point function for the ghost fields and, consequently, the counter terms (quantum corrections) should satisfy such a relation. The BRST invariance, in this way, is very fundamental in the study of gauge theories as far as renormalizability and gauge independence are concerned.

12.6 References

C. Becchi, A. Rouet and R. Stora, Communications in Mathematical Physics, **42**, 127 (1975).

K. Bleuler, Helv. Phys. Acta **23**, 567 (1950).

A. Das, *Finite Temperature Field Theory*, World Scientific Publishing (1997).

L. D. Faddeev and V. N. Popov, Physics Letters **25B**, 29 (1967).

V. N. Gribov, Nuclear Physics **B139**. 1 (1978).

S. N. Gupta, Proc. Phys. Soc. (London) **A63**, 681 (1950).

R. N. Mohapatra, Physical Review **D4**, 378 (1971).

CHAPTER 13

Anomalies

We have seen earlier that a continuous global symmetry in a quantum field theory leads to a current which is conserved and the Ward identities of the theory reflect this conservation law. Sometimes, however, the classical (tree level) conservation of the current is violated by quantum corrections (loop effects). This can happen, for example, if the regularization, used to define divergent amplitudes in a quantum field theory, does not respect the symmetry that leads to the conservation law. In this case, the divergence of the current density no longer vanishes and one says that there is an anomaly in the conservation law (or that the current is anomalous). In turn, this modifies the tree level Ward identities of the theory to new ones that are known as anomalous Ward identities of the theory (which reflect the modified conservation law). Anomalies are quite important from a physical point of view. Global anomalies such as the chiral anomaly have direct physical consequences. On the other hand, an anomaly in a gauge theory (namely, in the conservation of the gauge current) can render the theory inconsistent and, therefore, needs to be taken care of before one can make sense of such a theory. In this chapter, we will study some issues associated with this phenomenon.

13.1 Anomalous Ward identity

Let us consider a non-Abelian gauge theory interacting with massless fermions in $1 + 3$ dimensions. If we choose the gauge group to be $SU(3)$, then the fermions can be thought of as the quarks and the theory would describe quantum chromodynamics. For a general gauge group of $SU(n)$, the theory is described by the Lagrangian density

$$\mathcal{L} = -\frac{1}{4}F^a_{\mu\nu}F^{\mu\nu a} + i\bar{\psi}^i\gamma^\mu(D_\mu\psi)^i, \tag{13.1}$$

where $a = 1, 2, \cdots, n^2 - 1$, $i = 1, 2, \cdots, n$, and $\mu = 0, 1, 2, 3$. The (gauge) covariant derivative for the fermions is defined as

$$(D_\mu\psi)^i = \partial_\mu\psi^i + iA^a_\mu(T^a)^{ij}\psi^j. \tag{13.2}$$

Here $(T^a)^{ij}$ denote the generators of the gauge group $(SU(n))$ in the (fundamental) representation of the fermions and the gauge coupling has been absorbed into the gauge field. (Equivalently, we can say that the gauge coupling has been set to unity for simplicity.) Note that this theory in addition to having the infinitesimal non-Abelian local gauge invariance

$$\delta A^a_\mu(x) = (D_\mu\epsilon(x))^a,$$
$$\delta\psi^i(x) = -i\epsilon^a(x)(T^a\psi(x))^i,$$
$$\delta\bar{\psi}^i(x) = i\epsilon^a(x)(\bar{\psi}(x)T^a)^i, \tag{13.3}$$

where $\epsilon^a(x)$ is the real infinitesimal parameter of the gauge transformation, is also invariant under the infinitesimal global chiral transformations

$$\delta\psi^i(x) = -i\lambda\gamma_5\psi^i(x),$$
$$\delta\bar{\psi}^i(x) = -i\lambda\bar{\psi}^i(x)\gamma_5. \tag{13.4}$$

Here λ is an infinitesimal space-time independent real parameter and γ_5 denotes the pseudoscalar Dirac matrix (defined to be Hermitian). This global symmetry, which corresponds to an Abelian group of transformations $(U(1))$, is commonly known as the Abelian chiral symmetry of the theory. The invariances of the theory, of course, lead to conserved currents. Thus, for example, for the Abelian chiral symmetry, we have the conservation law

$$\partial_\mu J^\mu_5 = 0, \tag{13.5}$$

where the chiral (axial) current density is given by

$$J_5^\mu(x) = \bar{\psi}^i(x)\gamma_5\gamma^\mu\psi^i(x).$$ (13.6)

The chiral (axial) Ward identities are nothing but manifestations of the fact that the chiral current is conserved.

We note here that, if the quarks are massive, i.e., if there is a mass term for the fermions in the Lagrangian density of the form

$$\mathcal{L}_m = -m\bar{\psi}\psi = -m\bar{\psi}^i\psi^i,$$ (13.7)

then, the global chiral transformations in Eq. (13.4) will no longer be a symmetry of the theory and as a result, the chiral current in (13.6) would not be conserved. Rather, it will satisfy the relation

$$\partial_\mu J_5^\mu = -2im\bar{\psi}\gamma_5\psi.$$ (13.8)

For massive fermions, relation (13.8) would also lead to various relations between matrix elements known as the chiral (axial) Ward identities for the broken symmetry. However, when we calculate various matrix elements in a gauge theory (in either broken or unbroken case) regularized in a gauge invariant way, we find that the chiral Ward identities (Eq. (13.5) or Eq. (13.8)) are violated at the loop level. More specifically, the modifications to the tree level identities arise only at the one loop level.

There are various ways of understanding this violation. Let us begin with an analysis in the path integral formalism where we can write the generating functional as ($\hbar = 1$)

$$Z = \int \mathcal{D}A_\mu \mathcal{D}\bar{\psi}\mathcal{D}\psi \; e^{iS},$$ (13.9)

and here we are keeping the mass term in the action just for generality so that

$$S = \int \mathrm{d}^4x \left(-\frac{1}{4}F_{\mu\nu}^a F^{\mu\nu a} + i\bar{\psi}^i\gamma^\mu(D_\mu\psi)^i - m\bar{\psi}^i\psi^i \right).$$ (13.10)

Let us now make an infinitesimal local chiral field redefinition of the form (see Eq. (13.4))

$$\delta\psi^i(x) = -i\lambda(x)\gamma_5\psi^i(x),$$
$$\delta\bar{\psi}^i(x) = -i\lambda(x)\bar{\psi}^i(x)\gamma_5. \tag{13.11}$$

The generating functional should be invariant under such a field redefinition since we are integrating over all field configurations. Furthermore, for this analysis (namely, for evaluating the fermion path integral), we can treat the A_μ field as an external field and look at only the fermion part of the action. Namely, let us consider the fermion part of the path integral

$$Z_f[A] = \int \mathcal{D}\bar{\psi}\mathcal{D}\psi \; e^{iS_f}$$
$$= \int \mathcal{D}\bar{\psi}\mathcal{D}\psi \; e^{i\int \mathrm{d}^4x\left(i\bar{\psi}^i\gamma^\mu(D_\mu\psi)^i - m\bar{\psi}^i\psi^i\right)}, \tag{13.12}$$

so that under the redefinition in Eq. (13.11)

$$\delta Z_f[A] = \int \mathcal{D}\bar{\psi}\mathcal{D}\psi \left[\int \mathrm{d}^4x\left(-i(\partial_\mu\lambda)\bar{\psi}^i\gamma_5\gamma^\mu\psi - 2\lambda m\bar{\psi}^i\gamma_5\psi^i\right)\right] e^{iS_f}$$
$$= \int \mathcal{D}\bar{\psi}\mathcal{D}\psi \left[i\int \mathrm{d}^4x\lambda\left(\partial_\mu(\bar{\psi}^i\gamma_5\gamma^\mu\psi^i) + 2im\bar{\psi}^i\gamma_5\psi^i\right)\right] e^{iS_f}$$
$$= 0. \tag{13.13}$$

Here we have assumed that the path integral measure does not change under the infinitesimal field redefinition in Eq. (13.11). Equation (13.13) implies that

$$\left\langle \int \mathrm{d}^4x\left(\partial_\mu(\bar{\psi}^i(x)\gamma_5\gamma^\mu\psi^i(x)) + 2im\bar{\psi}^i(x)\gamma_5\psi^i(x)\right)\lambda(x)\right\rangle = 0,$$

or, $\quad \langle\partial_\mu J_5^\mu(x) + 2im\bar{\psi}^i(x)\gamma_5\psi^i(x)\rangle = 0, \tag{13.14}$

since this has to hold for any arbitrary (infinitesimal) $\lambda(x)$ at all coordinates. This is the naive (tree level) Ward identity (see Eq. (13.8)) associated with broken chiral symmetry and the natural question that arises is why does this fail at the one loop level. To understand this, let us note that in this derivation our only assumption has been that the path integral measure is insensitive to the chiral redefinition of the fields. Therefore, we should reanalyze the behavior of the functional measure under such a field redefinition more carefully to see if that may be the source of violation of the naive Ward identity.

To understand the question of the measure, let us assume that we can solve for the eigenstates of the gauge covariant Dirac operator,

$$i\not{D}\phi_n(x) = i\gamma^\mu D_\mu \phi_n(x) = \lambda_n \phi_n(x) \,. \tag{13.15}$$

Here $\phi_n(x)$'s represent the eigenstates of the Dirac operator and λ_n's are the corresponding eigenvalues. We have assumed that these are discrete states just for simplicity. Furthermore, since these sates define an orthonormal complete basis (the Dirac operator is a Hermitian operator), we have

$$\int d^4x \; \phi_n^\dagger(x)\phi_m(x) = \delta_{nm} \,,$$

$$\sum_n \phi_n(x)\phi_n^\dagger(y) = \delta^4(x-y) \,. \tag{13.16}$$

We can expand the fermion fields in the basis of these eigenfunctions as

$$\psi(x) = \sum_n a_n \phi_n(x) \,,$$

$$\bar{\psi}(x) = \sum_n \phi_n^\dagger(x) b_n \,, \tag{13.17}$$

where we are suppressing the index i for simplicity. For example, Eq. (13.17) can be thought of as a transformation from the wave function basis to the occupation number basis or any other discrete quantum number basis. In terms of bra and ket notations we can write

$$\psi(x) = \langle x|\psi\rangle = \sum_n \langle x|n\rangle\langle n|\psi\rangle$$

$$= \sum_n \phi_n(x)a_n = \sum_n a_n\phi_n(x)\,, \tag{13.18}$$

and so on. In this basis, the integration measure can be written as

$$\mathcal{D}\bar{\psi}(x)\mathcal{D}\psi(x) = \prod_n \mathrm{d}b_n\mathrm{d}a_n. \tag{13.19}$$

Let us now make an infinitesimal chiral field redefinition (see Eq. (13.11))

$$\psi(x) \to \psi'(x) = \psi(x) - i\lambda(x)\gamma_5\psi(x)\,,$$
$$\bar{\psi}(x) \to \bar{\psi}'(x) = \bar{\psi}(x) - i\lambda(x)\bar{\psi}(x)\gamma_5\,. \tag{13.20}$$

We can also expand the new variables $\psi'(x)$, $\bar{\psi}'(x)$ in the same eigenbasis. For example, we can write

$$\psi'(x) = \sum_n a'_n\phi_n(x) = \sum_{n,m} c_{nm}a_m\phi_n(x)\,, \tag{13.21}$$

where the matrix c_{nm} is given by

$$c_{nm} = \delta_{nm} - i\int \mathrm{d}^4x\,\lambda(x)\phi_n^\dagger(x)\gamma_5\phi_m(x)\,. \tag{13.22}$$

This can be seen easily in the following way

$$\psi'(x) = \langle x|\psi'\rangle = \sum_n \langle x|n\rangle\langle n|\psi'\rangle$$

$$= \sum_n a'_n\phi_n(x) = \sum_n \langle x|n\rangle\langle n|c|\psi\rangle\,, \tag{13.23}$$

where c denotes the infinitesimal chiral transformation matrix. Thus

$$\psi'(x) = \sum_n a'_n \phi_n(x) = \sum_n \langle x|n \rangle \langle n|c|\psi \rangle$$

$$= \sum_{n,m} \langle x|n \rangle \langle n|c|m \rangle \langle m|\psi \rangle = \sum_{n,m} c_{nm} a_m \phi_n(x) , \qquad (13.24)$$

where

$$c_{nm} = \langle n|c|m \rangle = \int d^4x d^4x' \langle n|x \rangle \langle x|c|x' \rangle \langle x'|m \rangle$$

$$= \int d^4x d^4x' \, \phi_n^\dagger(x) c(x,x') \phi_m(x') . \qquad (13.25)$$

We note that the chiral field redefinition in Eq. (13.11) is a local transformation so that $c(x, x')$ is a local function, i.e.,

$$\psi'(x) = c(x)\psi(x) = (\mathbb{1} - i\lambda(x)\gamma_5) \, \psi(x) . \qquad (13.26)$$

If we use this fact in Eq. (13.25), then we obtain

$$c_{nm} = \int d^4x d^4x' \phi_n^\dagger(x) c(x) \delta(x - x') \phi_m(x')$$

$$= \int d^4x \, \phi_n^\dagger(x) c(x) \phi_m(x)$$

$$= \int d^4x \, \phi_n^\dagger(x) (\mathbb{1} - i\lambda(x)\gamma_5) \, \phi_m(x)$$

$$= \delta_{nm} - i \int d^4x \, \lambda(x) \phi_n^\dagger(x) \gamma_5 \phi_m(x) . \qquad (13.27)$$

The above analysis shows that under the chiral field redefinition in Eq. (13.11)

$$\prod_n da_n \to \prod_n da'_n = (\det c_{nm})^{-1} \prod_n da_n , \qquad (13.28)$$

and we recognize that the inverse of the determinant (rather than the determinant) appears in this case simply because a_n's are fermionic in character (see Chapter 5). Similarly, under the field redefinition in Eq. (13.11)

$$\prod_n db_n \to \prod_n db'_n = (\det c_{nm})^{-1} \prod_n db_n. \qquad (13.29)$$

Therefore, it is clear that under the chiral field redefinition the fermion integration measure changes and contributes a nontrivial amount (unless the determinant is unity) given by

$$\mathcal{D}\bar{\psi}\mathcal{D}\psi \to (\det c_{nm})^{-2} \mathcal{D}\bar{\psi}\mathcal{D}\psi, \qquad (13.30)$$

where

$$\det c_{nm} = \det \left(\delta_{nm} - i \int d^4x \; \lambda(x)\phi_n^\dagger(x)\gamma_5\phi_m(x) \right)$$

$$= \exp \left(\text{Tr} \; \ln \left(\delta_{nm} - i \int d^4x \; \lambda(x)\phi_n^\dagger(x)\gamma_5\phi_m(x) \right) \right)$$

$$= \exp \left(\text{Tr} \; \left(-i \int d^4x \; \lambda(x)\phi_n^\dagger(x)\gamma_5\phi_m(x) \right) \right)$$

$$= \exp \left(-i \sum_n \int d^4x \; \lambda(x)\phi_n^\dagger(x)\gamma_5\phi_n(x) \right), \qquad (13.31)$$

where we have used the fact that $\lambda(x)$ is an infinitesimal parameter.

Since n takes infinitely many values, the sum in the exponent is ill defined. Therefore, we regularize the expression in the exponent as

$$\sum_n \int d^4x \; \lambda(x)\phi_n^\dagger(x)\gamma_5\phi_n(x)$$

$$= \lim_{M^2 \to \infty} \sum_n \int d^4x \; \lambda(x)\phi_n^\dagger(x)\gamma_5\phi_n(x)e^{-\left(\frac{\lambda_n}{M}\right)^2}$$

$$= \lim_{M^2 \to \infty} \sum_n \int d^4x \; \lambda(x)\phi_n^\dagger(x)\gamma_5 e^{-\left(\frac{i\not{D}}{M}\right)^2}\phi_n(x)$$

$$= \lim_{M^2 \to \infty} \sum_n \int d^4x \frac{d^4k}{(2\pi)^4} \frac{d^4k'}{(2\pi)^4} \lambda(x) \tilde{\phi}_n^\dagger(k') e^{ik' \cdot x}$$

$$\times \gamma_5 e^{-\left(\frac{i\not{D}}{M}\right)^2} e^{-ik \cdot x} \tilde{\phi}_n(k). \tag{13.32}$$

If we use the completeness relation for the basis functions in momentum space, namely,

$$\sum_n \tilde{\phi}_n(k) \tilde{\phi}_n^\dagger(k') = (2\pi)^4 \delta^4(k - k'), \tag{13.33}$$

then, Eq. (13.32) takes the form

$$\sum_n \int d^4x \, \lambda(x) \phi_n^\dagger(x) \gamma_5 \phi_n(x)$$

$$= \lim_{M^2 \to \infty} \int d^4x \frac{d^4k}{(2\pi)^4} \frac{d^4k'}{(2\pi)^4} (2\pi)^4 \delta^4(k - k')$$

$$\times \text{Tr} \, \lambda(x) \, e^{ik' \cdot x} \gamma_5 e^{-\left(\frac{i\not{D}}{M}\right)^2} e^{-ik \cdot x}$$

$$= \lim_{M^2 \to \infty} \int d^4x \frac{d^4k}{(2\pi)^4} \lambda(x) \, \text{Tr} \, e^{ik \cdot x} \gamma_5 e^{-\left(\frac{i\not{D}}{M}\right)^2} e^{-ik \cdot x}. \tag{13.34}$$

To simplify this expression, we note that using the properties of the Dirac matrices, we can write

$$(i\not{D})^2 = -\gamma^\mu \gamma^\nu D_\mu D_\nu$$

$$= -\left(\frac{1}{2}[\gamma^\mu, \gamma^\nu]_+ + \frac{1}{2}[\gamma^\mu, \gamma^\nu]\right) D_\mu D_\nu$$

$$= \left(-\eta^{\mu\nu} - \frac{1}{2}[\gamma^\mu, \gamma^\nu]\right) D_\mu D_\nu$$

$$= -D_\mu D^\mu - \gamma^\mu \gamma^\nu \frac{1}{2}[D_\mu, D_\nu]$$

$$= -D_\mu D^\mu - \frac{i}{2}\gamma^\mu \gamma^\nu F_{\mu\nu}, \tag{13.35}$$

where we are using the matrix representation for the gauge fields, namely, $A_\mu = A_\mu^a T^a, F_{\mu\nu} = F_{\mu\nu}^a T^a$ with the generators T^a in the representation of the fermions. It is clear, therefore, that the regularized exponent in (13.34) can be written as

$$\sum_n \int \mathrm{d}^4 x\, \lambda(x)\phi_n^\dagger(x)\gamma_5\phi_n(x)$$

$$= \lim_{M^2 \to \infty} \int \mathrm{d}^4 x \frac{\mathrm{d}^4 k}{(2\pi)^4} \lambda(x)\, \mathrm{Tr}\, e^{ik\cdot x}\gamma_5 e^{-\frac{1}{M^2}\left(-D_\mu D^\mu - \frac{i}{2}\gamma^\mu\gamma^\nu F_{\mu\nu}\right)}e^{-ik\cdot x}$$

$$= \lim_{M^2 \to \infty} \int \mathrm{d}^4 x \frac{\mathrm{d}^4 k}{(2\pi)^4} \lambda(x)\, \mathrm{Tr}\, \gamma_5 \frac{1}{2!}\left(\frac{i}{2M^2}\gamma^\mu\gamma^\nu F_{\mu\nu}\right)$$

$$\times \left(\frac{i}{2M^2}\gamma^\lambda\gamma^\rho F_{\lambda\rho}\right)e^{-\frac{k_\mu k^\mu}{M^2}}$$

$$= \lim_{M^2 \to \infty} \int \mathrm{d}^4 x \frac{\mathrm{d}^4 k}{(2\pi)^4} \lambda(x)\left(-\frac{1}{8M^4}\right)\, \mathrm{Tr}\, \left(\gamma_5\gamma^\mu\gamma^\nu\gamma^\lambda\gamma^\rho\right)$$

$$\times\, \mathrm{Tr}(F_{\mu\nu}F_{\lambda\rho})e^{-\frac{k^2}{M^2}}$$

$$= \lim_{M^2 \to \infty} \int \mathrm{d}^4 x\left(-\frac{1}{8M^4}\right)\lambda(x)\, 4i\epsilon^{\mu\nu\lambda\rho}\mathrm{Tr}(F_{\mu\nu}F_{\lambda\rho})\int \frac{\mathrm{d}^4 k}{(2\pi)^4}e^{-\frac{k^2}{M^2}}$$

$$= \lim_{M^2 \to \infty} \int \mathrm{d}^4 x\left(-\frac{i}{2M^4}\right)\lambda(x)\epsilon^{\mu\nu\lambda\rho}\, \mathrm{Tr}\, (F_{\mu\nu}F_{\lambda\rho})\frac{i\left(\sqrt{\pi M^2}\right)^4}{(2\pi)^4}$$

$$= \lim_{M^2 \to \infty} \int \mathrm{d}^4 x\left(\frac{1}{2M^4}\right)\lambda(x)\epsilon^{\mu\nu\lambda\rho}\, \mathrm{Tr}(F_{\mu\nu}(x)F_{\lambda\rho}(x))\frac{M^4}{16\pi^2}$$

$$= \frac{1}{32\pi^2}\int \mathrm{d}^4 x\, \lambda(x)\, \epsilon^{\mu\nu\lambda\rho}\, \mathrm{Tr}\, (F_{\mu\nu}(x)F_{\lambda\rho}(x)). \tag{13.36}$$

We note here that in the intermediate steps we have used the fact that the "Trace" is defined over the Dirac indices as well as the gauge indices and that in our discussions we have set the coupling constant for the gauge theory to unity. (We point out that the factor of i in evaluating the Gaussian integral arises from rotating to Euclidean space in order to do the integral carefully.) Therefore, we have determined that (with a gauge invariant regularization) the determinant is nontrivial and is given by

$$\det c_{nm} = \exp\left[-i\sum_n \int d^4x\ \lambda(x)\phi_n^\dagger(x)\gamma_5\phi_n(x)\right]$$

$$= \exp\left[-\frac{i}{32\pi^2}\int d^4x\ \lambda(x)\epsilon^{\mu\nu\lambda\rho}\ \mathrm{Tr}\left(F_{\mu\nu}(x)F_{\lambda\rho}(x)\right)\right]. \qquad (13.37)$$

In other words, we have determined that the functional integration measure changes nontrivially under the chiral redefinition of the fields and the change in the fermionic measure under an infinitesimal chiral transformation is given by (see also Eq. (13.30))

$$\delta\left(\mathcal{D}\bar\psi\mathcal{D}\psi\right) = \mathcal{D}\bar\psi'\mathcal{D}\psi' - \mathcal{D}\bar\psi\mathcal{D}\psi = \left((\det c_{nm})^{-2} - 1\right)\mathcal{D}\bar\psi\mathcal{D}\psi$$

$$= \mathcal{D}\bar\psi\mathcal{D}\psi\left[\frac{i}{16\pi^2}\int d^4x\ \lambda(x)\epsilon^{\mu\nu\lambda\rho}\ \mathrm{Tr}\left(F_{\mu\nu}(x)F_{\lambda\rho}(x)\right)\right]. \qquad (13.38)$$

As a result, we note that the change of the generating functional under an infinitesimal chiral field redefinition (see also Eq. (13.14)) is correctly given by

$$\delta Z_f[A] = 0 = \int \mathcal{D}\bar\psi\mathcal{D}\psi\left[i\int d^4x\ \lambda(x)\left\{\frac{1}{16\pi^2}\epsilon^{\mu\nu\lambda\rho}\ \mathrm{Tr}\left(F_{\mu\nu}F_{\lambda\rho}\right)\right.\right.$$

$$\left.\left. + \partial_\mu J_5^\mu + 2im\bar\psi\gamma_5\psi\right\}\right]e^{iS}, \qquad (13.39)$$

which leads to the relation

$$\langle\partial_\mu J_5^\mu(x) + 2im\bar\psi\gamma_5\psi\rangle + \frac{1}{16\pi^2}\epsilon^{\mu\nu\lambda\rho}\ \mathrm{Tr}\left(F_{\mu\nu}(x)F_{\lambda\rho}(x)\right) = 0. \qquad (13.40)$$

This corresponds to the operator identity

$$\partial_\mu J_5^\mu(x) = -2im\bar\psi\gamma_5\psi - \frac{1}{16\pi^2}\epsilon^{\mu\nu\lambda\rho}\ \mathrm{Tr}\left(F_{\mu\nu}(x)F_{\lambda\rho}(x)\right)$$

$$= -2im\bar\psi\gamma_5\psi - \frac{1}{16\pi^2}\ \mathrm{Tr}\left(F_{\mu\nu}(x)\tilde F^{\mu\nu}(x)\right), \qquad (13.41)$$

where the dual field strength tensor is defined as

$$\tilde{F}_{\mu\nu}(x) = \epsilon_{\mu\nu\lambda\rho} F^{\lambda\rho}(x) , \qquad\qquad (13.42)$$

and the last term in Eq. (13.41) on the right hand side is known as the anomaly associated with the chiral (axial) current conservation (in four space-time dimensions). Let us note here that sometimes in literature the right hand side of Eq. (13.41) may appear with the opposite sign and that is because in those works the chiral current is defined with an opposite sign. It is also worth pointing out here that this derivation clarifies two things. First, the regularization we used to define the infinite sum is gauge invariant (since the Dirac operator is gauge invariant) which does not respect chiral symmetry. The second point to note is that a determinant is normally associated with a one loop effect which clarifies why the anomaly is a one loop effect.

Anomalies have observable effects. For example, the correct pion life time can be obtained in a field theoretic calculation only if the chiral anomaly is taken into account. Furthermore, if we have a theory where there are both vectorial and axial vectorial couplings of the fermions to gauge fields (corresponding to local symmetries), the model cannot be renormalized unless the anomaly is cancelled. In grand unified theories where we treat both quarks and leptons as massless, they have precisely these kinds of couplings. Therefore, one cannot make a consistent grand unified theory unless one can cancel out the chiral anomaly (associated with the local chiral invariances) and this leads to the study of groups and representations which are anomaly free. In string theories, that are currently fashionable, similar anomaly free considerations (associated with gauge and gravitational anomalies) fix the gauge group uniquely to be $SO(32)$ or $E(8) \times E(8)$.

13.2 Schwinger model

The Schwinger model describes massless quantum electrodynamics (QED) in $1 + 1$ dimension. This is a model that can be exactly solved and the structure of the resulting effective gauge theory is directly related to the anomaly in the theory. Let us see how this arises. The Lagrangian density describing the interaction of massless charged fermions with an Abelian gauge field is given by

$$\mathcal{L} = -\frac{1}{4}F_{\mu\nu}F^{\mu\nu} + i\bar{\psi}\gamma^\mu\left(\partial_\mu + ieA_\mu\right)\psi, \quad \mu = 0, 1, \tag{13.43}$$

where e denotes the electric charge of the fermion. This theory is invariant under local $U(1)$ transformations as well as global chiral $U(1)$ transformations (since fermions are massless). We would examine the model in two different approaches. But first let us note some of the special features of two dimensional space time.

Note that in two dimensions there cannot be any transverse polarization and consequently the photon does not have any true (transverse) degrees of freedom. There is only an electric field associated with the photon since the only nontrivial component of the field strength tensor is F_{01}. Furthermore, the electromagnetic coupling in two dimensional QED is dimensional unlike in four dimensional QED where it is a dimensionless constant. This can be seen as follows. In natural units ($\hbar = c = 1$), the action is dimensionless

$$[S] = \left[\int \mathrm{d}^2x\, \mathcal{L}\right] = 0, \tag{13.44}$$

which implies that the Lagrangian density has the canonical dimension

$$[\mathcal{L}] = 2. \tag{13.45}$$

From the free part of $\mathcal{L}$ we determine the canonical dimension of the fields to be

$$[A_\mu] = 0, \quad [\psi] = [\bar{\psi}] = \frac{1}{2}. \tag{13.46}$$

With these, the interaction Lagrangian density leads to

$$[e] = 1. \tag{13.47}$$

Furthermore, by power counting (as well as gauge invariance) it is easily seen that this theory has only a mild divergence (at loop orders) so that this is a fairly well behaved theory.

Note also that in two dimensions if we choose our Dirac matrices to satisfy

$$(\gamma^0)^\dagger = \gamma^0, \quad (\gamma^0)^2 = \mathbb{1}, \tag{13.48}$$

$$(\gamma^1)^\dagger = -\gamma^1, \quad (\gamma^1)^2 = -\mathbb{1}, \tag{13.49}$$

and

$$\gamma_5 = \gamma^0 \gamma^1, \tag{13.50}$$

so that

$$\gamma_5^\dagger = \gamma^{1\dagger} \gamma^{0\dagger} = -\gamma^1 \gamma^0 = \gamma^0 \gamma^1 = \gamma_5 \,,$$

$$\gamma_5^2 = \gamma^0 \gamma^1 \gamma^0 \gamma^1 = -\left(\gamma^0\right)^2 \left(\gamma^1\right)^2 = \mathbb{1} \,, \tag{13.51}$$

then a possible representation for these matrices is given by

$$\gamma^0 = \begin{pmatrix} 0 & 1 \\ 1 & 0 \end{pmatrix} = \sigma_1,$$

$$\gamma^1 = \begin{pmatrix} 0 & -1 \\ 1 & 0 \end{pmatrix} = -i\sigma_2,$$

$$\gamma_5 = \begin{pmatrix} 1 & 0 \\ 0 & -1 \end{pmatrix} = \sigma_3, \tag{13.52}$$

where σ_i, $i = 1, 2, 3$ denote the three Pauli matrices. Furthermore, note that in addition to the symmetric second rank metric tensor which is diagonal and has the form

$$\eta^{\mu\nu} = (+, -) = \eta^{\nu\mu} \,, \tag{13.53}$$

in two dimensions, we also have a second rank anti-symmetric tensor given by

$$\epsilon^{\mu\nu} = -\epsilon^{\nu\mu} . \tag{13.54}$$

This is the Levi-Civita tensor in two dimensions which is completely anti-symmetric and is chosen as

$$\epsilon^{01} = 1 = -\epsilon^{10} . \tag{13.55}$$

Similarly, for the covariant Levi-Civita tensor, we have

$$\epsilon_{01} = -1 = -\epsilon_{10} . \tag{13.56}$$

Using the Levi-Civita tensor, we can write

$$\gamma_5 = -\frac{1}{2}\epsilon_{\mu\nu}\gamma^\mu\gamma^\nu = -\frac{1}{2}\epsilon^{\mu\nu}\gamma_\mu\gamma_\nu$$
$$= -\frac{1}{4}\epsilon^{\mu\nu}[\gamma_\mu,\gamma_\nu] = -\frac{1}{2}\epsilon^{\mu\nu}\sigma_{\mu\nu} , \tag{13.57}$$

where we have defined (in analogy with four dimensions)

$$\sigma_{\mu\nu} = \frac{1}{2}[\gamma_\mu,\gamma_\nu] . \tag{13.58}$$

This leads to

$$\epsilon^{\lambda\rho}\gamma_5 = -\frac{1}{2}\epsilon^{\lambda\rho}\epsilon^{\mu\nu}\sigma_{\mu\nu}$$
$$= -\frac{1}{2}\left(-\eta^{\lambda\mu}\eta^{\rho\nu} + \eta^{\lambda\nu}\eta^{\rho\mu}\right)\sigma_{\mu\nu},$$

or, $\quad \epsilon^{\lambda\rho}\gamma_5 = \sigma^{\lambda\rho},$

or, $\quad \sigma^{\mu\nu} = \epsilon^{\mu\nu}\gamma_5 . \tag{13.59}$

It is clear now that we can write

$$\gamma^\mu\gamma^\nu = \frac{1}{2}\{\gamma^\mu,\gamma^\nu\} + \frac{1}{2}[\gamma^\mu,\gamma^\nu] = \eta^{\mu\nu} + \sigma^{\mu\nu} = \eta^{\mu\nu} + \epsilon^{\mu\nu}\gamma_5 . \tag{13.60}$$

Furthermore, using Eqs. (13.58) and (13.60) we have ($\gamma^\nu \gamma^\mu \gamma_\nu = 0$ in two dimensions)

$$\sigma^{\mu\nu}\gamma_\nu = \epsilon^{\mu\nu}\gamma_5\gamma_\nu,$$

$$\text{or,}\quad \gamma^\mu = \epsilon^{\mu\nu}\gamma_5\gamma_\nu. \tag{13.61}$$

Conversely, we can write

$$\gamma_5\gamma^\mu = \epsilon^{\mu\nu}\gamma_\nu. \tag{13.62}$$

As a result of these identities, we note that in two dimensions, the vector and the axial vector current densities defined as

$$J^\mu = e\bar{\psi}\gamma^\mu\psi, \quad J_5^\mu = e\bar{\psi}\gamma_5\gamma^\mu\psi, \tag{13.63}$$

are related by the duality relations

$$J^\mu = \epsilon^{\mu\nu} J_{5\nu},$$

$$J_5^\mu = \epsilon^{\mu\nu} J_\nu. \tag{13.64}$$

The propagator for the (massless) free fermion ($A_\mu = 0$) can be related to the free propagator for the massless scalar field as

$$iS_F^{(0)}(x - x') = i\slashed{\partial}(iG_F^{(0)}(x - x')), \tag{13.65}$$

where

$$iS_F^{(0)}(x - x') = \langle T\left(\psi(x)\bar{\psi}(x')\right)\rangle,$$

$$iG_F^{(0)}(x - x') = \langle T\left(\phi(x)\phi(x')\right)\rangle, \tag{13.66}$$

and the free scalar propagator satisfies the equation

$$\Box(iG_F^{(0)}(x - x')) = -i\delta^2(x - x'). \tag{13.67}$$

To determine the free Green's functions let us note that the time ordered scalar Green's function can be written in terms of positive and negative frequency functions as

$$iG_F^{(0)}(x) = \theta(x^0)(-iG^{(+)}(x^0, x)) + \theta(-x^0)(iG^{(-)}(x^0, x)), \quad (13.68)$$

where

$$\mp iG^{(\pm)}(x^0, x) = \frac{1}{2(2\pi)} \int_{-\infty}^{\infty} \frac{dk}{|k|} e^{\mp i|k|x^0 + ikx}$$

$$= \frac{1}{2(2\pi)} \int_0^{\infty} \frac{dk}{k} e^{\mp ikx^0} \left(e^{ikx} + e^{-ikx} \right)$$

$$= \frac{1}{2\pi} \int_0^{\infty} \frac{dk}{k} e^{\mp ikx^0} \cos kx. \quad (13.69)$$

Therefore, from Eqs. (13.68) and (13.69) we have

$$iG_F^{(0)}(x^0, x) = \frac{1}{2\pi} \int_0^{\infty} \frac{dk}{k} e^{-ikx^0} \cos kx, \quad x^0 \geq 0, \quad (13.70)$$

and, for $x^0 \geq 0$, we have from Eq. (13.65)

$$iS_F^{(0)}(x^0, x) = i\gamma^\mu \partial_\mu \left(\frac{1}{2\pi} \int_0^{\infty} \frac{dk}{k} e^{-ikx^0} \cos kx \right)$$

$$= \frac{i}{2\pi} \int_0^{\infty} \frac{dk}{k} \left(-i\gamma^0 k \cos kx - \gamma^1 k \sin kx \right) e^{-ikx^0}$$

$$= -\frac{i\gamma^1}{2\pi} \int_0^{\infty} dk \left(-i\gamma^1 \gamma^0 \cos kx + \sin kx \right) e^{-ikx^0}$$

$$= -\frac{i\gamma^1}{2\pi} \int_0^{\infty} dk \left(i\gamma_5 \cos kx + \sin kx \right) e^{-ikx^0}$$

$$= \frac{\gamma^1 \gamma_5}{2\pi} \int\limits_0^\infty dk \, (\cos kx - i\gamma_5 \sin kx) \, e^{-ikx^0}$$

$$= \frac{\gamma^1 \gamma_5}{2\pi} \int\limits_0^\infty dk \, e^{-ikx^0 - i\gamma_5 kx} . \tag{13.71}$$

For $x^0 = 0$, this leads to

$$iS_F^{(0)}(0, x) \equiv iS_F^{(0)}(x) = \frac{\gamma^1 \gamma_5}{2\pi} \int\limits_0^\infty dk \, e^{-i\gamma_5 kx} = \frac{\gamma^1 \gamma_5}{2\pi} \frac{1}{i\gamma_5 x}$$

$$= \frac{\gamma^1}{2\pi i x} . \tag{13.72}$$

Let us now calculate the fermion propagator in the presence of an external electromagnetic field. Denoting by

$$iS_F^{(A)}(x, x') = \langle T \left(\psi(x)\bar{\psi}(x') \right) \rangle^A , \tag{13.73}$$

we note that this satisfies the differential equation

$$i\gamma^\mu \left(\partial_\mu + ieA_\mu \right) S_F^{(A)}(x, x') = \delta^2(x - x'). \tag{13.74}$$

To solve for the complete fermion Green's function (propagator) we note that the most general solution of Eq. (13.73) can be written as

$$S_F^{(A)}(x - x') = e^{-i(\phi(x) - \phi(x'))} S_F^{(0)}(x - x') , \tag{13.75}$$

where $S_F^{(0)}(x - x')$ is the free Green's function (already determined in Eq. (13.72) for equal times) satisfying

$$i\partial\!\!\!/ S_F^{(0)}(x - x') = \delta^2(x - x') , \tag{13.76}$$

and $\phi(x)$ is a 2×2 matrix function (in the Dirac space and not the scalar field) satisfying

$$i\gamma^\mu \left(-\partial_\mu(i\phi(x)) + ieA_\mu\right) = 0,$$

$$\text{or,} \quad \gamma^\mu \partial_\mu \phi(x) - e\gamma^\mu A_\mu = 0,$$

$$\text{or,} \quad \not{\partial}\phi(x) = e\gamma^\mu A_\mu,$$

$$\text{or,} \quad \Box\phi(x) = e\gamma^\mu\gamma^\nu \partial_\mu A_\nu = e\left(\eta^{\mu\nu} + \epsilon^{\mu\nu}\gamma_5\right)\partial_\mu A_\nu$$

$$= e\partial_\mu A^\mu + \frac{e}{2}\gamma_5\epsilon^{\mu\nu}F_{\mu\nu}, \tag{13.77}$$

so that

$$\Box\left(\text{Tr } \phi(x)\right) = 2e\partial_\mu A^\mu, \tag{13.78}$$

and

$$\Box\left(\text{Tr } \gamma_5\phi(x)\right) = e\epsilon^{\mu\nu}F_{\mu\nu}. \tag{13.79}$$

Let us note here that the whole point of this exercise is to show that in two dimensions one can always find a matrix function $\phi(x)$ such that (see Eq. (13.77))

$$\gamma^\mu A_\mu = \gamma^\mu \partial_\mu \phi(x). \tag{13.80}$$

That this is true can also be seen from the fact that in two dimensions we can always decompose a vector as

$$A_\mu = \partial_\mu \sigma + \epsilon_{\mu\nu}\partial^\nu \eta, \tag{13.81}$$

so that

$$\gamma^\mu A_\mu = \gamma^\mu \partial_\mu \sigma + \gamma^\mu \epsilon_{\mu\nu}\partial^\nu \eta$$

$$= \gamma^\mu \partial_\mu \sigma + \gamma^\mu \gamma_5 \partial_\mu \eta$$

$$= \gamma^\mu \partial_\mu \left(\sigma + \gamma_5\eta\right) = \gamma^\mu \partial_\mu \phi, \tag{13.82}$$

with

$$\phi(x) = \sigma(x) + \gamma_5 \eta(x) \,, \tag{13.83}$$

where we have used Eq. (13.62) in the intermediate step. Thus, using Eqs. (13.72) and (13.75) we determine that for equal times, $x^0 = x'^0$

$$iS_F^{(A)}(x - x') = e^{-i(\phi(x)-\phi(x'))} \frac{\gamma^1}{2\pi i(x - x')}. \tag{13.84}$$

The expectation value of the current is defined in a limiting gauge invariant manner as

$$\langle J^\mu(x) \rangle^A = \lim_{x \to x', \, x^0 = x'^0} \langle e\bar{\psi}(x')\gamma^\mu\psi(x)\rangle^A e^{ie \int_{x'}^{x} dx^\lambda A_\lambda}. \tag{13.85}$$

This leads explicitly to (since fermions anti-commute)

$$\langle J^\mu(x) \rangle^A = \lim_{x \to x', \, x^0 = x'^0} -e\text{Tr} \; \gamma^\mu \langle \psi(x)\bar{\psi}(x')\rangle^A e^{ie \int_{x'}^{x} dx^\lambda A_\lambda}$$

$$= \lim_{x \to x', \, x^0 = x'^0} -e\text{Tr} \; \gamma^\mu iS_F^{(A)}(x, x') e^{ie \int_{x'}^{x} dx^\lambda A_\lambda}$$

$$= \lim_{x \to x', \, x^0 = x'^0} -e\text{Tr} \left(\gamma^\mu \left[1 - i(x - x')\frac{\partial}{\partial x}\phi(x) \right] \frac{\gamma^1}{2\pi i(x - x')} \right.$$

$$\left. \times \; \left[1 + ie(x - x')A_1(x) \right] \right)$$

$$= \lim_{x \to x'} -\text{Tr} \left[\frac{e\gamma^\mu\gamma^1}{2\pi i(x - x')} - \frac{e}{2\pi}\gamma^\mu\partial_1\phi\gamma^1 + \frac{e^2}{2\pi}\gamma^\mu\gamma^1 A_1 \right]. \tag{13.86}$$

Neglecting the uninteresting infinite constant (coming from the first term), we determine

$$\langle J^\mu(x) \rangle^A = \frac{e}{2\pi} \; \text{Tr} \; \gamma^\mu \left(\partial_1\phi\gamma^1 - e\gamma^1 A_1 \right) \,. \tag{13.87}$$

Furthermore, we note from Eq. (13.83) that we can write

$$\phi(x) = \frac{1}{2} \, \text{Tr} \, (\phi(x)) + \frac{1}{2}\gamma_5 \, \text{Tr} \, (\gamma_5\phi(x)) \,, \tag{13.88}$$

so that Eq. (13.87) becomes

$$\langle J^\mu(x) \rangle^A = \frac{e}{2\pi} \, \text{Tr} \left[\left(\frac{1}{2}\gamma^\mu\gamma^1\partial_1 \text{Tr} \, \phi(x) - \frac{1}{2}\gamma^\mu\gamma^1\gamma_5\partial_1 \text{Tr} \, \gamma_5\phi(x) \right) \right.$$

$$\left. - e\gamma^\mu\gamma^1 A_1 \right]$$

$$= \frac{e}{2\pi} \left[\eta^{\mu 1}(\partial_1 \, (\text{Tr} \, \phi(x)) - 2A_1) - \epsilon^{\mu 1}\partial_1 \, \text{Tr} \, (\gamma_5\phi(x)) \right]. \tag{13.89}$$

On the other hand, since

$$\partial\!\!\!/\phi(x) = e\!\!\!/A,$$

$$\text{or,} \quad \partial\!\!\!/ \left[\frac{1}{2} \, \text{Tr} \, \phi(x) + \frac{1}{2}\gamma_5 \, \text{Tr} \, \gamma_5\phi(x) \right] = e\!\!\!/A, \tag{13.90}$$

we obtain

$$\text{Tr} \, \gamma^0\partial\!\!\!/ \left[\frac{1}{2} \, \text{Tr} \, \phi(x) + \frac{1}{2}\gamma_5 \, \text{Tr} \, \gamma_5\phi(x) \right] = \text{Tr} \, e\gamma^0 A\!\!\!/,$$

$$\text{or,} \quad \partial^0 \, \text{Tr} \, \phi(x) + \partial_1 \, \text{Tr} \, (\gamma_5\phi(x)) = 2eA^0, \tag{13.91}$$

and

$$\text{Tr} \, \gamma^1\partial\!\!\!/ \left[\frac{1}{2} \, \text{Tr} \, \phi(x) + \frac{1}{2}\gamma_5 \, \text{Tr} \, \gamma_5\phi(x) \right] = \text{Tr} \, e\gamma^1 A\!\!\!/,$$

$$\text{or,} \quad \partial^1 \, \text{Tr} \, \phi(x) - \partial_0 \, \text{Tr} \, (\gamma_5\phi(x)) = 2eA^1. \tag{13.92}$$

Using these relations as well as the definition in (13.89) we can write

$$\langle J^0(x)\rangle^A = -\frac{e}{2\pi}\partial_1 \text{ Tr } \gamma_5\phi(x)$$

$$= \frac{e}{2\pi}\left(\partial^0 \text{ Tr } \phi(x) - 2eA^0\right), \tag{13.93}$$

and

$$\langle J^1(x)\rangle^A = -\frac{e}{2\pi}\left(\partial_1 \text{ Tr } \phi(x) - 2eA_1\right)$$

$$= \frac{e}{2\pi}\left(\partial^1 \text{ Tr } \phi(x) - 2eA^1\right). \tag{13.94}$$

We can write the relations in Eqs. (13.93) and (13.94) in a covariant manner as

$$\langle J^\mu(x)\rangle^A = \frac{e}{2\pi}\left(\partial^\mu \text{ Tr } \phi(x) - 2eA^\mu\right). \tag{13.95}$$

Using the solution Eq. (13.78), namely,

$$\text{Tr } \phi(x) = 2e\,\square^{-1}\partial_\mu A^\mu, \tag{13.96}$$

in Eq. (13.95) we obtain

$$\langle J^\mu(x)\rangle^A = \frac{e}{2\pi}\left(2e\,\partial^\mu\square^{-1}\partial^\nu A_\nu - 2e\,A^\mu\right)$$

$$= -\frac{e^2}{\pi}\left(\eta^{\mu\nu} - \partial^\mu\square^{-1}\partial^\nu\right)A_\nu. \tag{13.97}$$

This is a gauge invariant current which is conserved as is readily seen from its transverse structure. However, it is also clear from this relation that

$$\partial_\mu\langle J_5^\mu(x)\rangle^A = \partial_\mu\epsilon^{\mu\nu}\langle J_\nu(x)\rangle^A = -\frac{e^2}{2\pi}\epsilon^{\mu\nu}F_{\mu\nu} \neq 0. \tag{13.98}$$

Namely, the chiral current is no longer conserved.

Let us next study the generating functional for the Schwinger model which has the form (see Eq. (13.9))

$$Z = \int \mathcal{D}A_\mu \mathcal{D}\bar{\psi}\mathcal{D}\psi \; e^{iS}, \tag{13.99}$$

where (see Eq. (13.43))

$$S = \int \mathrm{d}^2x \left(-\frac{1}{4}F_{\mu\nu}F^{\mu\nu} + i\bar{\psi}\gamma^\mu(\partial_\mu + ieA_\mu)\psi \right). \tag{13.100}$$

The integration over the fermion fields can be easily done (the action is quadratic in the variables) and the result is

$$Z = \int \mathcal{D}A_\mu \det\left(i\partial\!\!\!/ - eA\!\!\!/\right) e^{i\int \mathrm{d}^2x(-\frac{1}{4}F_{\mu\nu}F^{\mu\nu})}$$

$$= \int \mathcal{D}A_\mu \; e^{\mathrm{Tr}\,\ln(i\partial\!\!\!/ - eA\!\!\!/)} e^{i\int \mathrm{d}^2x(-\frac{1}{4}F_{\mu\nu}F^{\mu\nu})}$$

$$= \int \mathcal{D}A_\mu \; e^{iS_{\mathrm{eff}}}, \tag{13.101}$$

where

$$S_{\mathrm{eff}} = \int \mathrm{d}^2x \left(-\frac{1}{4}F_{\mu\nu}F^{\mu\nu} \right) - i\mathrm{Tr}\,\ln\left(i\partial\!\!\!/ - eA\!\!\!/\right)(x), \tag{13.102}$$

and "Tr" denotes trace over the Dirac indices as well as over the coordinate basis.

To understand the meaning of this effective action better, let us look at the Euler-Lagrange equation following from this (here "Tr" denotes trace over only the Dirac indices)

$$\partial_\mu \frac{\partial S_{\mathrm{eff}}}{\partial \partial_\mu A_\nu(x)} - \frac{\partial S_{\mathrm{eff}}}{\partial A_\nu(x)} = 0,$$

or, $\quad -\partial_\mu F^{\mu\nu}(x) - i\left(\mathrm{Tr}\,(i\partial\!\!\!/ - eA\!\!\!/)^{-1}(x,x)e\gamma^\nu\right) = 0,$

or, $\quad \partial_\mu F^{\mu\nu}(x) + e\,\mathrm{Tr}\,iS_F^{(A)}(x,x)\gamma^\nu = 0,$

or, $\quad \partial_\mu F^{\mu\nu}(x) - \langle J^\nu(x)\rangle^A = 0,$

or, $\quad \partial_\mu F^{\mu\nu}(x) + \frac{e^2}{\pi}\left(\eta^{\nu\lambda} - \partial^\nu \Box^{-1}\partial^\lambda\right)A_\lambda(x) = 0. \tag{13.103}$

Here we have used Eq. (13.97) in the last step. The effective theory now describes a massive photon with the mass of the photon given by

$$m_{\text{ph}}^2 = \frac{e^2}{\pi}. \tag{13.104}$$

This can be better seen by choosing a covariant gauge such as the Landau gauge

$$\partial_\mu A^\mu(x) = 0, \tag{13.105}$$

in which case the Euler-Lagrange equation (13.103) becomes

$$\partial_\mu F^{\mu\nu}(x) + m_{\text{ph}}^2 A^\nu(x) = 0,$$

$$\text{or,} \quad \Box A^\nu(x) + m_{\text{ph}}^2 A^\nu(x) = 0,$$

$$\text{or,} \quad \left(\Box + m_{\text{ph}}^2\right) A^\nu(x) = 0, \tag{13.106}$$

which is the equation for a massive gauge field. In this case the effective action in Eq. (13.102) for the photon field can be written as (not using the Landau gauge)

$$S_{\text{eff}} = \int d^2 x \left(-\frac{1}{4} F_{\mu\nu} F^{\mu\nu} + \frac{m_{\text{ph}}^2}{2} A_\mu \left(\eta^{\mu\nu} - \partial^\mu \Box^{-1} \partial^\nu \right) A_\nu \right), \tag{13.107}$$

which again shows that the photon in the effective theory has become massive.

Another way of seeing the same result is to look at the generating functional for the fermions

$$Z_f[A] = \int \mathcal{D}\bar{\psi}\mathcal{D}\psi \, e^{iS[\psi,\bar{\psi},A]}, \tag{13.108}$$

where

$$S[\psi, \bar{\psi}, A] = \int d^2x \; i\bar{\psi} \left(\partial\!\!\!/ + ieA\!\!\!/ \right) \psi .$$ (13.109)

It follows now that

$$\frac{\delta Z_f[A]}{\delta A_\mu(x)} = -i \int \mathcal{D}\bar{\psi}\mathcal{D}\psi \; (e\bar{\psi}(x)\gamma^\mu \psi(x)) \; e^{iS}$$

$$= -i \int \mathcal{D}\bar{\psi}\mathcal{D}\psi \; J^\mu(x) \; e^{iS} = -i\langle J^\mu(x)\rangle^A Z[A]$$

$$= \frac{ie^2}{\pi} \left(\eta^{\mu\nu} - \partial^\mu \Box^{-1} \partial^\nu \right) A_\nu \, Z[A] ,$$ (13.110)

where we have used the identification in Eq. (13.97). Integrating this, we obtain

$$Z_f[A] = e^{\frac{ie^2}{2\pi} \int d^2x \; A_\mu \left(\eta^{\mu\nu} - \partial^\mu \Box^{-1} \partial^\nu \right) A_\nu}$$

$$= e^{\frac{im_{\rm ph}^2}{2} \int d^2x \; A_\mu \left(\eta^{\mu\nu} - \partial^\mu \Box^{-1} \partial^\nu \right) A_\nu} .$$ (13.111)

Therefore, we can write the complete generating functional as

$$Z = \int \mathcal{D}A_\mu \mathcal{D}\bar{\psi}\mathcal{D}\psi \; e^{i \int d^2x \left(-\frac{1}{4} F_{\mu\nu} F^{\mu\nu} \right) + iS[\psi,\bar{\psi},A]}$$

$$= \int \mathcal{D}A_\mu \; e^{i \int d^2x \left(-\frac{1}{4} F_{\mu\nu} F^{\mu\nu} + \frac{m_{\rm ph}^2}{2} A_\mu \left(\eta^{\mu\nu} - \partial^\mu \Box^{-1} \partial^\nu \right) A_\nu \right)}$$

$$= \int \mathcal{D}A_\mu \; e^{iS_{\rm eff}} .$$ (13.112)

The phenomenon of the gauge field becoming massive through spontaneous breaking of a local symmetry is quite well understood through the Higgs mechanism (which involves a fundamental scalar field to break the symmetry). However, the novel feature here is that there is no fundamental scalar field in this theory. In this sense, it is called a dynamical symmetry breaking and the symmetry that is broken in this case is the chiral symmetry.

Note that the nonlocal effective Lagrangian density in Eq. (13.112) (or in Eq. (13.107))

$$\mathcal{L}_{\text{eff}} = -\frac{1}{4} F_{\mu\nu} F^{\mu\nu} + \frac{m_{\text{ph}}^2}{2} A_\mu \left(\eta^{\mu\nu} - \partial^\mu \Box^{-1} \partial^\nu \right) A_\nu \,, \tag{13.113}$$

can be written in a local manner by introducing a scalar field $B(x)$ as

$$\mathcal{L}_{\text{eff}} = -\frac{1}{4} F_{\mu\nu} F^{\mu\nu} + \frac{1}{2} \partial_\mu B(x) \partial^\mu B(x) - m_{\text{ph}} \epsilon^{\mu\nu} A_\mu \partial_\nu B(x) \,. \tag{13.114}$$

Integrating out the scalar field $B(x)$ (on which the action depends quadratically) we obtain the earlier form of the effective action. On the other hand, once we have this local form of the effective action, we can integrate out the A_μ field completely since the Lagrangian density depends on the gauge field at most quadratically. The resulting Lagrangian density in this case is given by

$$\mathcal{L}_{\text{eff}} = \frac{1}{2} \partial_\mu B(x) \partial^\mu B(x) - \frac{m_{\text{ph}}^2}{2} B(x) B(x) \,. \tag{13.115}$$

Namely, the complete Schwinger model including all quantum corrections is equivalent to a free, massive scalar field. This is often referred to as the bosonization of the model. The fermions have completely disappeared from the theory.

Let us now analyze the Schwinger model from a different point of view. The generating functional, as we have seen, is given by (the generating functional should be rigorously evaluated in the Euclidean space, but we will discuss it in the Minkowski space pointing out where care should be taken)

$$Z = \int \mathcal{D}A_\mu \mathcal{D}\bar{\psi} \mathcal{D}\psi \, e^{iS}, \tag{13.116}$$

where the action has the form

$$S = \int d^2x \left[-\frac{1}{4} F_{\mu\nu} F^{\mu\nu} + i\bar{\psi}\gamma^\mu \left(\partial_\mu + ieA_\mu \right) \psi \right]. \qquad (13.117)$$

Since the theory has a local gauge invariance let us choose a covariant gauge condition such as the Landau gauge condition

$$\partial_\mu A^\mu(x) = 0. \qquad (13.118)$$

Let us recall (see Eq. (13.81)) that in two dimensions we can write

$$A_\mu(x) = \partial_\mu \sigma(x) + \epsilon_{\mu\nu} \partial^\nu \eta(x). \qquad (13.119)$$

Therefore, the Landau gauge condition implies

$$\partial_\mu A^\mu(x) = 0 \quad \text{or,} \quad \sigma(x) = 0. \qquad (13.120)$$

In other words, in this gauge we can write $A_\mu(x) = \epsilon_{\mu\nu} \partial^\nu \eta(x)$ and the fermion Lagrangian density takes the form

$$\begin{aligned} \mathcal{L} &= i\bar{\psi}\gamma^\mu \left(\partial_\mu + ieA_\mu \right) \psi \\ &= i\bar{\psi}\gamma^\mu \left(\partial_\mu + ie\epsilon_{\mu\nu}(\partial^\nu \eta) \right) \psi \\ &= i\bar{\psi}\gamma^\mu \left(\partial_\mu + ie\gamma_5 (\partial_\mu \eta) \right) \psi, \qquad (13.121) \end{aligned}$$

where we have used Eq. (13.62).

Let us next note that if we make a local chiral field redefinition of the form

$$\begin{aligned} \psi(x) &\to \psi'(x) = e^{-ie\gamma_5 \alpha(x)} \psi(x), \\ \bar{\psi}(x) &\to \bar{\psi}'(x) = \bar{\psi}(x) e^{-ie\gamma_5 \alpha(x)}, \qquad (13.122) \end{aligned}$$

then the fermion Lagrangian density transforms as

$$\mathcal{L} \rightarrow i\bar{\psi}'(x)\gamma^\mu \left(\partial_\mu + ie\gamma_5(\partial_\mu \eta(x))\right)\psi'(x)$$

$$= i\bar{\psi}(x)e^{-ie\gamma_5\alpha(x)}\gamma^\mu \left(\partial_\mu + ie\gamma_5(\partial_\mu\eta)\right)e^{-ie\gamma_5\alpha(x)}\psi(x)$$

$$= i\bar{\psi}(x)\gamma^\mu \left(\partial_\mu - ie\gamma_5(\partial_\mu(\alpha(x) - \eta(x)))\right)\psi(x). \qquad (13.123)$$

Therefore, if we choose $\alpha(x) = \eta(x)$, the fermion field would decouple and the theory would appear to consist of only free photons and free massless fermions. This is quite different from what we have seen in the earlier analysis. The real story lies in the change in the functional measure under such a field redefinition.

In this two dimensional case the change in the measure for a finite field redefinition has to be calculated carefully. To do this, let us note that the finite field redefinition with a finite parameter $\alpha(x)$ can be obtained by making N successive infinitesimal field redefinitions with parameter $\epsilon(x)$ such that

$$\lim_{\epsilon\to0,\ N\to\infty} N\epsilon(x) = \alpha(x). \qquad (13.124)$$

An infinitesimal chiral field redefinition can be seen from Eq. (13.122) to be given by

$$\psi(x) \rightarrow \psi'(x) = \psi(x) - ie\epsilon(x)\gamma_5\psi(x),$$

$$\bar{\psi}(x) \rightarrow \bar{\psi}'(x) = \bar{\psi}(x) - ie\epsilon(x)\bar{\psi}(x)\gamma_5, \qquad (13.125)$$

Furthermore, let us assume that we have already made n successive infinitesimal transformations. The fermion Lagrangian density, after the n transformations, has the form (see Eq. (13.123))

$$\mathcal{L} = i\bar{\psi}(x)\gamma^\mu \left(\partial_\mu - ie\gamma_5(\partial_\mu(n\epsilon(x) - \eta(x)))\right)\psi(x). \qquad (13.126)$$

Let us define the eigenvalue equation

$$i\slashed{D}\phi_k(x) = i\gamma^\mu \left(\partial_\mu - ie\gamma_5(\partial_\mu(n\epsilon(x) - \eta(x)))\right)\phi_k(x)$$

$$= \lambda_k\phi_k(x). \qquad (13.127)$$

If the ϕ_k's form a complete set, then as discussed in Eq. (13.16), the eigenvectors satisfy

$$\int d^2x \; \phi_k^\dagger(x)\phi_m(x) = \delta_{km},$$

$$\sum_k \phi_k(x)\phi_k^\dagger(y) = \delta^2(x-y).$$ (13.128)

We can expand the fermion field variables ψ and $\bar{\psi}$ in this basis as

$$\psi(x) = \sum_n a_n\phi_n(x),$$

$$\bar{\psi}(x) = \sum_n b_n\phi_n^\dagger(x),$$ (13.129)

so that the functional integration measure (in the path integral) can be written as

$$\mathcal{D}\bar{\psi}\mathcal{D}\psi = \prod_n db_n da_n.$$ (13.130)

If we make an infinitesimal field redefinition as in Eq. (13.125), then as we have discussed in Eqs. (13.21)–(13.28), we can determine that

$$\psi(x) \to \psi'(x) = \sum_n a_n'\phi_n(x) = \sum_{n,m} c_{nm}a_m\phi_n(x),$$ (13.131)

where

$$c_{nm} = \delta_{nm} - ie \int d^2x \; \epsilon(x)\phi_n^\dagger(x)\gamma_5\phi_m(x).$$ (13.132)

Therefore,

$$\prod_n da_n \to \prod_n da_n' = (\det c_{nm})^{-1} \prod_n da_n.$$ (13.133)

Similarly,

$$\prod_n \mathrm{d}b_n \to \prod_n \mathrm{d}b'_n = (\det c_{nm})^{-1} \prod_n \mathrm{d}b_n \,, \qquad (13.134)$$

so that

$$\mathcal{D}\bar{\psi}\mathcal{D}\psi \to \mathcal{D}\bar{\psi}'\mathcal{D}\psi' = (\det c_{nm})^{-2}\,\mathcal{D}\bar{\psi}\mathcal{D}\psi. \qquad (13.135)$$

Let us, therefore, calculate the determinant

$$\begin{aligned}
\det c_{nm} &= \det\left(\delta_{nm} - ie\int \mathrm{d}^2x\; \epsilon(x)\phi_n^\dagger(x)\gamma_5\phi_m(x)\right) \\
&= \exp\left(\mathrm{Tr}\;\ln\left(\delta_{nm} - ie\int \mathrm{d}^2x\; \epsilon(x)\phi_n^\dagger(x)\gamma_5\phi_m(x)\right)\right) \\
&= \exp\left(-ie\sum_n \int \mathrm{d}^2x\; \epsilon(x)\phi_n^\dagger(x)\gamma_5\phi_n(x)\right). \qquad (13.136)
\end{aligned}$$

The exponent can be calculated in a regularized manner as before

$$\begin{aligned}
&-ie\sum_n \int \mathrm{d}^2x\; \epsilon(x)\phi_n^\dagger(x)\gamma_5\phi_n(x) \\
&= \lim_{M^2\to\infty} -ie\sum_n \int \mathrm{d}^2x\; \epsilon(x)\phi_n^\dagger(x)\gamma_5\phi_n(x)\, e^{-\frac{\lambda_n^2}{M^2}} \\
&= \lim_{M^2\to\infty} -ie\sum_n \int \mathrm{d}^2x\; \epsilon(x)\phi_n^\dagger(x)\gamma_5\, e^{\frac{\not{p}^2}{M^2}}\phi_n(x) \\
&= \lim_{M^2\to\infty} -ie\sum_n \int \mathrm{d}^2x\frac{\mathrm{d}^2k}{(2\pi)^2}\frac{\mathrm{d}^2k'}{(2\pi)^2}\; \epsilon(x) \\
&\qquad\qquad \times\; \tilde{\phi}_n^\dagger(k')e^{ik'\cdot x}\gamma_5 e^{\frac{\not{p}^2}{M^2}}e^{-ik\cdot x}\tilde{\phi}_n(k)\,. \qquad (13.137)
\end{aligned}$$

Using the completeness relation in the momentum space

$$\sum_n \tilde{\phi}_n(k)\tilde{\phi}_n^\dagger(k') = (2\pi)^2\delta^2(k-k')\,, \qquad (13.138)$$

the regularized exponent in the above relation takes the form

$$= \lim_{M^2 \to \infty} -ie \int d^2x \frac{d^2k}{(2\pi)^2} \, \epsilon(x) \, \text{Tr} \, e^{ik \cdot x} \gamma_5 \, e^{\frac{\rlap{/}D^2}{M^2}} e^{-ik \cdot x}. \qquad (13.139)$$

Let us note the two dimensional identity

$$\begin{aligned}
\rlap{/}D^2 &= \gamma^\mu D_\mu \gamma^\nu D_\nu \\
&= \gamma^\mu \gamma^\nu \left(\partial_\mu - ie\gamma_5(\partial_\mu(n\epsilon - \eta)) \right) \left(\partial_\nu - ie\gamma_5(\partial_\nu(n\epsilon - \eta)) \right) \\
&= (\eta^{\mu\nu} + \gamma_5 \epsilon^{\mu\nu}) \left(\partial_\mu - ie\gamma_5(\partial_\mu(n\epsilon - \eta)) \right) \left(\partial_\nu - ie\gamma_5(\partial_\nu(n\epsilon - \eta)) \right) \\
&= \partial^\mu \partial_\mu - e^2 (\partial^\mu(n\epsilon - \eta))(\partial_\mu(n\epsilon - \eta)) - 2ie\gamma_5(\partial^\mu(n\epsilon - \eta))\partial_\mu \\
&\quad - ie\gamma_5(\partial^\mu \partial_\mu(n\epsilon - \eta)) - 2iee^{\mu\nu}(\partial_\mu(n\epsilon - \eta))\partial_\nu \,. \qquad (13.140)
\end{aligned}$$

Using this in Eq. (13.139), we obtain (recall that $\text{Tr} \, \gamma_5 = 0$)

$$-ie \sum_n \int d^2x \, \epsilon(x)\phi^\dagger(x)\gamma_5\phi_n(x)$$

$$= \lim_{M^2 \to \infty} -ie \int d^2x \frac{d^2k}{(2\pi)^2} \epsilon(x) \left(-\frac{2ie}{M^2} \right) (\partial^\mu \partial_\mu(n\epsilon - \eta)) \, e^{-\frac{k^2}{M^2}}$$

$$= \lim_{M^2 \to \infty} -\frac{2e^2}{M^2} \int d^2x \, \epsilon(x)(\partial^\mu \partial_\mu(n\epsilon - \eta)) \times \frac{(-i\pi M^2)}{4\pi^2}$$

$$= \frac{ie^2}{2\pi} \int d^2x \, \epsilon(x)(\partial^\mu \partial_\mu(n\epsilon - \eta)) \,. \qquad (13.141)$$

Here the factor of $(-i)$ in the evaluation of the Gaussian integral arises from going to Euclidean space as we should have rigorously done.

Thus, we see that under this single infinitesimal transformation after n successive transformations, the measure changes as

$$\mathcal{D}\bar{\psi}\mathcal{D}\psi \to \mathcal{D}\bar{\psi}'\mathcal{D}\psi' = (\det c_{nm})^{-2}\,\mathcal{D}\bar{\psi}\mathcal{D}\psi$$

$$= \exp\left(-\frac{ie^2}{\pi}\int d^2x\ \epsilon(x)(\partial^\mu\partial_\mu(n\epsilon(x)-\eta))\right)\mathcal{D}\bar{\psi}\mathcal{D}\psi$$

$$= \left(1 - \frac{ie^2}{\pi}\int d^2x\ \epsilon(x)(\partial^\mu\partial_\mu(n\epsilon(x)-\eta))\right)\mathcal{D}\bar{\psi}\mathcal{D}\psi\,.$$

$$(13.142)$$

As we make N infinitesimal transformations, the measure changes as

$$\mathcal{D}\bar{\psi}\mathcal{D}\psi \to \prod_{n=0}^{N}\left(1 - \frac{ie^2}{\pi}\int d^2x\ \epsilon(x)(\partial^\mu\partial_\mu(n\epsilon-\eta))\right)\mathcal{D}\bar{\psi}\mathcal{D}\psi\,.$$

$$(13.143)$$

Using the result that if

$$X = \prod_{n=0}^{N}\left(1 + a\epsilon + nb\epsilon^2\right),\qquad\qquad (13.144)$$

then, for infinitesimal ϵ and large N such that $N\epsilon$ is finite, we have

$$\ln X = \sum_{n=0}^{N}\ln\left(1 + a\epsilon + nb\epsilon^2\right)$$

$$= \sum_{n=0}^{N}\left(a\epsilon + nb\epsilon^2\right)$$

$$= (N+1)a\epsilon + \frac{1}{2}N(N+1)b\epsilon^2\,.\qquad (13.145)$$

Therefore, in the limit $\epsilon \to 0, N \to \infty, N\epsilon = \alpha$, we obtain

$$\ln X = a\alpha + \frac{b}{2}\alpha^2,\quad \text{or,}\quad X = e^{a\alpha + \frac{b}{2}\alpha^2}\,.\qquad (13.146)$$

This gives the change in the measure for a finite chiral redefinition of fields to be

$$\mathcal{D}\bar{\psi}\mathcal{D}\psi \rightarrow \exp\left[-\frac{ie^2}{\pi}\int d^2x\ \alpha(x)\left(\partial^\mu\partial_\mu\left(\frac{1}{2}\alpha(x) - \eta\right)\right)\right]\mathcal{D}\bar{\psi}\mathcal{D}\psi.$$

(13.147)

Furthermore, if we choose $\alpha(x) = \eta(x)$, then we obtain

$$\mathcal{D}\bar{\psi}\mathcal{D}\psi \rightarrow \exp\left(-\frac{ie^2}{\pi}\int d^2x\ \eta(x)\left(\partial^\mu\partial_\mu\left(\frac{1}{2}\eta(x) - \eta(x)\right)\right)\right)\mathcal{D}\bar{\psi}\mathcal{D}\psi$$

$$= \exp\left(\frac{ie^2}{2\pi}\int d^2x\ \eta(x)(\partial^\mu\partial_\mu\eta(x))\right)\mathcal{D}\bar{\psi}\mathcal{D}\psi$$

$$= \exp\left(-\frac{ie^2}{2\pi}\int d^2x\ \partial^\mu\eta(x)\partial_\mu\eta(x)\right)\mathcal{D}\bar{\psi}\mathcal{D}\psi$$

$$= \exp\left(\frac{ie^2}{2\pi}\int d^2x\ \epsilon^{\mu\nu}\partial_\nu\eta(x)\epsilon_{\mu\lambda}\partial^\lambda\eta(x)\right)\mathcal{D}\bar{\psi}\mathcal{D}\psi$$

$$= \exp\left(\frac{ie^2}{2\pi}\int d^2x\ A^\mu(x)A_\mu(x)\right)\mathcal{D}\bar{\psi}\mathcal{D}\psi.$$

(13.148)

Thus we see that if we make a finite chiral field redefinition to decouple the fermions, there is a change in the functional measure leading to the generating functional

$$Z = \int \mathcal{D}A_\mu \mathcal{D}\bar{\psi}\mathcal{D}\psi\ e^{iS_{\text{eff}}},$$

(13.149)

where

$$S_{\text{eff}} = \int d^2x\left[-\frac{1}{4}F_{\mu\nu}F^{\mu\nu} + \frac{e^2}{2\pi}A^\mu A_\mu + i\bar{\psi}\gamma^\mu\partial_\mu\psi\right].$$

(13.150)

This shows again that the photon becomes massive in this theory and the fermions decouple completely from the spectrum. (This effective Lagrangian density is local because we have chosen to work in the

Landau gauge $\partial_\mu A^\mu = 0$ which sets the nonlocal terms to zero in Eq. (13.107).)

In this method, we see that the solubility of the model is closely related to the Jacobian of the chiral field redefinition that decouples the fermions. This, in turn, is connected with the anomaly of the system. This method can be extended to a very general class of two dimensional Abelian models which reduce to different known models in different limits.

13.3 References

S. Adler, Phys. Rev. **177**, 2426 (1969).

J. Bell and R. Jackiw, Nuovo Cimento **60 A**, 47 (1969).

A. Das and M. Hott, Zeit. Phys. **C 67**, 707 (1995).

A. Das and V. S. Mathur, Phys. Rev. **D 33**, 489 (1986).

K. Fujikawa, Phys. Rev. Lett. **42**, 1195 (1979); Phys. Rev. **D 21**, 2848 (1980).

R. Roskies and F. Schaposnik, Phys. Rev. **D23**, 558 (1981).

J. Schwinger, Phys. Rev. **128**, 2425 (1962).

CHAPTER 14

Systems at finite temperature

14.1 Statistical mechanics

Let us review very briefly various concepts from statistical mechanics. Let us consider not one quantum mechanical system, but a whole collection of identical quantum systems — an ensemble. Thus, for example, it can be an ensemble of oscillators or any other physical system. Let us further assume for simplicity that the physical system under consideration has discrete eigenvalues of energy. Each physical system in this ensemble can, of course, be in any eigenstate of energy and, therefore, there will be a distribution of the systems in the ensemble in different energy eigenstates. As a result, we can define p_n to represent the probability of finding a system in the ensemble to be in an energy eigenstate $|n\rangle$ with energy E_n. This is, of course, completely statistical in the sense that p_n can be identified simply with the number of physical systems in the state $|n\rangle$ divided by the total number of systems in the ensemble. Such a situation is quite physical as we know from our studies in statistical mechanics. Namely, we may have an ensemble of physical systems in thermal equilibrium with a heat bath described by a thermal probability distribution (Boltzmann distribution).

For any quantum mechanical ensemble with a given probability distribution p_n, the value of any observable quantity averaged over the entire ensemble (ensemble average) will take the form

$$\langle A \rangle = \bar{A} = \sum_n p_n \langle n|A|n \rangle = \sum_n p_n A_n, \qquad (14.1)$$

where we are assuming that the energy eigenstates are normalized and that

$$A_n = \langle n|A|n \rangle \,, \tag{14.2}$$

denotes the expectation value of the operator A in the quantum mechanical state $|n\rangle$. Thus, there are two kinds of averaging involved here. First, we have the average of the observable in a quantum state (expectation value) and second, we have the average of the expectation value with respect to the probability distribution of systems in the ensemble.

Being a probability, p_n has to satisfy certain conditions. Namely,

$$0 \le p_n \le 1, \quad \text{for all} \ \ n,$$

$$\sum_n p_n = 1 \,. \tag{14.3}$$

It is difficult to determine the probability distribution for a general ensemble. However, if we are dealing with a thermodynamic ensemble, namely, an ensemble interacting with a large heat bath, and if we allow sufficient time for it to achieve thermal equilibrium, then we know that the probability distribution, in this case, is given by the Maxwell-Boltzmann distribution. Namely, in this case, we can write

$$p_n = \frac{1}{Z} e^{-\frac{E_n}{kT}} \,. \tag{14.4}$$

Here E_n is the energy of the nth quantum state, k is the Boltzmann constant and T the equilibrium temperature of the system.

The normalization factor Z in Eq. (14.4) can be determined from the relations for the probabilities in Eq. (14.3) as

$$\sum_n p_n = 1,$$

$$\text{or,} \quad \frac{1}{Z} \sum_n e^{-\frac{E_n}{kT}} = 1,$$

$$\text{or,} \quad Z = \sum_n e^{-\frac{E_n}{kT}} = \sum_n e^{-\beta E_n} = \sum_n \langle n|e^{-\beta H}|n \rangle,$$

$$\text{or,} \quad Z(\beta) = \operatorname{Tr} e^{-\beta H} \,, \tag{14.5}$$

where we have defined

$$\beta = \frac{1}{kT}. \tag{14.6}$$

$Z(\beta)$ is known as the partition function of the system and plays a fundamental role in deriving the thermodynamic properties of the system.

For a statistical thermal ensemble, it is easy to see that the thermodynamic average (ensemble average) of any quantity defined in Eq. (14.1) will be given by

$$
\begin{aligned}
\langle A \rangle_\beta &= \sum_n p_n \langle n|A|n \rangle \\
&= \frac{1}{Z(\beta)} \sum_n e^{-\beta E_n} \langle n|A|n \rangle \\
&= \frac{1}{Z(\beta)} \sum_n \langle n|e^{-\beta H} A|n \rangle \\
&= \frac{1}{Z(\beta)} \operatorname{Tr}\left(e^{-\beta H} A\right) \\
&= \frac{\operatorname{Tr}\left(e^{-\beta H} A\right)}{\operatorname{Tr} e^{-\beta H}}.
\end{aligned}
\tag{14.7}
$$

This allows us to define an operator known as the density operator or the density matrix

$$\rho(\beta) = \frac{1}{Z(\beta)} e^{-\beta H}, \tag{14.8}$$

so that the thermal average of any operator can be written as

$$\langle A \rangle_\beta = \operatorname{Tr}\left(\rho(\beta)A\right) = \frac{1}{Z(\beta)} \operatorname{Tr}\left(e^{-\beta H} A\right). \tag{14.9}$$

In particular, the average energy associated with a system in the ensemble follows from Eq. (14.7) to be

$$\langle H \rangle_\beta = U = \frac{\mathrm{Tr}\left(e^{-\beta H} H\right)}{\mathrm{Tr}\, e^{-\beta H}}$$

$$= \frac{-\frac{\partial}{\partial \beta} \mathrm{Tr}\, e^{-\beta H}}{\mathrm{Tr}\, e^{-\beta H}}$$

$$= \frac{1}{Z(\beta)} \left(-\frac{\partial Z(\beta)}{\partial \beta}\right)$$

$$= -\frac{\partial \ln Z(\beta)}{\partial \beta}. \tag{14.10}$$

This is also known as the average internal energy associated with the ensemble.

The amount of order or the lack of it in an ensemble is defined through the entropy as

$$S = -\sum_n p_n \ln p_n = -\langle \ln p \rangle. \tag{14.11}$$

By definition, it is clear that the entropy is always positive semi-definite since $0 \leq p_n \leq 1$. Furthermore, its value is zero only for a pure ensemble for which

$$p_n = \delta_{nm}, \quad \text{for a fixed } m. \tag{14.12}$$

For such an ensemble, all the individual systems are in the same energy state and, therefore, it is an ordered ensemble. On the other hand, the larger the number of states over which the physical systems are distributed, the more disordered the ensemble becomes and the entropy increases. For a thermodynamic ensemble, as we have seen in Eq. (14.4),

$$p_n = \frac{1}{Z(\beta)} e^{-\beta E_n}. \tag{14.13}$$

Therefore, we can calculate the entropy of the ensemble to be

$$S = -\langle \ln p \rangle_\beta = -\sum_n p_n \ln p_n$$

$$= -\sum_n p_n (-\beta E_n - \ln Z(\beta))$$

$$= \beta \sum_n p_n E_n + \ln Z(\beta) \sum_n p_n$$

$$= \beta U + \ln Z(\beta)$$

$$= -\beta \frac{\partial \ln Z(\beta)}{\partial \beta} + \ln Z(\beta)$$

$$= -\beta^2 \frac{\partial}{\partial \beta} \left(\frac{1}{\beta} \ln Z(\beta) \right). \tag{14.14}$$

Here we have used Eqs. (14.10) and (14.3) in the intermediate steps.

Given the internal energy U and the entropy S, the free energy for an ensemble can be defined as

$$F(\beta) = U - \frac{S}{\beta}$$

$$= -\frac{\partial \ln Z(\beta)}{\partial \beta} + \frac{\partial \ln Z(\beta)}{\partial \beta} - \frac{1}{\beta} \ln Z(\beta)$$

$$= -\frac{1}{\beta} \ln Z(\beta). \tag{14.15}$$

In terms of the free energy we can define the other thermodynamical quantities as

$$U = -\frac{\partial \ln Z(\beta)}{\partial \beta} = \frac{\partial(\beta F)}{\partial \beta} = F + \beta \frac{\partial F}{\partial \beta},$$

$$S = -\beta^2 \frac{\partial}{\partial \beta} \left(\frac{1}{\beta} \ln Z(\beta) \right) = \beta^2 \frac{\partial F}{\partial \beta}. \tag{14.16}$$

It is also interesting to note from Eq. (14.15) that the partition function takes a particularly simple form when expressed in terms of the free energy. Namely, we can write

$$Z(\beta) = e^{-\beta F(\beta)}, \qquad (14.17)$$

and this is quite reminiscent of the relation between the generating functionals Z and W (in Euclidean space) that we have studied earlier. We have gone over some of these concepts in some detail in order to bring out the essential similarities with the concepts of path integral that we have been discussing so far.

One of the major interests in the study of statistical mechanics is the question of phase transitions in physical systems. Phase transitions are all too familiar to us from our studies of the different phases of water. Even in solids such as magnets, the hysteresis effect or the effect of spontaneous magnetization provides an example of a phase transition. Namely, we know that below the Curie temperature T_c, if a magnetic material is subjected to an external magnetic field, then the material develops a residual magnetization even when the external field is switched off. The amount of residual magnetization decreases as the temperature of the system approaches the Curie temperature and vanishes at T_c. For $T > T_c$, the system exhibits no spontaneous magnetization. Therefore, the temperature $T = T_c$ is a critical temperature separating the different phases of a magnetic material.

The behavior of physical systems near the critical point is of great significance. This can be studied from the point of view of statistical mechanics quite well. They can also be studied with equal ease using the concepts of path integrals. However, before we discuss the applications of path integrals to statistical mechanics, let us recapitulate how one uses statistical mechanics to study critical phenomena.

14.2 Critical exponents

Let us discuss the critical exponents in the context of a specific model which explains the properties of magnetization quite well. This model goes under the name of the planar Ising model or the Ising model in two dimensions (which we will study in the next chapter).

The crucial feature of this model is that it ascribes the magnetic properties of a material to its spin content. In fact, this is quite reasonable from our studies of microscopic systems where we know that elementary particles with a nontrivial spin (and charge) possess magnetic dipole moments

Let us consider a square lattice with equal spacing along both x and y directions. Let us also assume that at each lattice site labeled by $n = (n_1, n_2)$, there is a spin $S(n)$ which can either point up or down. Accordingly, we assume

$$S(n) = \begin{cases} 1 & \text{for spin up,} \\ -1 & \text{for spin down.} \end{cases} \tag{14.18}$$

Furthermore, let us assume that the spins interact as locally as is possible. In fact, the Hamiltonian for the planar Ising model is taken to be

$$H = -J \sum_{n,\hat{\mu}} S(n)S(n + \hat{\mu}), \tag{14.19}$$

where we have assumed a simplified coupling for the problem. Here $\hat{\mu}$ stands for either of the two unit vectors (along x or y axis) on the lattice. In simple language, then, the Ising model assumes nearest neighbor interaction between the spins which are supposed to be pointing only along the z-axis. The constant J measures the strength of the spin-spin interaction. It is clear that if its value is positive, then a minimum of the energy will be obtained when all the spins are pointing along the same direction — either up or down. Accordingly, such a coupling is known as a ferromagnetic coupling. Conversely, if J is negative, then the coupling is known as anti-ferromagnetic. It is worth pointing out here that the Hamiltonian for the Ising model has a discrete symmetry in the sense that if we flip all the spins of the system, then the Hamiltonian does not change.

Let us next subject this spin system to a constant external magnetic field B (this is like the external source we have discussed in connection with path integrals). In this case, the Hamiltonian becomes

$$H = -J \sum_{n,\hat{\mu}} S(n)S(n + \hat{\mu}) + B \sum_n S(n). \tag{14.20}$$

The partition function defined in Eqs. (14.5) and (14.17), for the present case, takes the form

$$Z(\beta, B) = e^{-\beta F(\beta, B)} = \text{Tr}\, e^{-\beta H} = \sum_{\text{config}} e^{-\beta E_{\text{config}}}. \tag{14.21}$$

The summation, here, is over all possible spin configurations on the lattice and E_{config} denotes the energy associated with each such spin configuration. Let us note that at every lattice site, the spin can take two possible values. Consequently, if N denotes the total number of lattice points (lattice sites), then there are 2^N possible spin configurations over which the summation in Eq. (14.21) has to be carried out. The true partition function, of course, has to be calculated in the thermodynamic limit when $N \to \infty$.

Let us note now from Eqs. (14.20) and (14.21) that, in this case, we have

$$\frac{\partial Z}{\partial B} = -\beta \frac{\partial F}{\partial B} Z = \text{Tr} \left(-\beta \left(\sum_n S(n) \right) e^{-\beta H} \right),$$

$$\text{or,} \quad \frac{\partial F}{\partial B} = \frac{1}{Z} \text{Tr} \left(\left(\sum_n S(n) \right) e^{-\beta H} \right) = \left\langle \sum_n S(n) \right\rangle_\beta$$

$$= \sum_n \langle S(n) \rangle_\beta. \tag{14.22}$$

Using the translation invariance of the theory (lattice), we can write

$$\langle S(n) \rangle_\beta = \langle S(0) \rangle_\beta, \tag{14.23}$$

so that we obtain

$$\frac{\partial F}{\partial B} = \sum_n \langle S(n) \rangle_\beta = N \langle S(0) \rangle_\beta. \tag{14.24}$$

This gives the total magnetization for the lattice and the mean magnetization per site can be obtained to be

$$M(\beta, B) = \frac{1}{N} \sum_n \langle S(n) \rangle_\beta = \langle S(0) \rangle_\beta = \frac{1}{N} \frac{\partial F}{\partial B}. \qquad (14.25)$$

This is, of course, a function of both the temperature and the applied magnetic field and its value can be calculated once we know the free energy or the partition function.

The magnetic susceptibility is proportional to the rate of change of magnetization with the applied field and is defined to be

$$\chi = -\left.\frac{\partial M}{\partial B}\right|_{B=0} = -\frac{1}{N} \left.\frac{\partial^2 F}{\partial B^2}\right|_{B=0}$$

$$= \frac{\beta}{N} \left(\sum_{n,m} \langle S(n)S(m) \rangle_\beta - \left(\sum_n \langle S(n) \rangle_\beta \right)^2 \right)_{B=0}$$

$$= \frac{\beta}{N} \left(N \sum_n \langle S(n)S(0) \rangle_\beta - N^2 (\langle S(0) \rangle_\beta)^2 \right)_{B=0}$$

$$= \beta \left(\sum_n \langle S(n)S(0) \rangle_\beta - N (\langle S(0) \rangle_\beta)^2 \right)_{B=0}, \qquad (14.26)$$

where we have used Eqs. (14.22) and (14.23). Thus, we see that the magnetic susceptibility is related to the fluctuations in the spin while the magnetization is given by the average value of spin. The susceptibility is large at those temperatures where the correlation between the spins is large. Note that if the system has no net magnetization, i.e., no spontaneous magnetization (when the external magnetic field is switched off), namely, if

$$\langle S(0) \rangle_\beta |_{B=0} = 0, \qquad (14.27)$$

then, the magnetic susceptibility is completely determined by the spin-spin correlation function. In other words, in this case we have

$$\chi = \beta \sum_n \langle S(n)S(0) \rangle_\beta \Big|_{B=0}. \qquad (14.28)$$

If, for some low temperature $T < T_c$ ($\beta > \beta_c$), we find in our spin system that

$$\langle S(0) \rangle_\beta |_{B=0} \neq 0 \,, \tag{14.29}$$

then, the system shows spontaneous magnetization or residual magnetization. In this case, we note that the discrete symmetry of the system is spontaneously broken. The spontaneous magnetization vanishes as we approach the critical temperature and for $T > T_c$ ($\beta < \beta_c$), the system will show no spontaneous magnetization simply because the random thermal motion will dominate. The behavior of spontaneous magnetization near the critical temperature, namely, how the magnetization vanishes as the temperature approaches the critical temperature

$$M|_{B=0} \sim (T - T_c)^\beta \,, \tag{14.30}$$

can be calculated once we know the partition function. Let us emphasize here that the parameter β in the exponent in Eq. (14.30) is not $\frac{1}{kT}$ which was defined earlier but represents a critical exponent. (The notation is unfortunate, but this is the convention.) The spontaneous magnetization defines an order parameter for the phase transition in the sense that its value separates the two different phases of the physical system.

We can similarly calculate the correlation length $\xi(T)$ between the spins at any temperature by analyzing the magnetic susceptibility. For very high temperatures, it is clear that the random thermal motion will not allow any appreciable correlation between the spins. However, as the temperature of the system is lowered to the critical temperature, the system may develop long range correlations and the behavior of the correlation length near the critical temperature is parameterized by another critical exponent of the form

$$\xi(T) \sim (T - T_c)^{-\nu} \,. \tag{14.31}$$

The magnetic susceptibility may similarly become large at this point and its behavior near the critical temperature is parameterized by yet another critical exponent as

$$\chi(T) \sim (T - T_c)^{-\gamma} . \tag{14.32}$$

Similarly, other thermodynamic quantities in the system such as the specific heat defined as

$$C = -T \frac{\partial^2 F}{\partial T^2} , \tag{14.33}$$

may also display a singular behavior at the critical point and all these can be calculated once we know the partition function or the free energy of the system.

14.3 Harmonic oscillator

The direct calculation of the partition function for the one dimensional quantum harmonic oscillator is quite straightforward. We know that for an oscillator with a natural frequency ω, the energy levels are given by

$$E_n = \omega \left(n + \frac{1}{2} \right), \quad n = 0, 1, 2, \cdots , \tag{14.34}$$

where we have set $\hbar = 1$. For this system then, the partition function can be derived using Eqs. (14.5) and (14.34) to be

$$Z(\beta) = \mathrm{Tr}\, e^{-\beta H} = \sum_{n=0}^{\infty} e^{-\beta E_n}$$

$$= \sum_{n=0}^{\infty} e^{-\beta \omega \left(n + \frac{1}{2} \right)} = e^{-\frac{\beta \omega}{2}} \left(\sum_{n=0}^{\infty} e^{-\beta \omega n} \right)$$

$$= e^{-\frac{\beta \omega}{2}} \left(\frac{1}{1 - e^{-\beta \omega}} \right) = \frac{1}{e^{\frac{\beta \omega}{2}} - e^{-\frac{\beta \omega}{2}}}$$

$$= \frac{1}{2 \sinh \frac{\beta \omega}{2}} . \tag{14.35}$$

Since we know the partition function, we can calculate the thermodynamic properties of the system.

Let us next see how we can calculate the partition function for the harmonic oscillator using the path integral method. Let us recall that we have already calculated the transition amplitude for the harmonic oscillator which has the form (see Eq. (3.71))

$$\langle x_f, T | x_i, 0 \rangle = \langle x_f | e^{-iHT} | x_i \rangle = N \int \mathcal{D}x \, e^{iS[x]},$$

$$\text{or,} \quad \langle x_f | e^{-iHT} | x_i \rangle = \left(\frac{m\omega}{2\pi i \sin \omega T} \right)^{\frac{1}{2}} e^{iS[x_{cl}]}, \tag{14.36}$$

and we have determined earlier that (see Eq. (3.91) with $J = 0$)

$$S[x_{cl}] = \frac{m\omega}{2 \sin \omega T} \left[(x_i^2 + x_f^2) \cos \omega T - 2x_i x_f \right]. \tag{14.37}$$

We recall that we have set $\hbar = 1$, $J = 0$ in the above equations and T here denotes the time interval between the initial and the final points of the trajectory.

On the other hand, from the definition of the partition function,

$$Z(\beta) = \text{Tr} \, e^{-\beta H}, \tag{14.38}$$

we note that the trace can be taken in any basis. In particular, if we choose the coordinate basis in the Schrödinger picture, then we can write

$$Z(\beta) = \int\limits_{-\infty}^{\infty} dx \, \langle x | e^{-\beta H} | x \rangle. \tag{14.39}$$

We now recognize the integrand in Eq. (14.39) merely as the transition amplitude (see Eq. (1.47)) (it is actually an expectation value as we have discussed earlier) for the harmonic oscillator with the identification

$$T = -i\beta,$$
$$x_f = x_i = x. \tag{14.40}$$

In other words, the integrand really is the transition amplitude between the same coordinate state in the Euclidean time with $\beta = \frac{1}{kT}$ (T is the temperature here) playing the role of the Euclidean time interval. Using this, then, we obtain from Eq. (14.39)

$$
Z(\beta) = \int\limits_{-\infty}^{\infty} dx \left(\frac{m\omega}{2\pi i(-i\sinh\beta\omega)} \right)^{\frac{1}{2}}
$$

$$
\times \; e^{\left(\frac{im\omega}{2(-i\sinh\beta\omega)} (2x^2\cosh\beta\omega - 2x^2) \right)}
$$

$$
= \left(\frac{m\omega}{2\pi\sinh\beta\omega} \right)^{\frac{1}{2}} \int\limits_{-\infty}^{\infty} dx \, e^{-\left(\frac{m\omega}{\sinh\beta\omega}(\cosh\beta\omega - 1)x^2 \right)}
$$

$$
= \left(\frac{m\omega}{2\pi\sinh\beta\omega} \right)^{\frac{1}{2}} \int\limits_{-\infty}^{\infty} dx \, e^{-(m\omega\tanh\frac{\beta\omega}{2})x^2}
$$

$$
= \left(\frac{m\omega}{2\pi\sinh\beta\omega} \right)^{\frac{1}{2}} \left(\frac{\pi}{m\omega\tanh\frac{\beta\omega}{2}} \right)^{\frac{1}{2}}
$$

$$
= \left(\frac{1}{2\sinh\beta\omega\tanh\frac{\beta\omega}{2}} \right)^{\frac{1}{2}}
$$

$$
= \left(\frac{1}{4\sinh^2\frac{\beta\omega}{2}} \right)^{\frac{1}{2}} = \frac{1}{2\sinh\frac{\beta\omega}{2}}. \tag{14.41}
$$

This is, of course, the partition function which we had found by a direct calculation in Eq. (14.35). Let us note next that since

$$
Z(\beta) = e^{-\beta F} = \frac{1}{2\sinh\frac{\beta\omega}{2}}, \tag{14.42}
$$

we obtain

$$
F = -\frac{1}{\beta}\ln Z = \frac{1}{\beta}\left(\ln 2 + \ln\sinh\frac{\beta\omega}{2} \right). \tag{14.43}
$$

Consequently, we note from Eq. (14.16) that

$$\langle H \rangle_\beta = U = \frac{\partial(\beta F)}{\partial \beta} = -\frac{\partial \ln Z}{\partial \beta}$$

$$= \frac{\omega}{2} \frac{\cosh \frac{\beta\omega}{2}}{\sinh \frac{\beta\omega}{2}} = \frac{\omega}{2} \frac{e^{\frac{\beta\omega}{2}} + e^{-\frac{\beta\omega}{2}}}{e^{\frac{\beta\omega}{2}} - e^{-\frac{\beta\omega}{2}}}$$

$$= \frac{\omega}{2} \frac{e^{\beta\omega} + 1}{e^{\beta\omega} - 1},$$

or,　　$$\langle H \rangle_\beta = U = \frac{\omega}{2} + \frac{\omega}{e^{\beta\omega} - 1}. \tag{14.44}$$

This is exactly what we would have obtained from Planck's law (remember that $\hbar = 1$). Among other things, it tells us that for low temperatures or large β, we have

$$\langle H \rangle_\beta = U \simeq \frac{\omega}{2}. \tag{14.45}$$

Namely, in such a case, the oscillators (in the ensemble) remain in the ground state. On the other hand, for very high temperatures or small β, we get

$$\langle H \rangle_\beta = U \simeq \frac{\omega}{2} + \frac{\omega}{\beta\omega} \simeq \frac{1}{\beta} = kT. \tag{14.46}$$

This is a manifestation of the equipartition of energy. (We expect the system to behave in a classical manner at very high temperature.)

　　This analysis of the derivation of the partition function for the harmonic oscillator from the path integral is quite instructive. It shows that the path integral for a $(D + 1)$-dimensional Euclidean quantum field theory can be related to the partition function of a D-dimensional quantum statistical system when the Euclidean time interval is identified with $\beta = \frac{1}{kT}$ as in Eq. (14.40). In fact, the relation between the two for a bosonic field theory can be written as

$$Z(\beta) = N \int_{\phi(0)=\phi(\beta)} \mathcal{D}\phi \, e^{-S_E}$$

$$= N \int_{\phi(0)=\phi(\beta)} \mathcal{D}\phi \, e^{-\int_0^\beta dt \int d^D x \mathcal{L}_E} . \tag{14.47}$$

Here we are assuming integration over the end points which is equivalent to taking the trace. Furthermore, we note that the field variables, in this path integral, are assumed to satisfy a periodic boundary condition

$$\phi(0) = \phi(\beta) . \tag{14.48}$$

This, as is clear, arises from the trace in the definition of the partition function. A careful analysis for the fermions shows that the partition function, in such a case, can be written exactly in the same manner but with anti-periodic boundary conditions. Namely, for fermions, we have

$$Z(\beta) = N \int \mathcal{D}\bar{\psi}\mathcal{D}\psi \, e^{-S_E[\psi,\bar{\psi}]}$$

$$= N \int \mathcal{D}\bar{\psi}\mathcal{D}\psi \, e^{-\int_0^\beta dt \int d^D x \, \mathcal{L}_E[\psi,\bar{\psi}]} , \tag{14.49}$$

with the boundary conditions

$$\psi(0) = -\psi(\beta),$$
$$\bar{\psi}(0) = -\bar{\psi}(\beta) . \tag{14.50}$$

These boundary conditions can be shown to be related to the question of quantum statistics associated with the different systems. This way of describing a quantum statistical system in equilibrium through a Euclidean path integral is known as the Matsubara formalism or the imaginary time formalism (since we rotate to imaginary time). Let us note here without going into details that there exist other formalisms which allow for the presence of both time and temperature in the theory simultaneously. These go under the name of real time formalisms.

The Matsubara formalism (the imaginary time formalism) also suggests that a $(D+1)$-dimensional bosonic Euclidean quantum field theory can be related to a $(D+1)$-dimensional classical statistical system in the following way. Let us consider a quantum mechanical system described by the Lagrangian

$$L = \frac{1}{2} m\dot{x}^2 - V(x). \tag{14.51}$$

Then, the generating functional for such a system will have the form (we use t to denote the time interval to avoid any confusion with temperature)

$$Z = N \int \mathcal{D}x\, e^{\frac{i}{\hbar} S[x]} = N \int \mathcal{D}x\, e^{\frac{i}{\hbar} \int\limits_0^t dt\, L}. \tag{14.52}$$

Here we have put back Planck's constant for reasons which will be clear shortly. If we rotate to Euclidean time, the generating functional takes the form

$$Z = N \int \mathcal{D}x\, e^{-\frac{1}{\hbar} \int\limits_0^t dt\left(\frac{1}{2} m\dot{x}^2 + V(x)\right)}. \tag{14.53}$$

This has precisely the form of a classical partition function if we identify

$$H = \int\limits_0^t dt \left(\frac{1}{2} m\dot{x}^2 + V(x)\right) \tag{14.54}$$

as governing the dynamics of the system and

$$\hbar = kT = \frac{1}{\beta}. \tag{14.55}$$

Here we should note that the variable, t, in this case should be treated as a space variable and not as a time. Let us also recall that the Planck's constant measures quantum fluctuations in a quantum mechanical system whereas temperature measures thermal fluctuations in a statistical system. The identification of the two above, therefore, relates the quantum fluctuations in a quantum mechanical system with the thermal fluctuations in a corresponding classical statistical system. This connection can be simply extended to a field theory where the Euclidean action would act as the Hamiltonian for the corresponding classical statistical system.

14.4 Fermionic oscillator

Let us next calculate the partition function for a fermionic oscillator with a natural frequency ω, using both the path integrals and the standard methods. We note from our discussion in Section 5.1 that the Hilbert space for the fermionic oscillator is quite simple. In fact, from Eqs. (5.13) and (5.14), we note that it is like a two level system with energy eigenvalues

$$E_0 = -\frac{\omega}{2},$$

$$E_1 = \frac{\omega}{2}. \tag{14.56}$$

Once again, we have set $\hbar = 1$ for simplicity here. Using Eq. (14.56) it follows from the definition of the partition function in Eq. (14.5) that for the fermionic oscillator we have

$$Z(\beta) = \mathrm{Tr}\, e^{-\beta H} = e^{-\beta E_0} + e^{-\beta E_1}$$

$$= e^{\frac{\beta\omega}{2}} + e^{-\frac{\beta\omega}{2}} = 2\cosh\frac{\beta\omega}{2}. \tag{14.57}$$

The evaluation of the partition function for the fermionic oscillator, in the path integral method, follows from the form of the transition amplitude derived in Eq. (5.99). Following our discussion in the last section, we note that in the case of fermions, we have to impose anti-periodic boundary conditions (see Eq. (14.50)), namely, for the calculation of the partition function, we require ($T \to -i\beta$)

$$\psi_f = -\psi_i \, ,$$

$$\bar{\psi}_f = -\bar{\psi}_i \, . \tag{14.58}$$

We can now calculate the partition function for the system with the identifications in Eqs. (14.40) and (14.58) as well as the result in Eq. (5.99) as

$$Z(\beta) = \int \mathrm{d}\bar{\psi}_i \mathrm{d}\psi_i \left(e^{\frac{\beta\omega}{2}} \, e^{\left(-e^{-\beta\omega} \bar{\psi}_i \psi_i - \bar{\psi}_i \psi_i \right)} \right)$$

$$= e^{\frac{\beta\omega}{2}} \int \mathrm{d}\bar{\psi}_i \mathrm{d}\psi_i \left(1 - (1 + e^{-\beta\omega}) \bar{\psi}_i \psi_i \right)$$

$$= e^{\frac{\beta\omega}{2}} \left(1 + e^{-\beta\omega} \right)$$

$$= 2 \cosh \frac{\beta\omega}{2} \, . \tag{14.59}$$

Here we have used the nilpotency properties of Grassmann variables (see Eq. (5.18)) as well as the integration rules given in Eqs. (5.27) and (5.28).

This is exactly the same result which we had obtained earlier in Eq. (14.57) for the partition function of the system. We note now from the definition in Eq. (14.15) that the free energy for the fermionic oscillator is given by

$$F(\beta) = -\frac{1}{\beta} \ln Z(\beta) = -\frac{1}{\beta} \left(\ln 2 + \ln \cosh \frac{\beta\omega}{2} \right) . \tag{14.60}$$

The average energy for the ensemble can now be calculated from Eq. (14.16) to be

$$\langle H \rangle_\beta = U = \frac{\partial(\beta F)}{\partial \beta}$$

$$= -\frac{\omega}{2} \frac{\sinh \frac{\beta\omega}{2}}{\cosh \frac{\beta\omega}{2}}$$

$$= -\frac{\omega}{2} \frac{e^{\frac{\beta\omega}{2}} - e^{-\frac{\beta\omega}{2}}}{e^{\frac{\beta\omega}{2}} + e^{-\frac{\beta\omega}{2}}} ,$$

$$= -\frac{\omega}{2} \frac{e^{\beta\omega} - 1}{e^{\beta\omega} + 1}$$

or, $\quad \langle H \rangle_\beta = U = -\dfrac{\omega}{2} + \dfrac{\omega}{e^{\beta\omega} + 1}$.   (14.61)

It now follows that for low temperatures or large β $(\beta = \frac{1}{kT})$, we have

$$\langle H \rangle_\beta = U \simeq -\frac{\omega}{2} .$$ (14.62)

Namely, the system likes to remain in the ground state for low temperatures whereas for high temperatures or small β, we obtain

$$\langle H \rangle_\beta = U \simeq -\frac{\omega}{2} + \frac{\omega}{2 + \beta\omega} = -\frac{\omega}{2} + \frac{\omega}{2}\left(1 + \frac{\beta\omega}{2}\right)^{-1}$$

$$\simeq -\frac{\omega}{2} + \frac{\omega}{2} - \frac{\beta\omega^2}{4}$$

$$= -\frac{\beta\omega^2}{4} = -\frac{\omega^2}{4kT} .$$ (14.63)

In this case, we see that the average energy of the system goes to zero inversely with the temperature which amounts to saying that the system tries to populate equally the two available energy states.

14.5 References

A. **Das**, *Finite Temperature Field Theory*, World Scientific Publishing.

K. **Huang**, *Statistical Mechanics*, John Wiley Publishing.

J. **Kogut**, Rev. Mod. Phys., **51**, 659 (1979).

T. **Matsubara**, Prog. Theor. Phys., **14**, 351 (1955).

J. **Schwinger**, J. Math. Physics **2** (1961) 407.

J. **Schwinger**, *Lecture Notes of Brandeis Summer Institute in Theoretical Physics* (1960).

H. **Umezawa** *et al.*, *Thermofield Dynamics and Condensed States*, North Holland Publishing.

CHAPTER 15

Ising model

15.1 One dimensional Ising model

Let us now apply some of the ideas of statistical mechanics which we have developed in the last chapter to the example of the one dimensional Ising model. The Hamiltonian for spins interacting through nearest neighbors on a one dimensional lattice (chain) is given by

$$
H = -J \sum_{i=1}^{N} s_i s_{i+1} + B \sum_{i=1}^{N} s_i
$$

$$
= -J \sum_{i=1}^{N} s_i s_{i+1} + B \sum_{i=1}^{N} \frac{1}{2}(s_i + s_{i+1}). \tag{15.1}
$$

Here we have assumed that the total number of lattice sites is N and that the spin system is being subjected to a constant external magnetic field B. Furthermore, we assume periodic condition on the lattice (cyclicity condition), namely,

$$
s_{N+1} = s_1. \tag{15.2}
$$

The spins take values $s_i = \pm 1$ representing the fact that they can either point "up" or "down". The classical partition function for this system is, by definition,

$$
Z(\beta) = \sum_{s_i = \pm 1} e^{-\beta H}
$$

$$
- \sum_{s_i = \pm 1} e^{\left(\beta J \sum_{i=1}^{N} s_i s_{i+1} - \beta B \sum_{i=1}^{N} \frac{1}{2}(s_i + s_{i+1}) \right)}. \tag{15.3}
$$

Let us ask whether there exists a quantum mechanical system whose Euclidean generating functional will give rise to the partition function for the one dimensional Ising model. Let us consider the quantum mechanical system described by the Hamiltonian

$$H_q = -\alpha\,\sigma_1 + \gamma\,\sigma_3\,, \tag{15.4}$$

where σ_1 and σ_3 are the two Pauli matrices and α and γ are two arbitrary real constant parameters. Let $|s\rangle$ denote the two component eigenstates of σ_3 such that (they correspond to the two dimensional column matrices $\begin{pmatrix} 1 \\ 0 \end{pmatrix}$ and $\begin{pmatrix} 0 \\ 1 \end{pmatrix}$)

$$\sigma_3|s\rangle = s|s\rangle\,, \quad s = \pm 1\,. \tag{15.5}$$

We can now calculate the Euclidean transition amplitude for the quantum system described by Eq. (15.4) between two eigenstates of σ_3, namely,

$$\langle s_{\text{fin}}|e^{-TH_q}|s_{\text{in}}\rangle\,. \tag{15.6}$$

Dividing the time interval T into N steps of infinitesimal length ϵ such that (for large N)

$$N\epsilon = T\,, \tag{15.7}$$

and introducing a complete set of eigenstates of σ_3 at every intermediate point, we obtain

$$\langle s_{\text{fin}}|e^{-TH_q}|s_{\text{in}}\rangle$$
$$= \sum_{s_i=\pm 1} \langle s_{N+1}|e^{-\epsilon H_q}|s_N\rangle \langle s_N|e^{-\epsilon H_q}|s_{N-1}\rangle \cdots \langle s_2|e^{-\epsilon H_q}|s_1\rangle\,, \tag{15.8}$$

where the intermediate sums are over the intermediate points $i = 2, 3, \ldots, N$. Furthermore, we have also identified

$$s_{\text{in}} = s_1, \quad s_{\text{fin}} = s_{N+1}. \tag{15.9}$$

Note that if ϵ is small, then we can expand the individual exponents in Eq. (15.8) and write

$$\langle s_{i+1}|e^{-\epsilon H_q}|s_i\rangle \simeq \langle s_{i+1}|(\mathbb{1} - \epsilon H_q))|s_i\rangle$$

$$= \langle s_{i+1}|(\mathbb{1} + \epsilon\alpha\sigma_1 - \epsilon\gamma\sigma_3)|s_i\rangle. \tag{15.10}$$

Using the relations,

$$\langle s_{i+1}|s_i\rangle = \left(\frac{1}{2}(s_i + s_{i+1})\right)^2 = \frac{1}{2}(1 + s_i s_{i+1}),$$

$$\langle s_{i+1}|\sigma_1|s_i\rangle = \left(\frac{1}{2}(s_i - s_{i+1})\right)^2 = \frac{1}{2}(1 - s_i s_{i+1}),$$

$$\langle s_{i+1}|\sigma_3|s_i\rangle = \frac{1}{2}(s_i + s_{i+1}), \tag{15.11}$$

which can be explicitly checked, we obtain

$$\langle s_{i+1}|e^{-\epsilon H_q}|s_i\rangle$$

$$\simeq \left(\frac{1}{2}(s_i + s_{i+1})\right)^2 + \epsilon\alpha\left(\frac{1}{2}(s_i - s_{i+1})\right)^2 - \epsilon\gamma\frac{1}{2}(s_i + s_{i+1}).$$

$$\tag{15.12}$$

Let us next see if the right hand side of Eq. (15.12) can be written as an exponential. From the fact that for any i, $s_i = \pm 1$, we also have the following identities.

$$\left(\frac{1}{2}(s_i + s_{i+1})\right)^{2n} = \left(\frac{1}{2}(s_i + s_{i+1})\right)^2, \quad \text{for } n \geq 1,$$

$$\left(\frac{1}{2}(s_i + s_{i+1})\right)^{2n+1} = \frac{1}{2}(s_i + s_{i+1}), \quad \text{for } n \geq 0,$$

$$\left(\frac{1}{2}(s_i - s_{i+1})\right)^{2n} = \left(\frac{1}{2}(s_i - s_{i+1})\right)^2, \quad \text{for } n \geq 1,$$

$$\left(\frac{1}{2}(s_i + s_{i+1})\right)^n \left(\frac{1}{2}(s_i - s_{i+1})\right)^m = 0, \quad \text{for } n, m \geq 1. \quad (15.13)$$

Using the relations in Eq. (15.13), then, we can obtain (for constant parameters δ and Δ)

$$e^{\left(\Delta\left(\frac{1}{2}(s_i - s_{i+1})\right)^2 + \delta \frac{1}{2}(s_i + s_{i+1})\right)}$$

$$= 1 + \sum_{n=1}^{\infty} \frac{1}{n!}\left[\Delta\left(\frac{1}{2}(s_i - s_{i+1})\right)^2 + \delta \frac{1}{2}(s_i + s_{i+1})\right]^n$$

$$= 1 + \sum_{n=1}^{\infty}\left[\frac{\Delta^n}{n!}\left(\frac{1}{2}(s_i - s_{i+1})\right)^{2n} + \frac{\delta^n}{n!}\left(\frac{1}{2}(s_i + s_{i+1})\right)^n\right]$$

$$= 1 + \sum_{n=1}^{\infty}\frac{\Delta^n}{n!}\left(\frac{1}{2}(s_i - s_{i+1})\right)^2$$

$$+ \sum_{n=0}^{\infty}\frac{\delta^{2n+1}}{(2n+1)!}\left(\frac{1}{2}(s_i + s_{i+1})\right)^{2n+1}$$

$$+ \sum_{n=1}^{\infty}\frac{\delta^{2n}}{(2n)!}\left(\frac{1}{2}(s_i + s_{i+1})\right)^{2n}$$

$$= 1 + (e^{\Delta} - 1)\left(\frac{1}{2}(s_i - s_{i+1})\right)^2 + \sum_{n=0}^{\infty}\frac{\delta^{2n+1}}{(2n+1)!}\frac{1}{2}(s_i + s_{i+1})$$

$$+ \sum_{n=1}^{\infty}\frac{\delta^{2n}}{(2n)!}\left(\frac{1}{2}(s_i + s_{i+1})\right)^2$$

$$= 1 + (e^\Delta - 1)\left(\frac{1}{2}(s_i - s_{i+1})\right)^2 + \sinh\delta\left(\frac{1}{2}(s_i + s_{i+1})\right)$$

$$+ (\cosh\delta - 1)\left(\frac{1}{2}(s_i + s_{i+1})\right)^2. \tag{15.14}$$

Let us now use the algebraic relation

$$\left(\frac{1}{2}(s_i + s_{i+1})\right)^2 + \left(\frac{1}{2}(s_i - s_{i+1})\right)^2 = 1, \tag{15.15}$$

which allows us to write Eq. (15.14) also as

$$e^{\left(\Delta\left(\frac{1}{2}(s_i - s_{i+1})\right)^2 + \delta\frac{1}{2}(s_i + s_{i+1})\right)}$$

$$= \cosh\delta\left(\frac{1}{2}(s_i + s_{i+1})\right)^2 + e^\Delta\left(\frac{1}{2}(s_i - s_{i+1})\right)^2$$

$$+ \sinh\delta\left(\frac{1}{2}(s_i + s_{i+1})\right). \tag{15.16}$$

We note that this has precisely the same form as the transition amplitude between two neighboring sites in Eq. (15.12) provided we make the identification

$$\cosh\delta = \frac{e^\delta + e^{-\delta}}{2} = 1,$$

$$e^\Delta = \epsilon\alpha,$$

$$\sinh\delta = \frac{e^\delta - e^{-\delta}}{2} = -\epsilon\gamma. \tag{15.17}$$

Equivalently, with the identification

$$e^\Delta = \epsilon\alpha,$$

$$e^\delta = 1 - \epsilon\gamma, \tag{15.18}$$

we can write

$$\langle s_{i+1}|e^{-\epsilon H_q}|s_i\rangle = e^{\left(\Delta\left(\frac{1}{2}(s_i-s_{i+1})\right)^2+\delta\frac{1}{2}(s_i+s_{i+1})\right)}. \tag{15.19}$$

Consequently, using this identification, we obtain (here summation over the end points is assumed)

$$\text{Tr}\,e^{-TH_q} = \sum_{s_i=\pm1}\langle s_1|e^{-\epsilon H_q}|s_N\rangle\langle s_N|e^{-\epsilon H_q}|s_{N-1}\rangle\cdots\langle s_2|e^{-\epsilon H_q}|s_1\rangle$$

$$= \sum_{s_i=\pm1} e^{\left(\Delta\sum_{i=1}^{N}\left(\frac{1}{2}(s_i-s_{i+1})\right)^2+\delta\sum_{i=1}^{N}\frac{1}{2}(s_i+s_{i+1})\right)}$$

$$= \sum_{s_i=\pm1} e^{\left(\Delta\sum_{i=1}^{N}\frac{1}{2}(1-s_is_{i+1})+\delta\sum_{i=1}^{N}\frac{1}{2}(s_i+s_{i+1})\right)}$$

$$= e^{\frac{N\Delta}{2}}\sum_{s_i=\pm1} e^{\left(-\frac{\Delta}{2}\sum_{i=1}^{N}s_is_{i+1}+\delta\sum_{i=1}^{N}\frac{1}{2}(s_i+s_{i+1})\right)}$$

$$= e^{\frac{N\Delta}{2}}\sum_{s_i=\pm1} e^{\left(\beta J\sum_{i=1}^{N}s_is_{i+1}-\beta B\sum_{i=1}^{N}\frac{1}{2}(s_i+s_{i+1})\right)}$$

$$= e^{\frac{N\Delta}{2}}\sum_{s_i=\pm1} e^{-\beta H}, \tag{15.20}$$

provided we identify

$$\Delta = -2\beta J, \quad \delta = -\beta B. \tag{15.21}$$

With these identifications, then we see that we can write

$$\text{Tr}\,e^{-TH_q} = e^{-N\beta J}Z(\beta), \tag{15.22}$$

where $Z(\beta)$ represents the partition function for the one dimensional ising model. Once again, this shows that the quantum fluctuations in a quantum theory can be related to the thermal fluctuations of a classical statistical system.

15.2 The partition function

To evaluate the partition function for the one dimensional Ising model explicitly, let us rewrite the exponent in Eq. (15.3) in the partition function in a way that is easy to use. Note that

$$e^{\left(\beta J s_i s_{i+1} - \beta B \frac{1}{2}(s_i + s_{i+1})\right)}$$

$$= e^{\left(-2\beta J \times \frac{1}{2}(-1 + 1 - s_i s_{i+1}) - \beta B \frac{1}{2}(s_i + s_{i+1})\right)}$$

$$= e^{\beta J} e^{\left(-2\beta J \left(\frac{1}{2}(s_i - s_{i+1})\right)^2 - \beta B \frac{1}{2}(s_i + s_{i+1})\right)}$$

$$= e^{\beta J} \left[\cosh \beta B \left(\frac{1}{2}(s_i + s_{i+1})\right)^2 + e^{-2\beta J} \left(\frac{1}{2}(s_i - s_{i+1})\right)^2 \right.$$

$$\left. - \sinh \beta B \frac{1}{2}(s_i + s_{i+1}) \right], \tag{15.23}$$

where we have used the identities in Eq. (15.13). It is now clear that if we define a matrix operator

$$K = e^{\beta J} \left[\cosh \beta B \mathbb{1} + e^{-2\beta J} \sigma_1 - \sinh \beta B \, \sigma_3 \right], \tag{15.24}$$

then, the matrix element of this operator between the eigenstates $|s_i\rangle$ and $|s_{i+1}\rangle$ of the σ_3 operator will be obtained using Eq. (15.11) to be

$$\langle s_{i+1} | K | s_i \rangle = e^{\left(\beta J s_i s_{i+1} - \beta B \frac{1}{2}(s_i + s_{i+1})\right)}. \tag{15.25}$$

Note that K is a 2×2 matrix and has the explicit form

$$K = \begin{pmatrix} e^{\beta J}(\cosh \beta B - \sinh \beta B) & e^{-\beta J} \\ e^{-\beta J} & e^{\beta J}(\cosh \beta B + \sinh \beta B) \end{pmatrix}$$

$$- \begin{pmatrix} e^{\beta(J-B)} & e^{-\beta J} \\ e^{-\beta J} & e^{\beta(J+B)} \end{pmatrix}. \tag{15.26}$$

From this analysis, it is clear that we can derive the partition function for the one dimensional Ising model explicitly as follows.

$$Z(\beta) = \sum_{s_i=\pm 1} e^{\left(\beta J \sum_{i=1}^{N} s_i s_{i+1} - \beta B \sum_{i=1}^{N} \frac{(s_i+s_{i+1})}{2} \right)}$$

$$= \sum_{s_i=\pm 1} \langle s_1|K|s_N \rangle \langle s_N|K|s_{N-1} \rangle \cdots \langle s_2|K|s_1 \rangle$$

$$= \mathrm{Tr}\, K^N = \lambda_1^N + \lambda_2^N, \tag{15.27}$$

where λ_1 and λ_2 are the two eigenvalues of the matrix K in Eq. (15.26).

The eigenvalues of the matrix K can be easily obtained from

$$\det(K - \lambda I) = 0,$$

$$\text{or,} \quad \det \begin{pmatrix} e^{\beta(J-B)} - \lambda & e^{-\beta J} \\ e^{-\beta J} & e^{\beta(J+B)} - \lambda \end{pmatrix} = 0,$$

$$\text{or,} \quad \lambda^2 - 2\lambda e^{\beta J} \cosh \beta B + e^{2\beta J} - e^{-2\beta J} = 0. \tag{15.28}$$

This is a quadratic equation whose solutions are easily obtained to be

$$\lambda = e^{\beta J} \cosh \beta B \pm \left(e^{2\beta J} \cosh^2 \beta B - e^{2\beta J} + e^{-2\beta J} \right)^{\frac{1}{2}}$$

$$= e^{\beta J} \cosh \beta B \pm \left(e^{2\beta J}(1 + \sinh^2 \beta B) - e^{2\beta J} + e^{-2\beta J} \right)^{\frac{1}{2}}$$

$$= e^{\beta J} \left(\cosh \beta B \pm \left(\sinh^2 \beta B + e^{-4\beta J} \right)^{\frac{1}{2}} \right). \tag{15.29}$$

If we identify the two eigenvalues as

$$\lambda_1 = e^{\beta J} \left(\cosh \beta B + \left(\sinh^2 \beta B + e^{-4\beta J} \right)^{\frac{1}{2}} \right),$$

$$\lambda_2 = e^{\beta J} \left(\cosh \beta B - \left(\sinh^2 \beta B + e^{-4\beta J} \right)^{\frac{1}{2}} \right), \qquad (15.30)$$

then, we note that since $\lambda_1 > \lambda_2$, for large N, we can approximately write

$$Z(\beta) = \text{Tr} \, K^N$$

$$= \lambda_1^N + \lambda_2^N \simeq \lambda_1^N$$

$$= \left[e^{\beta J} \left(\cosh \beta B + (\sinh^2 \beta B + e^{-4\beta J})^{\frac{1}{2}} \right) \right]^N. \qquad (15.31)$$

This method of evaluating the partition function is known as the matrix method and we recognize K as the transfer matrix for the system (see also Section 3.3).

We can now derive various quantities of thermodynamic interest. Let us note from Eq. (15.31) that we can write

$$\ln Z(\beta) = N \ln \left[e^{\beta J} \left(\cosh \beta B + (\sinh^2 \beta B + e^{-4\beta J})^{\frac{1}{2}} \right) \right]$$

$$= N \left[\beta J + \ln \left\{ \cosh \beta B + (\sinh^2 \beta B + e^{-4\beta J})^{\frac{1}{2}} \right\} \right]. \qquad (15.32)$$

Therefore, the average magnetization per site defined in Eq. (14.25) can now be derived from Eq. (15.32) to be

$$M = \frac{1}{N} \frac{\partial F}{\partial B} = \frac{1}{N} \frac{\partial}{\partial B} \left(-\frac{1}{\beta} \ln Z(\beta) \right) = -\frac{1}{N\beta} \frac{\partial \ln Z(\beta)}{\partial B}$$

$$= -\frac{1}{N\beta} \frac{N \left(\beta \sinh \beta B + \frac{1}{2} \frac{2\beta \sinh \beta B \cosh \beta B}{(\sinh^2 \beta B + e^{-4\beta J})^{\frac{1}{2}}} \right)}{\cosh \beta B + (\sinh^2 \beta B + e^{-4\beta J})^{\frac{1}{2}}}$$

$$= -\frac{\sinh \beta B}{\left(\sinh^2 \beta B + e^{-4\beta J} \right)^{\frac{1}{2}}}. \qquad (15.33)$$

It is interesting to note that when the external magnetic field is switched off, the magnetization vanishes. In this one dimensional system, therefore, there is no spontaneous magnetization and consequently, it cannot describe the properties of a magnet. The magnetic susceptibility for such a system can also be easily calculated (see Eq. (14.26)) and takes the form

$$\chi = - \left. \frac{\partial M}{\partial B} \right|_{B=0} = \beta\, e^{2\beta J}\,. \tag{15.34}$$

This shows that for $|2\beta J| < 1$, the susceptibility obeys Curie's law. Namely, in this case,

$$\chi \simeq \beta = \frac{1}{kT}. \tag{15.35}$$

The absence of spontaneous magnetization in the present system may appear puzzling because naively, we would expect the configurations where all the spins are "up" or "down" to correspond to minimum energy states (for $J > 0$). These, being ordered, we would have expected a nontrivial spontaneous magnetization for the system. The lack of magnetization can actually be understood through the instanton calculation which we discussed earlier. Let us recall that for the double-well potential (see Section 7.4 as well as Chapter 8) the two naive ground states would give

$$\langle x \rangle = \pm a\,. \tag{15.36}$$

On the other hand, as we have seen earlier in Eq. (7.59), the true ground state is a mixture of these two states (the symmetric state) such that

$$\langle x \rangle_{\text{true}} = 0\,. \tag{15.37}$$

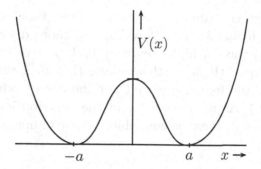

Figure 15.1: Double well potential.

In this case, we showed explicitly that the tunneling or the presence of instanton states contributes significantly leading to the mixing of the states and restoring the symmetry.

In the one dimensional Ising spin system, we can correspondingly think of the following two spin configurations

$$\uparrow\uparrow\uparrow\uparrow\uparrow\uparrow\uparrow\uparrow\uparrow,$$

$$\downarrow\downarrow\downarrow\downarrow\downarrow\downarrow\downarrow\downarrow\downarrow,$$

as denoting the two ground states for which the magnetization is nonzero

$$\langle M \rangle_\beta \neq 0. \tag{15.38}$$

However, in the present case, there are other spin configurations such as

$$\uparrow\uparrow\uparrow\uparrow\downarrow\downarrow\downarrow\downarrow \quad \text{—one kink or one instanton,}$$

$$\uparrow\uparrow\uparrow\downarrow\downarrow\uparrow\uparrow\uparrow \quad \text{—two kinks or one instanton-anti-instanton,}$$

and so on which contribute significantly. Even though these configurations have higher energy, they are also more disordered. Consequently, they have a higher entropy and as a result can have a lower free energy. (It is worth recalling that in a thermodynamic ensemble, it is the free energy and not the energy which plays the dominant role.) As a consequence of the contributions from these spin configurations, the true ensemble average of magnetization vanishes. Namely,

$$\langle M \rangle_\beta^{\text{true}} = 0 \,. \tag{15.39}$$

This qualitative discussion can actually be made more precise through the use of path integrals.

Finally, let us note here that as we have discussed earlier, path integrals are defined by discretizing space-time variables. In fact, space-time lattices are often used to define a regularized quantum field theory. The continuum theory is, of course, obtained in the limit when the lattice spacing goes to zero. Viewed in this way, the spin Hamiltonian is seen to have the continuum limit

$$H = -J \sum_i s_i s_{i+1} + B \sum_i s_i$$

$$= \frac{J}{2} \sum_i (s_{i+1} - s_i)^2 + \frac{B}{2} \sum_i (s_i + s_{i+1}) - NJ$$

$$\xrightarrow{\text{continuum}} \int \mathrm{d}x \left(\frac{\alpha}{2} (\partial s(x))^2 + \gamma s(x) \right) + \text{constant.} \tag{15.40}$$

where α and γ are two constants. Namely, we can think of the one dimensional Ising model as corresponding to a one dimensional free scalar field theory interacting with a constant external source in the continuum limit.

15.3 Two dimensional Ising model

Let us next consider a two dimensional array of spins on a square lattice interacting through nearest neighbors. Once again, let us use periodic boundary conditions along both the axes so that

$$s_{1,i_2} = s_{N+1,i_2}, \quad s_{i_1,1} = s_{i_1,N+1}, \tag{15.41}$$

where we are using the notation that $i = (i_1, i_2)$ denotes a point on the two dimensional lattice and we are assuming that N denotes the total number of lattice sites along any axis.

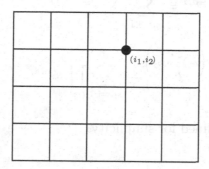

The total number of points on the lattice is then obtained to be

$$n = N^2. \tag{15.42}$$

The spins are assumed to take only the values ± 1, namely,

$$s_i = s_{i_1,i_2} = \pm 1, \quad \text{for all } i_1, i_2, \tag{15.43}$$

corresponding to the fact that they can only point "up" or "down". The Hamiltonian describing the interaction of the spins is given by (see also Eq. (14.20))

$$H = -J \sum_{\langle ij \rangle} s_i s_j = -J \sum_{i_1,i_2=1}^{N} \left(s_{i_1,i_2} s_{i_1+1,i_2} + s_{i_1,i_2} s_{i_1,i_2+1} \right). \tag{15.44}$$

The symbol $\langle ij \rangle$ is introduced as a short hand for sites which are nearest neighbors and the coupling J is assumed to be positive. We can also think of the sum in Eq. (15.44) as being taken over all the links of the lattice. (Remember that a link connects two nearest neighbors on a lattice.)

The partition function for the system described by the Hamiltonian in Eq. (15.44) can now be defined to be

$$
Z(\beta) = \sum_{s_i = \pm 1} e^{-\beta H} = \sum_{s_i = \pm 1} e^{\left(\beta J \sum_{\langle ij \rangle} s_i s_j \right)}
$$

$$
= \sum_{s_i = \pm 1} e^{\left(\kappa \sum_{\langle ij \rangle} s_i s_j \right)} = \sum_{s_i = \pm 1} \prod_{\langle ij \rangle} e^{\kappa s_i s_j} , \tag{15.45}
$$

where we have defined for simplicity

$$
\kappa = \beta J . \tag{15.46}
$$

Note that we are discussing the simpler case where the spin system is not interacting with an external magnetic field. This partition function, as it stands, appears to be only slightly more complicated than that for the one dimensional case in Eq. (15.3). However, as we will see, this partition function is much more difficult to evaluate in closed form. Before going into the actual evaluation of this partition function, let us discuss some of the symmetries associated with this system.

15.4 Duality

Let us note that since

$$
s_i = \pm 1 , \tag{15.47}
$$

we can expand the exponent in the partition function in Eq. (15.45) to obtain

$$
e^{\kappa s_i s_j} = \cosh \kappa + s_i s_j \sinh \kappa
$$

$$
= \cosh \kappa \left(1 + s_i s_j \tanh \kappa \right) . \tag{15.48}
$$

Therefore, we can also write

$$Z(\beta) = \sum_{s_i=\pm 1} \prod_{\langle ij \rangle} e^{\kappa s_i s_j}$$

$$= \sum_{s_i=\pm 1} \prod_{\langle ij \rangle} \cosh \kappa \, (1 + s_i s_j \tanh \kappa)$$

$$= (\cosh \kappa)^{2n} \sum_{s_i=\pm 1} \prod_{\langle ij \rangle} (1 + s_i s_j \tanh \kappa)$$

$$= (\cosh \kappa)^{2n} \sum_{s_i=\pm 1} \prod_{\langle ij \rangle} \sum_{l=0}^{1} (s_i s_j \tanh \kappa)^l$$

$$= (\cosh \kappa)^{2n} \sum_{s_i=\pm 1} \prod_{\langle ij \rangle} \sum_{l=0}^{1} (\tanh \kappa)^l (s_i s_j)^l . \tag{15.49}$$

We see that we can simplify this expression by assigning a (link) number $l_k = l_{ij} = l_{ji} = (0, 1)$ to each link between the sites i and j and rewriting

$$Z(\beta) = (\cosh \kappa)^{2n} \sum_{l_k} (\tanh \kappa)^{l_1 + l_2 + \cdots} \sum_{s_i=\pm 1} \prod_{\langle ij \rangle} (s_i s_j)^{l_{ij}} . \tag{15.50}$$

Let us next note that the product on the right hand side in Eq. (15.50) can simply be understood as the product of the spins at each lattice site with an exponent corresponding to the sum of the link numbers for the four links meeting at that site. Namely, for nearest neighbors j

$$\sum_{s_i=\pm 1} \prod_{\langle ij \rangle} (s_i s_j)^{l_{ij}} = \sum_{s_i=\pm 1} \prod_i (s_i)^{\sum_j l_{ij}} = \sum_{s_i=\pm 1} \prod_i (s_i)^{n_i} , \tag{15.51}$$

where we have defined, for nearest neighbors j,

$$n_i = \sum_j l_{ij} . \tag{15.52}$$

The sum over the spin values can now be done to give

$$\sum_{s_i=\pm 1} \prod_{\langle ij \rangle} (s_i s_j)^{l_{ij}} = \prod_i \sum_{s_i=\pm 1} (s_i)^{n_i} = \prod_i \left(1 + (-1)^{n_i}\right). \qquad (15.53)$$

It is clear that the expression in Eq. (15.53) vanishes when n_i is odd. For even n_i, on the other hand, it has the value

$$\sum_{s_i=\pm 1} \prod_{\langle ij \rangle} (s_i s_j)^{l_{ij}} = 2^n. \qquad (15.54)$$

Putting everything back in Eq. (15.50), we obtain

$$Z(\beta) = (2 \cosh^2 \kappa)^n \sum_{l_k} (\tanh \kappa)^{l_1 + l_2 + \cdots} = Z(\kappa). \qquad (15.55)$$

The constraint here is that the l_k's in Eq. (15.55) must satisfy

$$\sum_j l_{ij} = 0, \quad \mod 2, \qquad (15.56)$$

for any four links joining at a site. In other words, if l_1, l_2, l_3 and l_4 denote the link numbers for four links meeting at a common site, then

$$l_1 + l_2 + l_3 + l_4 = 0, \quad \mod 2. \qquad (15.57)$$

Let us next consider the dual lattice associated with our original lattice. It is constructed by placing a lattice site at the center of each plaquette of the original lattice.

Thus, each plaquette of the dual lattice encloses a given site of the original lattice and intersects the four links originating from that site. Let us also define a dual variable σ_i at each site of the dual lattice and assume that it takes values ± 1. Denoting by $(1, 2, 3, 4)$ the sites of the dual lattice which enclose the point k of the original lattice, we note that for every link that is intersected by a dual link, we can define

Figure 15.2: A plaquette in the dual lattice.

$$l_1 = \frac{1}{2}(1 - \sigma_1\sigma_2),$$

$$l_2 = \frac{1}{2}(1 - \sigma_2\sigma_3),$$

$$l_3 = \frac{1}{2}(1 - \sigma_3\sigma_4),$$

$$l_4 = \frac{1}{2}(1 - \sigma_4\sigma_1). \tag{15.58}$$

We see that each of the l_k's have the value 0 or 1 as required. Furthermore, we also have

$$l_1 + l_2 + l_3 + l_4 = \frac{1}{2}(4 - \sigma_1\sigma_2 - \sigma_2\sigma_3 - \sigma_3\sigma_4 - \sigma_4\sigma_1)$$

$$= \frac{1}{2}\left(4 - (\sigma_1 + \sigma_3)(\sigma_2 + \sigma_4)\right)$$

$$= 0, \quad \mathrm{mod}\ 2. \tag{15.59}$$

In other words, the constraint equation in Eq. (15.57) can be naturally solved through the dual lattice variables.

Going back to the expression for the partition function derived in Eq. (15.55), we note that using Eq. (15.58) we can write

$$(\tanh \kappa)^{l_1} = (\tanh \kappa)^{\frac{1}{2}(1-\sigma_1\sigma_2)} = e^{-\kappa^*(1-\sigma_1\sigma_2)}, \tag{15.60}$$

where we have defined

$$\tanh \kappa = e^{-2\kappa^*} . \tag{15.61}$$

We note here that when $\kappa = \beta J$ is large, κ^* is small and *vice versa*. Substituting this back into Eq. (15.55), we obtain

$$Z(\kappa) = (2\cosh^2 \kappa)^n \sum_{\sigma_i=\pm 1} e^{\left(-\kappa^* \sum_{\langle ij \rangle} (1-\sigma_i\sigma_j)\right)}$$

$$= (2\cosh^2 \kappa)^n e^{-2n\kappa^*} \sum_{\sigma_i=\pm 1} e^{\left(\kappa^* \sum_{\langle ij \rangle} \sigma_i\sigma_j\right)},$$

$$\text{or,} \quad \frac{Z(\kappa)}{(2\cosh^2 \kappa)^n} = e^{-2n\kappa^*} \sum_{\sigma_i=\pm 1} e^{\left(\kappa^* \sum_{\langle ij \rangle} \sigma_i\sigma_j\right)} = \frac{Z(\kappa^*)}{(e^{2\kappa^*})^n}. \tag{15.62}$$

Using trigonometric identities, this can also be written as

$$\frac{Z(\kappa)}{(\sinh 2\kappa)^{\frac{n}{2}}} = \frac{Z(\kappa^*)}{(\sinh 2\kappa^*)^{\frac{n}{2}}}. \tag{15.63}$$

The relation in Eq. (15.62) (or Eq. (15.63)) is quite interesting in that the relation

$$\tanh \kappa = e^{-2\kappa^*} , \tag{15.64}$$

which can also be written as

$$\sinh 2\kappa \sinh 2\kappa^* = 1 , \tag{15.65}$$

defines a transformation between strong and weak couplings (or high and low temperatures (see Eq. (15.46))). And we find that the corresponding partition functions are related as well. Consequently, if there exists a single phase transition in this model (which was known from earlier general arguments due to Peirels), it must occur at a unique point where (note that κ or κ_c is positive as is clear from Eq. (15.64))

$$\kappa = \kappa_c = \kappa_c^*,$$

or, $\quad \sinh^2 2\kappa_c = 1,$

or, $\quad \sinh 2\kappa_c = 1,$

or, $\quad e^{2\kappa_c} - e^{-2\kappa_c} = 2,$

or, $\quad e^{2\kappa_c} = \sqrt{2} + 1,$

or, $\quad \kappa_c = \beta_c J = \frac{1}{2} \ln\left(\sqrt{2} + 1\right),$

or, $\quad \beta_c = \frac{1}{2J} \ln\left(\sqrt{2} + 1\right).$ $\hfill (15.66)$

15.5 High and low temperature expansions

Quite often in statistical mechanics, the partition function cannot be evaluated exactly for an arbitrary temperature. In such a case, we study ensemble averages of observables associated with the system as an expansion in temperature at very high temperatures as well as at very low temperatures to see if any meaningful conclusion regarding the system can be obtained. In the language of field theory, we have seen in Eq. (14.55) that temperature can be related to Planck's constant which in some sense measures the quantum coupling. Therefore, high and low temperature expansions are also known as strong and weak coupling expansions (or approximations).

Let us go back to the partition function for the two dimensional Ising model

$$Z(\kappa) = \sum_{s_i = \pm 1} e^{\left(\kappa \sum_{\langle ij \rangle} s_i s_j\right)}, \hfill (15.67)$$

where, as in Eq. (15.46), we have identified

$$\kappa = \beta J = \frac{J}{kT} \,.$$

(15.68)

We have seen in Eq. (15.55) that we can write

$$\frac{Z(\kappa)}{\left(2\cosh^2\kappa\right)^n} = \sum_{l_k=0,1} (\tanh\kappa)^{l_1+l_2+\cdots} \,,$$

(15.69)

where the link numbers are assumed to satisfy

$$l_1 + l_2 + l_3 + l_4 = 0, \quad \mathrm{mod}\, 2 \,,$$

(15.70)

for any four links meeting at a lattice site. If the temperature is high enough, then κ is small and so is $\tanh\kappa$ and the right hand side of Eq. (15.69) can be expanded in a power series in $\tanh\kappa$. To do this, let us note that the link numbers, l_k's, can only take values 0 or 1. Accordingly, let us postulate the graphical rule that if $l_k = 0$ for a particular link, then we will not draw a bond for the link connecting the two lattice sites whereas if $l_k = 1$, then we will draw a bond. With this rule then, the constraint on the link numbers in Eq. (15.70) simply says that there must be an even number of bonds meeting at a given lattice site. Consequently, we note that the first term of the expansion on the right hand side will correspond to the case where there are no bonds on the lattice giving a unit contribution (since this corresponds to the case for which all link numbers vanish).

The next term in the series will have the form

 $\longrightarrow (\tanh\kappa)^4.$

In other words, the first nontrivial term in the series will correspond to the product of the weight factor $\tanh\kappa$ over a single plaquette. The plaquette can be drawn in n-different ways on the lattice (recall the periodic boundary condition) and hence this term will come with a multiplicity of n. (It is clear that the constraint in Eq. (15.70) at every site would lead to closed diagrams of connected links.)

The next term in the series will represent the product of the weight factor $\tanh\kappa$ over a plaquette involving two lattice lengths

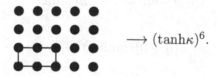 $\longrightarrow (\tanh\kappa)^6.$

It is not hard to see that such a diagram can be drawn in $2n$ different ways and hence this term will come with a multiplicity of $2n$. At the next order the diagrams that will contribute are

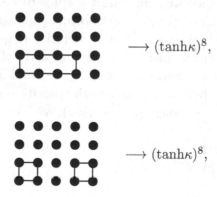 $\longrightarrow (\tanh\kappa)^8,$

$\longrightarrow (\tanh\kappa)^8,$

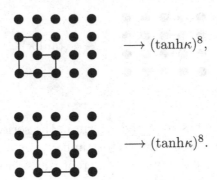

$\longrightarrow (\tanh \kappa)^8,$

$\longrightarrow (\tanh \kappa)^8.$

The combinatorics can be worked out in a straightforward manner for these graphs $(2n+\frac{1}{2}n(n-5)+4n+n)$ so that the high temperature expansion of the partition function will have the form

$$\frac{Z(\kappa)}{(2\cosh^2 \kappa)^n} = 1 + n(\tanh \kappa)^4 + 2n(\tanh \kappa)^6$$

$$+ \frac{1}{2}n(n+9)(\tanh \kappa)^8 + \cdots . \tag{15.71}$$

To obtain the low temperature expansion, let us note that when T is small,

$$\kappa = \frac{J}{kT}, \tag{15.72}$$

is large. If $T = 0$, then we expect all the spins to be frozen along one axis, say up. Therefore, the low temperature expansion would merely measure the deviation from such an ordered configuration. Namely, the low temperature expansion will be a measure of how many spins flip as T becomes nonzero but small. Thus, dividing the partition function by $e^{2n\kappa}$ (which is the value of the partition function when all the spins are pointing along one direction), we have from Eq. (15.67)

$$\frac{Z(\kappa)}{e^{2n\kappa}} = \sum_{s_i=\pm 1} e^{\left(\kappa \sum_{\langle ij \rangle} (-1+s_i s_j) \right)}. \tag{15.73}$$

To develop the right hand side diagrammatically, let us draw a cross on the lattice site to represent a flipped spin. Thus, the first term on the right hand side of Eq. (15.73) will correspond to a diagram without any flipped spin of the form

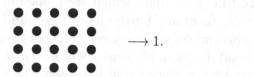 $\longrightarrow 1.$

The next term in the series will correspond to the case where one of the spins on the lattice has flipped and will represent a diagram of the form

 $\longrightarrow e^{-8\kappa}.$

In other words, a single flipped spin will interact with each of the nearest neighbors with weight $e^{-2\kappa}$ and since there are four nearest neighbors for any site, the term would lead to a contribution of $e^{-8\kappa}$. Furthermore, the flipped spin can occur at any lattice site and hence this term will come with a multiplicity of n.

The next term in the series will correspond to two flipped spins. Interestingly enough, this leads to two possibilities. Namely, the flipped spins can be nearest neighbors or they need not be. Diagrammatically, the two possibilities can be represented as

$\longrightarrow e^{-12\kappa},$

$\longrightarrow e^{-16\kappa}.$

In other words, in the first configuration, the interaction between the two nearest neighbor spins which are flipped does not contribute to the partition function. Furthermore, the number of ways a pair of flipped spins can occur as nearest neighbors is $2n$. The multiplicity of the second diagram, obviously, will be $\frac{1}{2}n(n-5)$ (this is because once a site with a flipped spin is chosen, the four nearest neighbor sites cannot have another flipped spin). However, that is not the only kind of diagram which contributes an amount $e^{-16\kappa}$. In fact, there are other classes of diagrams which also contribute the same amount, namely, ones where there are three or four flipped spins which are nearest neighbors (with multiplicities $2n$, $4n$ and n respectively)

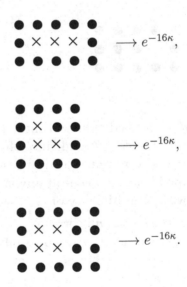

Thus, we can consistently derive a low temperature expansion of the partition function which has the form

$$\frac{Z(\kappa)}{e^{2n\kappa}} = 1 + ne^{-8\kappa} + 2ne^{-12\kappa} + \frac{1}{2}n(n+9)e^{-16\kappa} + \cdots . \quad (15.74)$$

It is clear now that if we denote, for low temperatures,

$$\kappa = \kappa^*, \tag{15.75}$$

then, we can write the low temperature expansion also as

$$\frac{Z(\kappa^*)}{e^{2n\kappa^*}} = 1 + ne^{-8\kappa^*} + 2ne^{-12\kappa^*} + \frac{1}{2}n(n+9)e^{-16\kappa^*} + \cdots . \tag{15.76}$$

Thus, comparing Eqs. (15.71) and (15.76) we see that under the mapping

$$\tanh \kappa = e^{-2\kappa^*}, \tag{15.77}$$

we have

$$\frac{Z(\kappa)}{(2\cosh^2 \kappa)^n} = \frac{Z(\kappa^*)}{e^{2n\kappa^*}} . \tag{15.78}$$

This is, of course, the duality relation that we had derived earlier in Eq. (15.62). Here we see explicitly that the duality mapping (transformation) really takes us from the high temperature expansion to the low temperature expansion and *vice versa*.

15.6 Quantum mechanical model

Before finding the correspondence between the two dimensional Ising model and a quantum mechanical model, let us derive the transfer matrix for the system. Let us begin by writing the Hamiltonian for the system in a way that is better suited for our manipulations. Let us label the sites on a given row by $1 \leq i \leq N$ and the rows by $1 \leq m \leq N$.

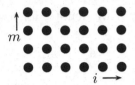

Then, we can write the interaction energy between the spins on a row m as

$$H(m) = -J \sum_{i=1}^{N} s_i(m)s_{i+1}(m) \, . \tag{15.79}$$

Similarly, the interaction energy between two adjacent rows, say m and $(m+1)$, can be written as

$$H(m, m+1) = -J \sum_{i=1}^{N} s_i(m)s_i(m+1) \, . \tag{15.80}$$

Given this, we can write (see Eq. (15.44)) the total energy of the system as

$$H = \sum_{m=1}^{N} \left(H(m) + H(m, m+1) \right)$$

$$= -J \sum_{m=1}^{N} \sum_{i=1}^{N} \left(s_i(m)s_{i+1}(m) + s_i(m)s_i(m+1) \right) \, . \tag{15.81}$$

If we desire, we can also introduce an external magnetic field at this point. However, let us ignore it for simplicity.

The partition function will involve the exponent

$$e^{-\beta H} = e^{\left(-\beta \sum\limits_{m=1}^{N} (H(m)+H(m,m+1))\right)}. \tag{15.82}$$

Let us, for simplicity, concentrate only on a single factor of this exponent. Namely, let us look at

$$e^{-\beta(H(m)+H(m,m+1))}$$

$$= e^{\beta J\left(\sum\limits_{i=1}^{N}(s_i(m)s_{i+1}(m)+s_i(m)s_i(m+1))\right)}$$

$$= \prod_{i=1}^{N} e^{\beta J s_i(m)s_{i+1}(m)} e^{\beta J s_i(m)s_i(m+1)}. \tag{15.83}$$

Let us next note that on every row there are N sites and if we introduce a two component eigenvector of σ_3 at every site, then we can define a 2^N dimensional vector space on every row through a direct product as

$$|s(m)\rangle = |s_1(m)\rangle \otimes |s_2(m)\rangle \otimes \cdots |s_N(m)\rangle$$

$$= |s_1, s_2, \ldots, s_N\rangle, \tag{15.84}$$

where s_i represents the eigenvalue of σ_3 for the state $|s_i(m)\rangle$. We can define an inner product on such states as

$$\langle s(m+1)|s(m)\rangle = \langle s_1', s_2', \ldots, s_N' | s_1, s_2, \ldots, s_N\rangle$$

$$= \delta_{s_1 s_1'} \delta_{s_2 s_2'} \cdots \delta_{s_N s_N'}. \tag{15.85}$$

Similarly, the completeness relation will take the form

$$\sum_{s_i=\pm 1} |s(m)\rangle\langle s(m)| = \mathbb{1}, \tag{15.86}$$

where $\mathbb{1}$ denotes the $2^N \times 2^N$ identity matrix.

With these preliminaries, let us now introduce the following set of $2^N \times 2^N$ matrices (there will be N of each)

$$\sigma_1(i) = \mathbb{1} \otimes \mathbb{1} \otimes \cdots \otimes \sigma_1 \otimes \mathbb{1} \cdots \otimes \mathbb{1},$$

$$\sigma_3(i) = \mathbb{1} \otimes \mathbb{1} \otimes \cdots \otimes \sigma_3 \otimes \mathbb{1} \cdots \otimes \mathbb{1}. \tag{15.87}$$

Namely, all the entries in the above expression correspond to the trivial 2×2 identity matrix except at the i th entry. Let us also record here the product formula for matrices defined through direct products, namely,

$$(A \otimes B)(C \otimes D) = AC \otimes BD. \tag{15.88}$$

We note that the σ_1 in the ith place acts on the vectors $|s_i\rangle$ and, therefore, as a 2×2 matrix, we can write

$$\langle s_i'|\alpha e^{\theta \sigma_1}|s_i\rangle = \langle s_i'|\alpha \cosh \theta + \alpha \sigma_1 \sinh \theta|s_i\rangle$$

$$= \begin{pmatrix} \alpha \cosh \theta & \alpha \sinh \theta \\ \alpha \sinh \theta & \alpha \cosh \theta \end{pmatrix}, \tag{15.89}$$

where α is a constant. On the other hand, we note that a term such as

$$e^{\beta J s_i(m) s_i(m+1)}, \tag{15.90}$$

can also be written as a 2×2 matrix of the form

$$e^{\beta J s_i(m) s_i(m+1)} = \begin{pmatrix} e^{\beta J} & e^{-\beta J} \\ e^{-\beta J} & e^{\beta J} \end{pmatrix}. \tag{15.91}$$

Therefore, comparing the two relations in Eqs. (15.89) and (15.91), we note that we can identify

$$\langle s_i'|\alpha e^{\theta \sigma_1}|s_i\rangle = e^{\beta J s_i(m) s_i(m+1)}, \tag{15.92}$$

provided the following relations are true, namely,

$$\alpha \cosh \theta = e^{\beta J},$$

$$\alpha \sinh \theta = e^{-\beta J}. \tag{15.93}$$

Equivalently, we can make the above identification provided

$$\tanh \theta = e^{-2\beta J},$$

$$\alpha = (2 \sinh 2\beta J)^{\frac{1}{2}}. \tag{15.94}$$

Consequently, it is clear that if we define a $2^N \times 2^N$ matrix as

$$K_1 = \alpha^N e^{\theta \sigma_1(1)} \times e^{\theta \sigma_1(2)} \cdots \times e^{\theta \sigma_1(N)}$$

$$= (2 \sinh 2\beta J)^{\frac{N}{2}} e^{\theta \sum_{i=1}^{N} \sigma_1(i)}, \tag{15.95}$$

with

$$\tanh \theta = e^{-2\beta J}, \tag{15.96}$$

then, we can write

$$\langle s(m+1)|K_1|s(m)\rangle = e^{\beta J \sum_{i} s_i(m) s_i(m+1)}. \tag{15.97}$$

Similarly, let us also note that

$$\langle s(m+1)|e^{\left(\tilde{\theta}\sigma_3(i+1)\sigma_3(i)\right)}|s(m)\rangle$$

$$= \langle s(m+1)| \cosh \tilde{\theta} + \sigma_3(i+1)\sigma_3(i) \sinh \tilde{\theta}|s(m)\rangle$$

$$= \langle s(m+1)|s(m)\rangle \left(\cosh \tilde{\theta} + s_{i+1}(m)s_i(m) \sinh \tilde{\theta}\right)$$

$$= \langle s(m+1)|s(m)\rangle \, e^{\tilde{\theta} s_{i+1}(m)s_i(m)}$$

$$= \langle s(m+1)|s(m)\rangle \, e^{\beta J s_i(m)s_{i+1}(m)}, \tag{15.98}$$

provided we make the identification

$$\tilde{\theta} = \beta J . \tag{15.99}$$

Thus, defining

$$K_2 = e^{\beta J \sum_{i=1}^{N} \sigma_3(i+1)\sigma_3(i)} , \tag{15.100}$$

we note that we can write

$$K = K_1 K_2$$

$$= (2\sinh 2\beta J)^{\frac{N}{2}} e^{\theta \sum_{i=1}^{N} \sigma_1(i)} \times e^{\beta J \sum_{i=1}^{N} \sigma_3(i+1)\sigma_3(i)} , \tag{15.101}$$

which defines the transfer matrix for the system. Namely,

$$\langle s(m+1)|K|s(m)\rangle = \langle s(m+1)|K_1 K_2|s(m)\rangle$$

$$= \langle s(m+1)|K_1|s(m)\rangle \, e^{\beta J \sum_{i=1}^{N} s_i(m)s_{i+1}(m)}$$

$$= e^{\beta J \sum_{i=1}^{N} s_i(m)s_i(m+1)} \, e^{\beta J \sum_{i=1}^{N} s_i(m)s_{i+1}(m)}$$

$$= e^{-\beta(H(m)+H(m,m+1))} . \tag{15.102}$$

The partition function can now be written as

$$Z(\beta) = \sum_{s_i=\pm 1} e^{-\beta H}$$

$$= \sum_{s_i=\pm 1} \langle s(1)|K|s(N)\rangle\langle s(N)|K|s(N-1)\rangle \cdots \langle s(2)|K|s(1)\rangle$$

$$= \operatorname{Tr} K^N . \tag{15.103}$$

This, therefore, is the starting point for the Onsager solution of the two dimensional Ising model.

In field theory language, we are looking for a quantum Hamiltonian whose Euclidean transition amplitude will yield the partition function. Furthermore, in field theory, even if we are dealing with a theory on the lattice, we would prefer the time variable to be continuous. Thus, let us identify one of the axes, say the vertical one, to correspond to time and we choose the separation between the rows to be ϵ, a very small quantity.

The continuum time limit, of course, will be obtained by choosing $\epsilon \to 0$. Let us note that we are only changing the spacing among the rows. The spacing along a row, of course, is unchanged. At first sight, this may appear bothersome. But, let us recall that if there is a critical point in the theory, then the correlation lengths become quite large in this limit and in such a limit the lattice structure becomes quite irrelevant. Let us also note here that by making the lattice asymmetric, we have actually destroyed the isotropy of the system and, consequently, the couplings along different axes can, in principle, be different. Thus, allowing for different couplings along the two axes, we can write

$$H = \sum_{m=1}^{N} \sum_{i=1}^{N} \left(-J s_i(m) s_{i+1}(m) - J' s_i(m) s_i(m+1) \right)$$

$$= \sum_{m=1}^{N} \left(H(m) + H(m, m+1) \right), \tag{15.104}$$

where

$$H(m) = -J \sum_{i=1}^{N} s_i(m)s_{i+1}(m),$$

$$H(m, m+1) = -J' \sum_{i=1}^{N} s_i(m)s_i(m+1). \tag{15.105}$$

In this case, the partition function will have the form

$$Z(\beta) = \sum_{s_i=\pm 1} e^{-\beta H}$$

$$= \sum_{s_i=\pm 1} e^{-\beta \sum_{m=1}^{N} (H(m)+H(m,m+1))}$$

$$= \sum_{s_i=\pm 1} \prod_{m=1}^{N} e^{\beta J \sum_i s_i(m)s_{i+1}(m) + \beta J' \sum_i s_i(m)s_i(m+1)}. \tag{15.106}$$

In the quantum field theory language, we write the Euclidean time interval as

$$T_{\text{Eucl.}} = N\epsilon, \tag{15.107}$$

and assume that there exists a quantum Hamiltonian H_q such that we can write

$$Z(\beta) = \text{Tr} \, e^{-T_{Eucl.}H_q} = \text{Tr} \, e^{-N\epsilon H_q}$$

$$= \sum_{s_i=\pm 1} \langle s(1)|e^{-\epsilon H_q}|s(N)\rangle \cdots \langle s(2)|e^{-\epsilon H_q}|s(1)\rangle$$

$$= \sum_{s_i=\pm 1} \prod_{m=1}^{N} \langle s(m+1)|e^{-\epsilon H_q}|s(m)\rangle. \tag{15.108}$$

Thus, comparing Eqs. (15.106) and (15.108), we recognize that the two can be identified if

$$\langle s(m+1)|e^{-\epsilon H_q}|s(m)\rangle = e^{\beta \sum_i (Js_i(m)s_{i+1}(m)+J's_i(m)s_i(m+1))}.$$

$$(15.109)$$

From our discussion of the transfer matrix in Eqs. (15.101) and (15.102), we immediately conclude that

$$H_q = -\sum_i (\sigma_1(i) + \lambda\sigma_3(i+1)\sigma_3(i)), \qquad (15.110)$$

where λ is a constant parameter and as before, we can identify

$$\epsilon\lambda = \beta J,$$

$$\tanh\epsilon \simeq \epsilon = e^{-2\beta J'}. \qquad (15.111)$$

This relation is quite interesting in the sense that it expresses the coupling strengths as functions of the lattice spacing. In particular, we note that as we make the spacing between the rows smaller, the corresponding coupling between the rows becomes stronger. This is renormalization group behavior of the couplings in its crudest form.

15.7 Duality in the quantum system

We have been able to relate the 2-d Ising model to a one dimensional quantum mechanical system with a Hamiltonian

$$H_q = -\sum_{i=1}^{N} (\sigma_1(i) + \lambda\sigma_3(i+1)\sigma_3(i)). \qquad (15.112)$$

Let us next consider the dual lattice corresponding to this one dimensional lattice and define the dual operators on the dual lattice as

$$\mu_1(i) = \sigma_3(i+1)\sigma_3(i),$$

$$\mu_3(i) = \prod_{j=1}^{i} \sigma_1(j)\,. \tag{15.113}$$

It is easy to see that

$$\mu_1^2(i) = (\sigma_3(i+1)\sigma_3(i))^2 = I,$$

$$\mu_3^2(i) = \left(\prod_{j=1}^{i} \sigma_j\right)^2 = I\,. \tag{15.114}$$

These results can be shown to follow from the basic commutation relations of the Pauli matrices, namely,

$$[\sigma_1(i), \sigma_1(j)] = 0 = [\sigma_3(i), \sigma_3(j)], \quad \text{if } i \neq j,$$

$$\sigma_1^2(i) = \mathbb{1} = \sigma_3^2(i),$$

$$[\sigma_1(i), \sigma_3(i)]_+ = 0\,. \tag{15.115}$$

Using these, we can also derive that for $i \neq j$,

$$[\mu_1(i), \mu_1(j)] = [\sigma_3(i+1)\sigma_3(i), \sigma_3(j+1)\sigma_3(j)] = 0,$$

$$[\mu_3(i), \mu_3(j)] = \left[\prod_{k=1}^{i} \sigma_1(k), \prod_{l=1}^{j} \sigma_1(l)\right] = 0\,. \tag{15.116}$$

On the other hand,

$$[\mu_1(i), \mu_3(i)]_+ = \left[\sigma_3(i+1)\sigma_3(i), \prod_{j=1}^{i}\sigma_1(j)\right]_+$$

$$= \sigma_3(i+1)\prod_{j=1}^{i-1}[\sigma_3(i), \sigma_1(i)]_+\sigma_1(j)$$

$$= 0. \tag{15.117}$$

Thus, $\mu_1(i)$ and $\mu_3(i)$ also have the same algebraic properties as the Pauli matrices on the original lattice. Furthermore, let us note that by definition,

$$\mu_3(i+1)\mu_3(i) = \sigma_1(i+1). \tag{15.118}$$

Using these, then, we note that we can write

$$H_q(\lambda) = -\sum_{i=1}^{N}(\sigma_1(i) + \lambda\sigma_3(i+1)\sigma_3(i))$$

$$= -\sum_{i=1}^{N}(\mu_3(i+1)\mu_3(i) + \lambda\mu_1(i))$$

$$= -\lambda\sum_{i=1}^{N}\left(\mu_1(i) + \lambda^{-1}\mu_3(i+1)\mu_3(i)\right)$$

$$= \lambda H_q(\lambda^{-1}). \tag{15.119}$$

This is the self duality relation for this system. Namely, it maps the strong coupling properties of the system to its weak coupling properties. This shows, in particular, that the energy eigenvalues of this system must also satisfy the relation

$$E(\lambda) = \lambda E(\lambda^{-1}). \tag{15.120}$$

For some finite value of λ, if there is a phase transition such that the correlation lengths become infinite or that some energy eigenvalue becomes zero (zero mode), then the above duality relation implies that this must happen at $\lambda = 1$. This is precisely how we had determined the critical temperature for the 2-d Ising model in Eq. (15.66). Let us also note here that since we have a quantum mechanical description of the 2-d Ising model, we can also develop a perturbation theory in the standard manner.

15.8 References

K. Huang, *Statistical Mechanics*, John Wiley Publishing.

J. Kogut, Rev. Mod. Phys., **51**, 659 (1979).

R. Savit, Rev. Mod. Phys., **52**, 453 (1980).

CHAPTER 16

Proper time formalism

The idea of proper time method goes back to Fock who had noticed that sometimes it is useful to express the propagator or the Green's function as an integral over an additional auxiliary variable. Schwinger used this observation and developed the idea fully to what is now known as the proper time method or the proper time formalism. This is quite useful in studying many properties of quantum field theories, in particular, gauge theories since it, equivalently, introduces a gauge invariant regularization. It can be used at zero temperature as well as at finite temperature. The proper time formalism was the precursor of the zeta function regularization and the heat kernel method which we will discuss in the next chapter. These latter methods are quite useful in the covariant study of quantum field theories on nontrivial backgrounds (with or without boundaries) as well as in the study of Casimir energy.

16.1 Scalar propagator in D dimensions

To introduce the idea of the proper time method, let us consider a free, massive scalar field theory in D (flat) space-time dimensions described by a Lagrangian density

$$\mathcal{L} = \frac{1}{2} \partial_\mu \phi(x) \partial^\mu \phi(x) - \frac{m^2}{2} \phi^2(x) = \frac{1}{2} \phi(x) \hat{H} \phi(x), \tag{16.1}$$

where x denotes the coordinates in D space-time dimensions, m represents the mass of the field (particles) and we have identified (after an integration by parts)

$$\hat{H} = -(\Box + m^2), \qquad \Box = \eta^{\mu\nu} \partial_\mu \partial_\nu. \tag{16.2}$$

Here $\eta^{\mu\nu}$ denotes the Minkowski metric in D dimensions. (We parenthetically note here that the operator in the quadratic terms of the Lagrangian density is generally denoted by $\hat{O}$. However, we have denoted the operator here as $\hat{H}$ for reasons to become clear shortly.) The tree level Green's function (propagator), for the massive scalar field, satisfies the equation (see, for example, Eq. (3.75) in one dimension)

$$(\Box + m^2)G(x - x') = -\delta^D(x - x'). \tag{16.3}$$

Here $G(x - x')$ denotes the Green's function of the theory (the propagator is given by $iG(x - x')$) and can be evaluated in the following way. (There are various ways of determining the Green's function. The present method introduces the proper time method which is useful in various other studies in quantum field theory.) Equation (16.3) can be written in the operator form as ($\mathbb{1}$ denotes the identity operator in D dimensions)

$$(\Box + m^2)\hat{G} = -\mathbb{1}, \qquad \langle x|\hat{G}|x'\rangle = G(x - x'), \tag{16.4}$$

so that we have (see Eq. (16.2))

$$
\begin{aligned}
\hat{G} = \frac{1}{\hat{H}} &= \lim_{\epsilon \to 0^+} -i \int_0^\infty d\tau\, e^{i\tau(\hat{H}+i\epsilon)} \\
&= \lim_{\epsilon \to 0^+} -i \int_0^\infty d\tau\, e^{i\tau(-(\Box+m^2)+i\epsilon)} \\
&= \lim_{\epsilon \to 0^+} -i \int_0^\infty d\tau\, e^{i\tau(\hat{p}_\mu \hat{p}^\mu - m^2 + i\epsilon)} \\
&= \lim_{\epsilon \to 0^+} -i \int_0^\infty d\tau\, e^{-\epsilon\tau}\, e^{i\tau(\hat{p}^2 - m^2)},
\end{aligned} \tag{16.5}
$$

where we have identified ($\hbar = 1$)

$$\hat{p}_\mu = i\partial_\mu, \qquad [\hat{x}^\mu, \hat{p}_\nu] = -i\delta^\mu_\nu\, \mathbb{1}, \qquad \hat{p}^2 = \hat{p}_\mu \hat{p}^\mu. \tag{16.6}$$

We note here that τ in Eq. (16.5) is an auxiliary variable, commonly known as the proper time for reasons to be clear shortly. The "$i\epsilon$"

prescription (which leads to the Feynman prescription for the Green's function) is needed for convergence of the integral at the upper limit. (We point out here that sometimes the proper time is also denoted by s, but we use τ for closer analogy with (proper) time.)

Let $|x\rangle$ denote the eigenstates of the coordinate operator in the Schrödinger picture, namely,

$$\hat{x}^\mu |x\rangle = x^\mu |x\rangle, \quad \langle x|\hat{x}^\mu = x^\mu \langle x|, \tag{16.7}$$

and let us define new (time translated) coordinate states

$$|x,\tau\rangle = e^{-i\tau\hat{H}}|x\rangle = e^{-i\tau\hat{H}}|x,0\rangle,$$

$$\langle x,\tau| = \langle x|e^{i\tau\hat{H}} = \langle x,0|e^{i\tau\hat{H}}, \tag{16.8}$$

where we have used the fact that $\hat{H}$ in Eq. (16.2) is Hermitian and we have identified $|x\rangle = |x,0\rangle$. These coordinate states are eigenstates of the coordinate operator

$$\hat{x}^\mu(\tau) = e^{-i\tau\hat{H}}\hat{x}^\mu e^{i\tau\hat{H}}, \quad \hat{x}^\mu(0) = \hat{x}^\mu, \tag{16.9}$$

so that

$$\hat{x}^\mu(\tau)|x,\tau\rangle = e^{-i\tau\hat{H}}\hat{x}^\mu e^{i\tau\hat{H}}e^{-i\tau\hat{H}}|x\rangle$$

$$= x^\mu e^{-i\tau\hat{H}}|x\rangle = x^\mu|x,\tau\rangle,$$

$$\langle x,\tau|\hat{x}^\mu(\tau) = \langle x|e^{i\tau\hat{H}}e^{-i\tau\hat{H}}\hat{x}^\mu e^{i\tau\hat{H}}$$

$$= x^\mu\langle x|e^{i\tau\hat{H}} = x^\mu\langle x,\tau|, \tag{16.10}$$

where we have used Eq. (16.7). Similarly, we have (see also Eq. (16.6))

$$\hat{p}_\mu(\tau) = e^{-i\tau\hat{H}}\hat{p}_\mu e^{i\tau\hat{H}}, \quad [\hat{x}^\mu(\tau), \hat{p}_\nu(\tau)] = -i\delta^\mu_\nu \mathbb{1}. \tag{16.11}$$

Equations (16.9) and (16.11) lead to the (differential) time evolution equations for the operators as (this is the reason why we have denoted the quadratic operator as $\hat{H}$, namely, it corresponds to the Hamiltonian for proper time evolution of the system)

$$\frac{\mathrm{d}\hat{p}_\mu(\tau)}{\mathrm{d}\tau} = i[\hat{p}_\mu(\tau), \hat{H}] = i[\hat{p}_\mu(\tau), \hat{p}^2 - m^2] = 0,$$

$$\frac{\mathrm{d}\hat{x}^\mu(\tau)}{\mathrm{d}\tau} = i[\hat{x}^\mu(\tau), \hat{p}^2 - m^2] = i(-2i\hat{p}^\mu(\tau)) = 2\hat{p}^\mu(\tau), \qquad (16.12)$$

where we have used the commutation relations from Eqs. (16.6) and (16.11). These equations can be easily solved to yield

$$\hat{p}_\mu(\tau) = \hat{p}_\mu(0) = \hat{p}_\mu,$$

$$\hat{x}^\mu(\tau) = \hat{x}^\mu(0) + 2\hat{p}^\mu\tau = \hat{x}^\mu + 2\hat{p}^\mu\tau, \qquad (16.13)$$

leading to

$$\hat{p}^\mu = \frac{\hat{x}^\mu(\tau) - \hat{x}^\mu(0)}{2\tau} = \frac{\hat{x}^\mu(\tau) - \hat{x}^\mu}{2\tau}. \qquad (16.14)$$

As a result, we obtain

$$\hat{H} = \hat{p}^\mu\hat{p}_\mu - m^2 = \frac{1}{4\tau^2}(\hat{x}^\mu(\tau) - \hat{x}^\mu)(\hat{x}_\mu(\tau) - \hat{x}_\mu) - m^2$$

$$= \frac{1}{4\tau^2}(\hat{x}^\mu(\tau)\hat{x}_\mu(\tau) - 2\hat{x}^\mu(\tau)\hat{x}_\mu + \hat{x}^\mu\hat{x}_\mu + 2iD\tau) - m^2, \qquad (16.15)$$

where D denotes the number of space-time dimensions and we have used Eq. (16.13) in the intermediate step, namely, (see Eq. (16.6) and the identity operator, $\mathbb{1}$, is understood here)

$$[\hat{x}^\mu, \hat{x}^\nu(\tau)] = [\hat{x}^\mu, \hat{x}^\nu + 2\hat{p}^\nu\tau] = -2i\eta^{\mu\nu}\tau. \qquad (16.16)$$

From the definition in Eq. (16.4), we see that we can write the Green's function as (see also Eq. (16.5))

$$G(x - x') = \langle x|\hat{G}|x'\rangle = -i \int_0^\infty d\tau\, e^{-\epsilon\tau} \langle x|e^{i\hat{H}\tau}|x'\rangle$$

$$= -i \int_0^\infty d\tau\, e^{-\epsilon\tau} \langle x|e^{i\hat{H}\tau}|x'\rangle$$

$$= -i \int_0^\infty d\tau\, e^{-\epsilon\tau} \langle x, \tau|x', 0\rangle, \tag{16.17}$$

where the limit $\epsilon \to 0$ is understood and we have used the identification in Eqs. (16.8). We note that the transition amplitude

$$\langle x, \tau|x', 0\rangle = \langle x, 0|e^{i\hat{H}\tau}|x', 0\rangle = \langle x|e^{i\hat{H}\tau}|x'\rangle, \tag{16.18}$$

represents the probability that an initial coordinate state $|x'\rangle = |x', 0\rangle$ would propagate to the state $|x, \tau\rangle$ in a given (proper) time τ. This transition amplitude plays an important role in the proper time method as well as in the heat kernel method (to be discussed in the next chapter). It follows from Eq. (16.18) that

$$\partial_\tau \langle x, \tau|x', 0\rangle = i\langle x|e^{i\hat{H}\tau}\hat{H}|x'\rangle = i\langle x, \tau|\hat{H}|x', 0\rangle$$

$$= \left(\frac{i}{4\tau^2}(x - x')^2 - \frac{D}{2\tau} - im^2 \right) \langle x, \tau|x', 0\rangle. \tag{16.19}$$

Here we have used the form of the Hamiltonian in Eq. (16.15) as well as the eigenstate conditions in Eqs. (16.7) and (16.10). Equation (16.19) can be trivially integrated to determine

$$\langle x, \tau|x', 0\rangle = \frac{C}{\tau^{\frac{D}{2}}}\, e^{-\frac{i}{4\tau}(x-x')^2 - im^2\tau}, \tag{16.20}$$

where C is a constant determined from the initial (normalization) condition

$$\langle x, \tau|x', 0\rangle \xrightarrow{\tau \to 0} \langle x|x'\rangle = \delta^D(x - x'). \tag{16.21}$$

There are various ways of determining the constant C. We discuss only two (equivalent) methods to bring out some subtleties. We recall that the integral of the one dimensional Gaussian

$$\int_{-\infty}^{\infty} dx \, e^{-\alpha x^2} = \sqrt{\frac{\pi}{\alpha}}, \quad \alpha > 0, \tag{16.22}$$

leads to the well known Gaussian representation for the Dirac delta function as (area under the delta function has to be unity)

$$\sqrt{\frac{\alpha}{\pi}} \, e^{-\alpha x^2} \xrightarrow{\alpha \to \infty} \delta(x). \tag{16.23}$$

The integral in Eq. (16.22) can be easily generalized to the D-dimensional Euclidean space as

$$\int_{-\infty}^{\infty} d^D x \, e^{-\alpha \mathbf{x} \cdot \mathbf{x}} = \left(\frac{\pi}{\alpha} \right)^{\frac{D}{2}}, \quad \alpha > 0, \tag{16.24}$$

similarly leading to

$$\left(\frac{\alpha}{\pi} \right)^{\frac{D}{2}} e^{-\alpha \mathbf{x} \cdot \mathbf{x}} \xrightarrow{\alpha \to \infty} \delta^D(x). \tag{16.25}$$

On the other hand, for an oscillatory Gaussian in one dimension of the form $e^{\pm i \alpha x^2}, \alpha > 0$, the normalization can be obtained in one of two ways as follows. First, we can rotate to "Euclidean space" (for either of the signs in the exponent) as

$$x \to \sqrt{\pm i} \, \bar{x}, \tag{16.26}$$

leading to

$$\int_{-\infty}^{\infty} dx \, e^{\pm i \alpha x^2} \to \int_{-\infty}^{\infty} \sqrt{\pm i} d\bar{x} \, e^{\pm i \alpha (\pm i \bar{x}^2)}$$

$$= \sqrt{\pm i} \int_{-\infty}^{\infty} d\bar{x} \, e^{-\alpha \bar{x}^2} = \sqrt{\frac{\pm i \pi}{\alpha}}. \tag{16.27}$$

This leads to a representation of the one dimensional Dirac delta function in terms of the one dimensional oscillatory Gaussian of the form

$$\sqrt{\frac{\mp i\alpha}{\pi}}\, e^{\pm i\alpha x^2} \xrightarrow{\alpha \to \infty} \delta(x). \tag{16.28}$$

Equation (16.27) (and, therefore, Eq. (16.28)) can also be obtained in an alternative manner by introducing a regularization parameter. Namely, we note that we can write (ϵ is a regularization parameter different from the one introduced earlier)

$$\int \mathrm{d}x\, e^{\pm i\alpha x^2} = \lim_{\epsilon \to 0+} \int \mathrm{d}x\, e^{-(\epsilon \mp i\alpha)x^2} = \lim_{\epsilon \to 0} \sqrt{\frac{\pi}{\epsilon \mp i\alpha}}$$

$$= \sqrt{\frac{\pm i\pi}{\alpha}}, \tag{16.29}$$

which coincides with Eq. (16.27). This can be generalized to D-dimensional Euclidean space to give

$$\left(\frac{\mp i\alpha}{\pi}\right)^{\frac{D}{2}} e^{\pm i\alpha \mathbf{x} \cdot \mathbf{x}} \xrightarrow{\alpha \to \infty} \delta^D(x). \tag{16.30}$$

In a D-dimensional Minkowski space, the invariant length of the form

$$x^2 = (x^0)^2 - (x^1)^2 - \cdots - (x^{D-1})^2, \tag{16.31}$$

on the other hand, has one positive signature and $D-1$ negative signatures coming from the structure of the metric. Correspondingly, using Eq. (16.27) (or Eq. (16.28)), we conclude that in such a case (we restrict to the negative exponent here which coincides with Eq. (16.20)),

$$\left(\frac{i\alpha}{\pi}\right)^{\frac{1}{2}} \left(\frac{-i\alpha}{\pi}\right)^{\frac{D-1}{2}} e^{-i\alpha x^2}$$

$$= (-i)^{\frac{D}{2}-1} \left(\frac{\alpha}{\pi}\right)^{\frac{D}{2}} e^{-i\alpha x^2} \xrightarrow{\alpha \to \infty} \delta^D(x). \tag{16.32}$$

We can now compare Eqs. (16.20) and (16.21) with Eq. (16.32) (with $\alpha = \frac{1}{4\tau}$) which determines ($\frac{1}{\tau^{\frac{D}{2}}}$ has already been factored out in Eq. (16.20))

$$C = \frac{(-i)^{\frac{D}{2}-1}}{(4\pi)^{\frac{D}{2}}}. \tag{16.33}$$

As a result, we can write the transition amplitude as (see Eq. (16.20))

$$\langle x, \tau | x', 0 \rangle = \frac{(-i)^{\frac{D}{2}-1}}{(4\pi\tau)^{\frac{D}{2}}} e^{-\frac{i}{4\tau}(x-x')^2 - i(m^2 - i\epsilon)\tau}. \tag{16.34}$$

This allows us to write the scalar Green's function in D-dimensions as (see Eq. (16.17))

$$G(x - x') = \lim_{\epsilon \to 0^+} -i \int_0^\infty d\tau\, e^{-\epsilon\tau} \langle x, \tau | x', 0 \rangle$$

$$= \lim_{\epsilon \to 0^+} \left(\frac{-i}{4\pi} \right)^{\frac{D}{2}} \int_0^\infty \frac{d\tau}{\tau^{\frac{D}{2}}} e^{-\frac{i}{4\tau}((x-x')^2 - i\epsilon) - i(m^2 - i\epsilon)\tau}. \tag{16.35}$$

We note here that we have introduced a regularization parameter for the integral to be convergent at the lower limit $\tau = 0$ in addition to the one that was already there to make the integral well behaved at the upper limit. In principle, the new parameter can be distinct, however, we have chosen them to be the same for simplicity. The integral in Eq. (16.35) can be done with the help of the tables (see, for example, Gradshteyn and Ryzhik 3.471.9) and leads to

$$G(x - x') = \lim_{\epsilon \to 0^+} \left(\frac{-i}{2\pi} \right)^{\frac{D}{2}} \left(\frac{m^2 - i\epsilon}{(x - x')^2 - i\epsilon} \right)^{\frac{D-2}{4}}$$

$$\times K_{\frac{D}{2}-1}(i\sqrt{(m^2 - i\epsilon)((x - x')^2 - i\epsilon)}), \tag{16.36}$$

where we have used the property of the modified Bessel function, $K_{-\nu} = K_\nu$. This is the massive scalar propagator in D space-time dimensions. We can obtain the massless limit of Eq. (16.36) using the Bessel function identity

$$K_\nu(z) \xrightarrow{z \to 0} \frac{1}{2} \Gamma(\nu) \left(\frac{z}{2}\right)^{-\nu}, \tag{16.37}$$

which leads to the massless Green's function (in D dimensions) of the form

$$G(x - x') = \lim_{\epsilon \to 0^+} \frac{(-i)^{D-1}}{4\pi^{\frac{D}{2}}} \frac{\Gamma(\frac{D}{2} - 1)}{((x - x')^2 - i\epsilon)^{\frac{D}{2} - 1}}. \tag{16.38}$$

In particular, for $D = 4$, this leads to the well known coordinate space result

$$G(x - x') = \lim_{\epsilon \to 0^+} \frac{i}{4\pi^2} \frac{1}{(x - x')^2 - i\epsilon}, \tag{16.39}$$

while, for $D = 3$, we have

$$G(x - x') = \lim_{\epsilon \to 0^+} -\frac{1}{4\pi} \frac{1}{\sqrt{(x - x')^2 - i\epsilon}}. \tag{16.40}$$

Both of these results can be readily checked with the Fourier transforms of the corresponding momentum space Feynman Green's functions.

16.2 Evaluating determinants

Determinants arise quite regularly in quantum theories, as we have seen earlier. For example, if we have a quadratic Lagrangian density, as in Eq. (16.1), and we integrate out the field variables in the path integral formalism, then it leads to the determinant $(\det \hat{H})^{-\frac{1}{2}}$ (see, for example, Eq. (4.3)) which is a constant. If the quantum mechanical system is in equilibrium at a finite temperature, then this determinant is a temperature dependent quantity (constant for a given temperature) leading to the partition function (free energy) of the theory. When we have a quantum field theory which is quadratic in the basic field variables, but is interacting with an external field nontrivially, then integrating out the basic field variables leads to a determinant which is a functional of the external fields. Such a determinant leads naturally to an effective action for the external field. We will talk about determining the effective action in the proper time formalism in the next section. Here we indicate how the proper time formalism can be used to calculate the partition functions of various quantum mechanical systems.

16.2.1 Bosonic oscillator

Let us consider a bosonic oscillator which is a 0+1 dimensional quantum field theory and which we have studied in detail in Chapters **3** and **4** (we have also calculated the partition function for the oscillator in Chapter **14**). The partition function is best studied in the imaginary time (Euclidean space) (see, for example, Eqs. (4.8) and (14.40)) where the Lagrangian for the oscillator has the form (see Eq. (4.9) with $\hbar = 1 = m$)

$$
L = \frac{1}{2}\left(\left(\frac{dx}{dt}\right)^2 - \omega^2 x^2\right) = -\frac{1}{2}\,x\left(\frac{d^2}{dt^2} + \omega^2\right)x
$$

$$
\rightarrow -\frac{1}{2}\,x\left(-\frac{d^2}{dt^2} + \omega^2\right)x = -L_E, \tag{16.41}
$$

where ω denotes the natural frequency of the oscillator. (Here we have neglected a total derivative term, rotated to imaginary time $t \rightarrow -it$ and do not use τ to denote the imaginary time as in Eq. (4.8) since we are using τ for the proper time in this chapter. Therefore, t in the following is really the imaginary time.) Integrating out the field variable $x(t)$ in the path integral leads to the partition function for the bosonic oscillator

$$
Z(\beta) = \left[\det\left(-\frac{d^2}{dt^2} + \omega^2\right)\right]^{-\frac{1}{2}}
$$

$$
= \exp\left\{-\frac{1}{2}\operatorname{Tr}\ln\left(-\frac{d^2}{dt^2} + \omega^2\right)\right\}, \tag{16.42}
$$

where $\beta = \frac{1}{kT}$ with k denoting the Boltzmann constant (see Eq. (14.6)) and "Tr" stands for the trace to be taken in any complete basis (remember also that $\det A = e^{\operatorname{Tr}\ln A}$). The determinant has to be evaluated in the space of periodic functions with a time interval β (partition function involves a trace, see Eq. (14.5)). We point out here that Eq. (16.42) incorporates two results derived earlier. Namely, the generating functional for an action quadratic in the dynamical variables is proportional to the determinant of the quadratic operator in the action (see Eq. (4.19)) as well as the fact that the partition function is the generating functional of the system evaluated on the periodic space with the end points identified and integrated (see Eq. (14.47)).

We recall from the definition of the Gamma function (s is assumed to be a real parameter)

$$\int_0^\infty d\tau\, \tau^{s-1} e^{-\alpha\tau} = \alpha^{-s}\Gamma(s), \quad \alpha > 0,\ s > 0, \tag{16.43}$$

that, in the limit $s \to 0$, the right hand side of Eq. (16.43) leads to

$$\lim_{s\to 0} \alpha^{-s}\Gamma(s) = \lim_{s\to 0} (1 - s\ln\alpha + O(s^2)) \left(\frac{1}{s} + O(s^0)\right)$$

$$= C - \ln\alpha, \tag{16.44}$$

where C is an uninteresting divergent constant. Therefore, up to an uninteresting, additive constant, we can write

$$\ln\alpha = -\int_0^\infty \frac{d\tau}{\tau} e^{-\alpha\tau}, \tag{16.45}$$

which can also be generalized to any positive operator as

$$\ln\hat{H} = -\int_0^\infty \frac{d\tau}{\tau} e^{-\tau\hat{H}}. \tag{16.46}$$

Therefore, we can write

$$\text{Tr}\,\ln\hat{H} = -\int_0^\infty \frac{d\tau}{\tau}\,\text{Tr}\,e^{-\tau\hat{H}}, \tag{16.47}$$

where the "trace" can be evaluated in any basis. In particular, in the coordinate basis, we can write

$$\text{Tr}\, e^{-\tau \hat{H}} = \int dt\, \langle t | e^{-\tau \hat{H}} | t \rangle, \tag{16.48}$$

where the limits of integration depend on the particular problem being studied (for example, zero temperature or finite temperature).

To determine the partition function in Eq. (16.42), let us identify

$$\hat{H} = \left(-\frac{d^2}{dt^2} + \omega^2 \right) = (\hat{p}^2 + \omega^2), \tag{16.49}$$

with (see Eq. (16.6))

$$\hat{p} = i\frac{d}{dt}, \quad [\hat{t}, \hat{p}] = -i, \tag{16.50}$$

Furthermore, we define as before (see Eq. (16.8) with $\tau \to i\tau$)

$$|t, \tau\rangle = e^{\tau \hat{H}} |t, 0\rangle = e^{\tau \hat{H}} |t\rangle, \quad \langle t, \tau | = \langle t, 0 | e^{-\tau \hat{H}} = \langle t | e^{-\tau \hat{H}}, \tag{16.51}$$

where $\hat{t}|t\rangle = t|t\rangle$. This allows us to write (see Eqs. (16.9), (16.11) and (16.12))

$$\hat{t}(\tau) = e^{\tau \hat{H}}\, \hat{t}\, e^{-\tau \hat{H}}, \quad \hat{p}(\tau) = e^{\tau \hat{H}}\, \hat{p}\, e^{-\tau \hat{H}}, \quad [\hat{t}(\tau), \hat{p}(\tau)] = -i, \tag{16.52}$$

with $\hat{t}(0) = \hat{t}$, $\hat{p}(0) = \hat{p}$ and it follows that

$$\hat{t}(\tau)|t, \tau\rangle = t|t, \tau\rangle, \quad \langle t, \tau|\hat{t}(\tau) = t\,\langle t, \tau|. \tag{16.53}$$

Therefore, we can write

$$\langle t, \tau|t', 0\rangle = \langle t|e^{-\tau \hat{H}}|t'\rangle, \quad \partial_\tau \langle t, \tau|t', 0\rangle = -\langle t, \tau|\hat{H}|t', 0\rangle. \tag{16.54}$$

As before, the transition amplitude can be determined by solving the time evolution equations (see Eq. (16.52))

$$\frac{d\hat{p}(\tau)}{d\tau} = -[\hat{p}(\tau), \hat{H}] = -[\hat{p}(\tau), \hat{p}^2 + \omega^2] = 0,$$

$$\frac{d\hat{t}(\tau)}{d\tau} = -[\hat{t}(\tau), \hat{H}] = -[\hat{t}(\tau), \hat{p}^2 + \omega^2] = 2i\hat{p}(\tau) = 2i\hat{p}, \quad (16.55)$$

which lead to

$$\hat{p}(\tau) = \hat{p}, \quad \hat{t}(\tau) = \hat{t} + 2i\hat{p}\tau. \tag{16.56}$$

This, in turn, determines

$$\hat{p} = \frac{\hat{t}(\tau) - \hat{t}}{2i\tau}, \quad [\hat{t}, \hat{t}(\tau)] = 2\tau. \tag{16.57}$$

Using these, as discussed in the last section, we now obtain

$$\hat{H} = \frac{(\hat{t}(\tau) - \hat{t})^2}{-4\tau^2} + \omega^2$$

$$= -\frac{1}{4\tau^2}((\hat{t}(\tau))^2 - 2\hat{t}(\tau)\hat{t} + (\hat{t})^2 - 2\tau) + \omega^2. \tag{16.58}$$

Substituting this into Eq. (16.54) we obtain

$$\partial_\tau \langle t, \tau | t', 0 \rangle = \left[\frac{1}{4\tau^2}(t - t')^2 - \frac{1}{2\tau} - \omega^2 \right] \langle t, \tau | t', 0 \rangle, \tag{16.59}$$

which can be integrated to give (together with a normalizing constant discussed in the last section)

$$\langle t, \tau | t', 0 \rangle = K(t - t', \tau) = \frac{1}{\sqrt{4\pi\tau}} e^{-\frac{(t-t')^2}{4\tau} - \omega^2\tau}. \tag{16.60}$$

This can be compared with Eq. (16.34) for $D = 1$ and with appropriate analytic continuation of τ (namely, $\tau \to i\tau$) as can be seen from Eq. (16.51) as well as $m^2 \to -\omega^2$.

The transition amplitude in Eq. (16.60) has been derived without any reference to the periodicity condition that we need at finite temperature. A periodic finite temperature transition amplitude can be obtained from Eq. (16.60) through the linear superposition

$$\langle t, \tau | t', 0 \rangle^{(\beta)} = K^{(\beta)}(t - t', \tau) = \sum_{n=-\infty}^{\infty} K(t - t' - n\beta, \tau)$$

$$= K^{(\beta)}(t - t' - \beta, \tau). \tag{16.61}$$

Therefore, on the space of periodic functions, we can write (see Eq. (16.48))

$$\mathrm{Tr}\,\ln \hat{H} = -\int_0^\infty \frac{\mathrm{d}\tau}{\tau}\, \mathrm{Tr}\, e^{-\tau \hat{H}}$$

$$= -\int_0^\infty \frac{\mathrm{d}\tau}{\tau} \int_0^\beta \mathrm{d}t\, \langle t | e^{-\tau \hat{H}} | t \rangle^{(\beta)}$$

$$= -\int_0^\infty \frac{\mathrm{d}\tau}{\tau} \int_0^\beta \mathrm{d}t\, \langle t, \tau | t, 0 \rangle^{(\beta)}$$

$$= -\int_0^\infty \frac{\mathrm{d}\tau}{\tau} \int_0^\beta \mathrm{d}t\, \sum_{n=-\infty}^{\infty} K(-n\beta, \tau)$$

$$= -\beta \sum_{n=-\infty}^{\infty} \int_0^\infty \frac{\mathrm{d}\tau}{\tau}\, K(-n\beta, \tau), \tag{16.62}$$

where the integral over t is trivial (since there is no t dependence in the integrand) and we have also interchanged the summation and integration in the last step. Using the form of K from Eq. (16.60), we have

$$\mathrm{Tr}\,\ln \hat{H} = -\frac{\beta}{\sqrt{4\pi}} \sum_{n=-\infty}^{\infty} \int_0^\infty \mathrm{d}\tau\, \tau^{-\frac{3}{2}}\, e^{-\frac{n^2\beta^2}{4\tau} - \omega^2\tau}$$

$$= -\frac{2\beta}{\sqrt{4\pi}} \sum_{n=1}^{\infty} \int_0^\infty \mathrm{d}\tau\, \tau^{-\frac{3}{2}}\, e^{-\frac{n^2\beta^2}{4\tau} - \omega^2\tau}$$

$$\qquad - \frac{\beta}{\sqrt{4\pi}} \int_0^\infty \mathrm{d}\tau\, \tau^{-\frac{3}{2}}\, e^{-\omega^2\tau}. \tag{16.63}$$

The first of these integrals gives the modified Bessel function (see, for example, Gradshteyn and Ryzhik 3.471.9) while the second is a Gamma function leading to

$$\mathrm{Tr}\ln\hat{H} = -\frac{2\beta}{\sqrt{4\pi}}\sum_{n=1}^{\infty}2\left(\frac{n\beta}{2\omega}\right)^{-\frac{1}{2}}K_{\frac{1}{2}}(n\beta\omega) - \frac{\beta}{\sqrt{4\pi}}\omega\Gamma\left(-\frac{1}{2}\right)$$

$$= -2\sum_{n=1}^{\infty}\frac{1}{n}e^{-n\beta\omega} + \beta\omega. \tag{16.64}$$

Here we have used the particular form of $K_{\frac{1}{2}}(z)$ as well as the properties of the Gamma function.

The sum in the first term in Eq. (16.64) can be evaluated in a simple manner as follows.

$$\sum_{n=1}^{\infty}\frac{1}{n}e^{-n\beta\omega} = \sum_{n=1}^{\infty}\int_{\beta\omega}^{\infty}dz\, e^{-nz} = \int_{\beta\omega}^{\infty}dz\sum_{n=1}^{\infty}e^{-nz}$$

$$= \int_{\beta\omega}^{\infty}dz\,\frac{e^{-z}}{1-e^{-z}} = \ln(1-e^{-z})\big|_{\beta\omega}^{\infty}$$

$$= -\ln(1-e^{-\beta\omega}) = -\ln\left(2\sinh\frac{\beta\omega}{2}\right) + \frac{\beta\omega}{2}. \tag{16.65}$$

Substituting this result into Eq. (16.64), we obtain

$$\mathrm{Tr}\ln\hat{H} = 2\ln\left(2\sinh\frac{\beta\omega}{2}\right). \tag{16.66}$$

The partition function now follows from Eq. (16.42) to have the form

$$Z(\beta) = e^{-\frac{1}{2}\mathrm{Tr}\ln\hat{H}} = \left(2\sinh\frac{\beta\omega}{2}\right)^{-1} = \frac{1}{2\sinh\frac{\beta\omega}{2}}, \tag{16.67}$$

which is the well known partition function for the bosonic oscillator (also derived in Eqs. (14.35) and (14.41) in different ways).

16.2.2 Fermionic oscillator

The fermionic oscillator, as we have seen in Chapter **5**, is a first order system which is described by the Euclidean Lagrangian (see Eq. (5.45))

$$L_E = \bar{\psi} \left(\frac{\mathrm{d}}{\mathrm{d}t} + \omega \right) \psi, \quad \omega > 0, \tag{16.68}$$

where t denotes the imaginary (Euclidean) time. If we naively identify the operator in the quadratic term to be

$$\hat{H} = \frac{\mathrm{d}}{\mathrm{d}t} + \omega, \tag{16.69}$$

then, we will conclude that the partition function is given by (we have a fermionic system so that the power of the determinant is positive and there are two degrees of freedom making the power unity)

$$Z_F(\beta) = \det \left(\frac{\mathrm{d}}{\mathrm{d}t} + \omega \right) = e^{\mathrm{Tr}\, \ln\left(\frac{\mathrm{d}}{\mathrm{d}t} + \omega \right)}. \tag{16.70}$$

This would lead to proper time evolution equations (see, for example, Eq. (16.55)) with the solutions

$$p(\tau) = p(0) = p, \quad t(\tau) = t(0) + \tau, \tag{16.71}$$

which do not allow us to express p in terms of t. As a result, the evaluation of the transition amplitude becomes difficult. Therefore, we will try another method to evaluate the determinant (partition function).

As we have emphasized in Chapter **5**, a fermionic (Grassmann) system is inherently a matrix system and we can combine the two dynamical variables into the two component matrices (see Eq. (5.61))

$$\Psi = \begin{pmatrix} \psi \\ \bar{\psi} \end{pmatrix}, \qquad \bar{\Psi} = \Psi^\dagger \sigma_3 = \begin{pmatrix} \bar{\psi} & -\psi \end{pmatrix}. \tag{16.72}$$

Here σ_3 denotes the diagonal Pauli matrix. With this, the Hermitian Lagrangian Eq. (5.44) (see also Eq. (5.53)) rotates to imaginary time as

$$L_E = \frac{1}{2}\bar{\Psi}\left(\sigma_3\frac{d}{dt} + \omega\right)\Psi,$$ (16.73)

which can also be seen from Eq. (5.64) (with the source $\Theta = 0$). As a result, the operator in the quadratic terms can be identified with the matrix operator

$$\hat{H} = \sigma_3\frac{d}{dt} + \omega.$$ (16.74)

Correspondingly, the partition function can be written as (the system has a single fermionic matrix degree of freedom, but the quadratic operator is a 2×2 matrix)

$$Z(\beta) = \left[\det\left(\sigma_3\frac{d}{dt} + \omega\right)\right]^{\frac{1}{2}} = \left[\det\left(-\frac{d^2}{dt^2} + \omega^2\right)\right]^{\frac{1}{2}}$$

$$= e^{\frac{1}{2}\text{Tr} \ln(-\frac{d^2}{dt^2} + \omega^2)}.$$ (16.75)

Here we have used the fact that the determinant involves both a coordinate part as well as a matrix part and the matrix part of the determinant is trivial since the two matrices (σ_3 and $\mathbb{1}$) are diagonal. What remains, therefore, is only a determinant over any complete basis states (such as coordinate states).

There are several things to note from Eq. (16.75). First, the operator in the exponent is what we had dealt with earlier in the bosonic oscillator case (see Eq. (16.49)) so that the analysis of proper time evolution and the determination of the transition amplitude can be carried over completely from Eqs. (16.49)–(16.60). There is a sign difference in the exponent in Eq. (16.75) (compared with Eq. (16.42)) which results from the fermionic nature of the system and should be kept in mind. Most importantly, we should remember that at finite temperature fermions obey anti-periodicity conditions (see Eq. (14.50)) so that the determinant (and, therefore, the transition amplitude) in this case has to be evaluated over anti-periodic functions. We can determine the transition amplitude over the space of anti-periodic functions as (compare with Eq. (16.61))

$$\langle t, \tau | t', 0 \rangle^{(\beta)} = \bar{K}^{(\beta)}(t - t', \tau) = \sum_{n=-\infty}^{\infty} (-1)^n K(t - t' - n\beta, \tau)$$

$$= -\bar{K}^{(\beta)}(t - t' - \beta, \tau), \tag{16.76}$$

where $K(t - t', \tau)$ is already determined in Eq. (16.60). Therefore, in the space of anti-periodic functions, we have (compare with Eqs. (16.62) and (16.63))

$$\operatorname{Tr} \ln \left(-\frac{d^2}{dt^2} + \omega^2 \right) = -\beta \sum_{n=-\infty}^{\infty} (-1)^n \int_0^{\infty} \frac{d\tau}{\tau} K(-n\beta, \tau)$$

$$= -\frac{2\beta}{\sqrt{4\pi}} \sum_{n=1}^{\infty} (-1)^n \int_0^{\infty} d\tau\, \tau^{-\frac{3}{2}} e^{-\frac{n^2\beta^2}{4\tau} - \omega^2 \tau}$$

$$- \frac{\beta}{\sqrt{4\pi}} \int_0^{\infty} d\tau\, \tau^{-\frac{3}{2}} e^{-\omega^2 \tau}. \tag{16.77}$$

The τ integral can be evaluated as in Eq. (16.64) and leads to

$$\operatorname{Tr} \ln \left(-\frac{d^2}{dt^2} + \omega^2 \right) = 2 \sum_{n=1}^{\infty} \frac{(-1)^{n+1}}{n} e^{-n\beta\omega} + \beta\omega. \tag{16.78}$$

As in Eq. (16.65), we can write

$$\sum_{n=1}^{\infty} \frac{(-1)^{n+1}}{n} e^{-n\beta\omega} = \sum_{n=1}^{\infty} (-1)^{n+1} \int_{\beta\omega}^{\infty} dz\, e^{-nz}$$

$$= \int_{\beta\omega}^{\infty} dz \sum_{n=1}^{\infty} (-1)^{n+1} e^{-nz} = \int_{\beta\omega}^{\infty} dz \frac{e^{-z}}{1 + e^{-z}}$$

$$= -\ln(1 + e^{-z})\big|_{\beta\omega}^{\infty} = \ln \left(2 \cosh \frac{\beta\omega}{2} \right) - \frac{\beta\omega}{2}. \tag{16.79}$$

Substituting this back into Eq. (16.78), we obtain

$$\operatorname{Tr} \ln \left(-\frac{d^2}{dt^2} + \omega^2 \right) = 2 \ln \left(2 \cosh \frac{\beta\omega}{2} \right), \tag{16.80}$$

so that the partition function (see Eq. (16.75)) can be determined to have the form

$$Z(\beta) = e^{\frac{1}{2}\mathrm{Tr}\,\ln(-\frac{d^2}{dt^2}+\omega^2)} = 2\cosh\frac{\beta\omega}{2}, \qquad (16.81)$$

which coincides with Eqs. (14.57) and (14.59).

16.3 Effective actions

In the last section, we evaluated the partition functions (determinants with periodicity) for two simple systems, namely, the bosonic and the fermionic oscillator. The Lagrangians for these two systems were quadratic in the dynamical variables with a constant quadratic operator which is why they could be exactly evaluated. However, when a dynamical system, say, a fermion, is interacting with a space-time dependent external field (say, an external gauge field), even though the system is quadratic in the dynamical variables, the determinant becomes a functional of the external field and leads to an effective action for the external field. We may or may not be able to determine the determinant (and, therefore, the effective action) exactly in a closed form in such a case. When we cannot evaluate it exactly, we resort to a perturbative determinantion of the effective action. In this section, we explain this idea (within the context of proper time formalism) with two examples at zero temperature and one example at finite temperature.

First, let us recall some definitions. The Lagrangian density for a fermion interacting with an external gauge field in an arbitrary dimension is given by

$$\mathcal{L} = \bar{\psi}(\gamma^\mu(i\partial_\mu - A_\mu(x)) - m)\psi, \quad \mu = 0, 1, 2, \cdots, D-1, \quad (16.82)$$

where $D > 1$ denotes the number of space-time dimensions and we have set the coupling constant to unity for simplicity. (We will treat the $0+1$ dimensional case separately in the following.) This Lagrangian density is quadratic in the dynamical field variables and, therefore, the generating functional can be formally determined to be ($\hbar = 1$)

$$Z[A] = e^{iW[A]} = N\left[\det(i\not\partial - m - \not\! A)\right], \qquad (16.83)$$

where N is a normalization constant (normally divergent beyond $0 + 1$ dimension). This determines the generating functional for the connected Green's functions (or the effective action) to be

$$W[A] = -i\ln Z[A] = -i\ln[N\det(i\not\partial - m - \not\! A)]. \qquad (16.84)$$

Subtracting out the value at $A_\mu = 0$, we obtain the normalized effective action to be

$$\begin{aligned}
\Gamma[A] = W[A] - W[A = 0] &= -i\ln\frac{[\det(i\not\partial - m - \not\! A)]}{[\det(i\not\partial - m)]} \\
&= -i\ln\det\left(\mathbb{1} - S(i\not\partial)\not\! A\right) \\
&= -i\mathrm{Tr}\,\ln\left(\mathbb{1} - S(i\not\partial)\not\! A\right), \qquad (16.85)
\end{aligned}$$

where $S(i\not\partial) = (i\not\partial - m)^{-1}$ is the free fermion Green's function operator with momentum vector replaced by $i\partial_\mu$. Clearly, this effective action is normalized so as to vanish when the external source is set to zero, namely, $\Gamma[A = 0] = 0$. At this point, we can Taylor expand the logarithm and evaluate the trace using the coordinate basis. This would lead to the one loop effective action in the usual perturbation theory. However, we will describe how to evaluate the effective action through the use of the proper time formalism.

Using Eq. (16.46), we can give an integral representation to $\Gamma[A]$

$$\begin{aligned}
\Gamma[A] &= i\int_0^\infty \frac{d\tau}{\tau}\,\mathrm{Tr}\,e^{-\tau(\mathbb{1} - S(i\not\partial)\not\! A)} \\
&= i\int_0^\infty \frac{d\tau}{\tau}\,e^{-\tau}\,\mathrm{Tr}\,e^{-\tau\hat{H}}, \qquad (16.86)
\end{aligned}$$

where we have used the identity

$$e^{-\tau\mathbb{1}} = e^{-\tau}\,\mathbb{1}.$$

Furthermore, we have denoted

$$\hat{H} = -S(i\partial)\slashed{A} = -S(\hat{p})\slashed{A}(\hat{x}), \tag{16.87}$$

where we have identified, as in Eq. (16.6), $\hat{p}_\mu = i\partial_\mu$ and the free (operator) Green's function for a massive fermion (in D dimensions) has the form

$$S(\hat{p}) = (\hat{\slashed{p}} - m)^{-1} = \frac{\hat{\slashed{p}} + m}{\hat{p}^2 - m^2}. \tag{16.88}$$

The Feynman prescription is understood here although we do not write it explicitly for simplicity. Choosing the trace (trace has two parts, a matrix trace and a coordinate trace) to be evaluated in the coordinate basis states we can write

$$\Gamma[A] = i \int_0^\infty \frac{d\tau}{\tau} e^{-\tau} \int d^D x \, \text{tr} \, \langle x|e^{-\tau\hat{H}}|x\rangle, \tag{16.89}$$

where "tr" stands for trace over Dirac (or other) matrix indices. This one loop effective action is regularized in a gauge invariant manner through the use of the proper time method (regularization).

Unlike the other cases which we have studied earlier, $\hat{H}$ now contains both coordinates and momenta (in a mixed manner). As a result, the proper time Hamiltonian equations (see, for example, Eq. (16.12) or (16.55)) can not be solved in a simple manner. Therefore, we try to solve the (proper) time evolution equation for the transition amplitude directly (see, for example, Eq. (16.54) for $0 + 1$ dimension and the discussions in that section), namely,

$$\partial_\tau \langle x, \tau|x', 0\rangle = -\langle x, \tau|\hat{H}|x', 0\rangle = \langle x, \tau|S(\hat{p})\slashed{A}(\hat{x})|x', 0\rangle, \tag{16.90}$$

where we have used the identification in Eq. (16.87). If this equation will have a closed form solution, there will be a corresponding exact (closed form) effective action. If not, then the solution of this equation will lead to a perturbative solution for the effective action (which, of course, completely agrees with the form obtained from a perturbative diagrammatic calculation). We will now illustrate how this method leads to exact effective actions in two cases at zero temperature.

16.3.1 Zero temperature

To start with, let us analyze the $0 + 1$ dimensional fermion interacting with an external electromagnetic field through minimal coupling (coupling constant set to unity). The minimally coupled Lagrangian has the form (see also Eq. (16.68) and remember that $e = 1$)

$$L = \bar{\psi}((i\partial_t - A(t)) - m)\psi, \quad m > 0, \tag{16.91}$$

where ψ and $\bar{\psi}$ are two fermion fields. Here the external gauge field $A(t)$ is an arbitrary function of t. This system can be described also as a two component matrix system as in Eq. (16.72) and would lead to equivalent results, but we will continue with Eq. (16.91) for simplicity. Following Eqs. (16.85) and (16.86), we can write

$$\begin{aligned}
\Gamma[A] &= -i \ln \frac{\det(i\partial_t - m - A)}{\det(i\partial_t - m)} \\
&= -i \ln \det(1 - S(i\partial_t)A(t)) \\
&= -i \operatorname{Tr} \ln(1 - S(i\partial_t)A(t)) \\
&= i \int_0^\infty \frac{d\tau}{\tau} e^{-\tau} \operatorname{Tr} e^{\tau S(i\partial_t)A(t)} \\
&= i \int_0^\infty \frac{d\tau}{\tau} e^{-\tau} \operatorname{Tr} e^{-\tau \hat{H}},
\end{aligned} \tag{16.92}$$

where we have identified

$$\hat{H} = -S(i\partial_t)A(\hat{t}) = -S(\hat{p})A(\hat{t}), \tag{16.93}$$

with the identification, as in Eq. (16.6), $\hat{p} = i\partial_t$. The free Green's function, in this case, is given by

$$S(\hat{p}) = (\hat{p} - m)^{-1} = \frac{1}{\hat{p} - m + i\epsilon}. \tag{16.94}$$

In this case, Eq. (16.90) takes the form

$$\partial_\tau \langle t, \tau | t', 0 \rangle = -\langle t, \tau | \hat{H} | t', 0 \rangle = \langle t, \tau | S(\hat{p}) A(\hat{t}) | t', 0 \rangle$$
$$= \langle t, \tau | S(\hat{p}) | t', 0 \rangle A(t'). \tag{16.95}$$

For later use, let us calculate the matrix element

$$\langle t, 0 | S(\hat{p}) | t', 0 \rangle = \int dp \, \langle t, 0 | p \rangle \langle p | S(\hat{p}) | t', 0 \rangle$$

$$= \int dp \, \frac{1}{p - m + i\epsilon} \, \langle t, 0 | p \rangle \langle p | t', 0 \rangle$$

$$= \int \frac{dp}{2\pi} \frac{1}{p - m + i\epsilon} \, e^{-ip(t-t')}$$

$$= -i\theta(t - t') \, e^{-im(t-t')}. \tag{16.96}$$

Namely, we see that in 0+1 dimension, the Feynman Green's function coincides with the retarded Green's function. This can be understood simply from the fact that, in non-relativistic quantum mechanics, time only flows forward. Using this as well as introducing a complete set of coordinate states, we can write Eq. (16.95) as

$$\partial_\tau \langle t, \tau | t', 0 \rangle = \int dt_1 \, \langle t, \tau | t_1, 0 \rangle \langle t_1, 0 | S(\hat{p}) | t', 0 \rangle A(t')$$

$$= -i \int dt_1 \, \langle t, \tau | t_1, 0 \rangle \theta(t_1 - t') A(t') e^{-im(t_1-t')}. \tag{16.97}$$

This equation can be formally integrated leading to

$$\langle t, \tau | t', 0 \rangle = \delta(t - t')$$

$$- i \int_0^\tau d\tau' \int dt_1 \, \langle t, \tau' | t_1, 0 \rangle \theta(t_1 - t') A(t') e^{-im(t_1-t')}. \tag{16.98}$$

Here the homogeneous term on the right hand side is needed to ensure the normalization condition (see Eq. (16.21))

$$\langle t, \tau | t', 0 \rangle \xrightarrow{\tau \to 0} \delta(t - t').$$

The integral equation in Eq. (16.98) can be solved iteratively and may not have a closed form in general. However, in the present case it does have a closed form expression which can be seen as follows

$$\langle t, \tau | t', 0 \rangle = \delta(t - t') - i \int_0^\tau d\tau' \theta(t - t') A(t') e^{-im(t-t')}$$

$$+ (-i)^2 \int_0^\tau d\tau' \int_0^{\tau'} d\tau'' \int dt_1$$

$$\times \theta(t - t_1) \theta(t_1 - t') A(t_1) A(t') e^{-im(t-t')} + \cdots$$

$$= \delta(t - t') - i\tau \theta(t - t') A(t') e^{-im(t-t')}$$

$$+ \frac{(-i\tau)^2}{2} \int dt_1 \theta(t - t_1) \theta(t_1 - t') A(t_1) A(t') e^{-im(t-t')}$$

$$+ \cdots . \tag{16.99}$$

In taking the trace (see Eq. (16.92)), we have to identify $t' = t$ and integrate over t. The first term in Eq. (16.99) gives an unimportant constant which can be neglected. Furthermore, the theta function in the second term leads to $\theta(0) = \frac{1}{2}$. The higher order terms all vanish because of opposing theta functions, namely,

$$\theta(t - t_1) \theta(t_1 - t_2) \cdots \theta(t_n - t) = 0. \tag{16.100}$$

Consequently, in this case, we can determine the effective action exactly (see Eq. (16.92))

$$\Gamma[A] = i \int_0^\infty \frac{d\tau}{\tau} \int dt \, e^{-\tau} \left(-\frac{i\tau}{2} \right) A(t) = \frac{1}{2} \int dt \, A(t). \tag{16.101}$$

This indeed coincides with the perturbative calculation of the effective action for this system at zero temperature.

The second model, which we would like to discuss, is the theory of massless fermions interacting with an external gauge field in $1 + 1$ dimensions (namely, it is the Schwinger model, discussed extensively in Section 13.2 from various points of view, without the kinetic term for the gauge field). It is an exactly soluble model at zero temperature implying that the effective action can be obtained in exact closed form. Here we derive the exact effective action at zero temperature in the proper time formalism. The Lagrangian density is given by (see Eq. (16.82) with $m = 0$ and recall that we have set the coupling strength $e = 1$)

$$\mathcal{L} = \bar{\psi}\gamma^\mu(i\partial_\mu - A_\mu)\psi, \quad \mu = 0, 1, \tag{16.102}$$

where a standard representation for the two dimensional Dirac matrices is given in Eq. (13.52). Exploiting the gauge invariance of the theory (Schwinger model), we choose the gauge condition

$$\partial_\mu A^\mu = 0, \tag{16.103}$$

for simplicity which allows us to write the gauge field, in $1 + 1$ dimensions, as (see, for example, Eq. (13.62))

$$A_\mu = \epsilon_{\mu\nu}\partial^\nu\phi, \quad \slashed{A} = \gamma^\mu A_\mu = -\gamma_5\slashed{\partial}\phi, \tag{16.104}$$

where ϕ denotes a scalar field. Furthermore, the free fermion propagator, in this massless case, can be written as (see also Eq. (16.88))

$$S(\hat{p}) = \frac{1}{\hat{\slashed{p}}} = \frac{\hat{\slashed{p}}}{\hat{p}^2}, \tag{16.105}$$

where the Feynman prescription is understood.

Introducing a complete set of coordinate and momentum basis states, Eq. (16.90) takes the form

$$\partial_\tau \langle x, \tau | x', 0 \rangle = \int d^2x_1 d^2p_1 \langle x, \tau | x_1, 0 \rangle \langle x_1, 0 | p_1 \rangle \left\langle p_1 \left| \frac{\hat{\slashed{p}}}{\hat{p}^2} \slashed{A}(\hat{x}) \right| x', 0 \right\rangle$$

$$= \int d^2x_1 \frac{d^2p_1}{(2\pi)^2} \langle x, \tau | x_1, 0 \rangle \frac{\slashed{p}_1}{p_1^2} \slashed{A}(x') e^{-ip_1 \cdot (x_1 - x')}. \tag{16.106}$$

This can be integrated to give

$$\langle x, \tau | x', 0 \rangle = \delta^2(x - x') + \int_0^\tau d\tau' \int d^2 x_1 \frac{d^2 p_1}{(2\pi)^2} \langle x, \tau | x_1, 0 \rangle$$

$$\times \frac{\not{p}_1 \not{A}(x')}{p_1^2} e^{-ip_1 \cdot (x_1 - x')}, \qquad (16.107)$$

where the first homogeneous term is needed for the normalization of the state (see Eq. (16.21)). As before, this integral equation can be iterated to give

$$\langle x, \tau | x', 0 \rangle = \delta^2(x - x') + \int_0^\tau d\tau' \int \frac{d^2 p_1}{(2\pi)^2} \frac{\not{p}_1 \not{A}(x')}{p_1^2} e^{-ip_1 \cdot (x - x')}$$

$$+ \int_0^\tau d\tau' \int_0^{\tau'} d\tau'' \int d^2 x_1 \frac{d^2 p_1}{(2\pi)^2} \frac{d^2 p_2}{(2\pi)^2} \frac{\not{p}_2 \not{A}(x_1)}{p_2^2} \frac{\not{p}_1 \not{A}(x')}{p_1^2}$$

$$\times e^{-ip_2 \cdot (x - x_1) - ip_1 \cdot (x_1 - x')} + \cdots . \qquad (16.108)$$

We note here that at the end, we have to take a trace which involves a Dirac trace as well as setting $x = x'$ and integrating over x. In this limit, the first term gives an unimportant constant which can be neglected. The linear term (in A) vanishes because the integrand is odd in p_1. The terms higher than quadratic (in A) can also be shown to vanish so that the only meaningful nontrivial term on the right hand side is the term quadratic in the fields. Using the properties of the Dirac matrices in two dimensions (see Section **13.2**), we can evaluate the Dirac trace (remember also to use the gauge condition Eq. (16.103)) and carrying out the momentum integral, we finally obtain

$$\text{Tr} \langle x, \tau | x, 0 \rangle = -\frac{i\tau^2}{\pi} \int d^2 x \, A_\mu(x) A^\mu(x), \qquad (16.109)$$

Therefore, the effective action, Eq. (16.89), is determined to be

$$\Gamma[A] = i \int_0^\infty \frac{d\tau}{\tau} e^{-\tau} \left(-\frac{i\tau^2}{\pi} \right) \int d^2 x \, A_\mu(x) A^\mu(x)$$

$$= \frac{1}{2\pi} \int d^2 x \, A_\mu(x) A^\mu(x). \qquad (16.110)$$

This is exactly the effective action for the Schwinger model (in gauge Eq. (16.103) and with $e = 1$) evaluated in, see for example, Eq. (13.111). (If we had not used the gauge condition Eq. (16.103) in evaluating the trace in Eq. (16.108), we would have obtained exactly the same gauge invariant effective action as in Eq. (13.111).)

The simplest way to see that the higher order terms in A in Eq. (16.108) vanish is as follows. Let us recall from Eq. (16.104) that, in the gauge Eq. (16.103), we can write (remember Eq. (16.6))

$$A(\hat{x}) = -\gamma_5 \hat{\partial}\phi(\hat{x}) = i\gamma_5[\hat{p}, \phi(\hat{x})]. \tag{16.111}$$

As a result, we can write (remember that $\frac{\hat{p}}{\hat{p}^2} = \frac{1}{\hat{p}}$)

$$\frac{1}{\hat{p}}A(\hat{x}) = -i\gamma_5 \frac{1}{\hat{p}}[\hat{p}, \phi(\hat{x})] = -i\gamma_5 \left(\phi(\hat{x}) - \frac{1}{\hat{p}}\phi(\hat{x})\hat{p}\right), \tag{16.112}$$

where we have used the fact that γ_5 and $\gamma^\mu, \mu = 0, 1$ anti-commute. This leads to

$$\frac{1}{\hat{p}}A(\hat{x})\frac{1}{\hat{p}}A(\hat{x}) = -i\gamma_5(\phi(\hat{x}) - \frac{1}{\hat{p}}phi(\hat{x})\hat{p}) \frac{1}{\hat{p}}A(\hat{x}),$$

$$= -i\gamma_5(\phi(\hat{x})\frac{1}{\hat{p}}A(\hat{x}) - \frac{1}{\hat{p}}\phi(\hat{x})A(\hat{x}))$$

$$= -i\gamma_5[\phi(\hat{x}), \frac{1}{\hat{p}}A(\hat{x})], \tag{16.113}$$

where, in the last line, we have used the fact that $\phi(\hat{x})$ and $A(\hat{x})$ commute (since both of them are functions of $\hat{x}$). Using this, we can now show that

$$\mathrm{Tr}\left(\frac{1}{\hat{p}}A(\hat{x})\right)^{n+1} = \mathrm{Tr}\left(-\frac{i\gamma_5}{n}\left[\phi(\hat{x}), \left(\frac{1}{\hat{p}}A(\hat{x})\right)^n\right]\right), \tag{16.114}$$

where $n > 0$. For $n > 1$, the momentum integrals in the trace are finite and, then, the trace vanishes because of cyclicity. For $n = 1$, the momentum integrals are divergent and one can not use the cyclicity of the trace and, therefore, the terms for $n = 0, 1$ need to be evaluated explicitly which we have done in arriving at Eq. (16.110).

16.3.2 Finite temperature

At finite temperature, the Green's functions (boson/fermion) become temperature dependent and more complicated in order to incorporate the periodicity/anti-periodicity properties. That makes the proper time evaluation of the effective action (for interacting theories) quite difficult and one way to evaluate the effective action in Eq. (16.85) is through a Taylor expansion of the logarithm which coincides with the perturbative calculation. Here we will describe an alternative method for evaluation of effective actions at finite temperature. (It is also worth noting here that the proper time formalism naturally introduces a gauge invariant ultraviolet regularization at zero temperature. At finite temperature, on the other hand, the temperature dependent part of the amplitudes are ultraviolet finite and, therefore, there is no need for regularization.)

We recall from discussions in Chapter **14** that there are two major formalisms to describe field theories at finite temperature. The imaginary time formalism (also known as the Matsubara formalism), where time is rotated to the imaginary axis and the time (interval) is traded for temperature, is quite useful in calculating partition functions (as we have done in Section **16.2** (as well as in Chapter **14**)). However, they are not so useful in calculating effective actions. The reason is simple. The effective actions give rise to Feynman (time ordered) correlation functions whereas the imaginary time formalism leads naturally to retarded and advanced Green's functions. As a result, we need to use the real time formalism where both time and temperature are present. The price one has to pay for this is that the (dynamical) field degrees of freedom in the theory are doubled. We will use the closed time path formalism in our discussion for its simplicity and versatility. The formalism is characterized by a path in the complex time plane as shown in Fig. 16.1, with the original fields on the C_+ branch and the doubled fields on the C_- branch. The vertical branch $C_\perp$ decouples from the correlation functions.

We recall from Eq. (16.85) that since

$$\Gamma[A] = -i \operatorname{Tr} \ln(i\slashed{\partial} - m - \slashed{A}) - i \operatorname{Tr} \ln(i\slashed{\partial} - m), \qquad (16.115)$$

taking the functional derivative with respect to the external field, we obtain (remember that the coupling constant has been set to unity)

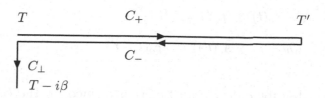

Figure 16.1: The closed time path contour in the complex t plane. Here $T \to -\infty$, while $T' \to \infty$ and β denotes the inverse temperature (in units of the Boltzmann constant k).

$$\frac{\delta \Gamma[A]}{\delta A_\mu(t, \mathbf{x})} = i \operatorname{tr}(\gamma^\mu S(t, \mathbf{x}; t, \mathbf{x})),$$

$$\frac{\delta \Gamma[A]}{\delta A_\mu(t, -\mathbf{p})} = i \operatorname{tr}(\gamma^\mu S(t, t; \mathbf{p})). \tag{16.116}$$

Here "tr" denotes the Dirac trace in higher dimensions (in $0 + 1$ dimension, as we see from Eq. (16.91), there is no Dirac trace since there are no gamma matrices) and the quantity on the right hand side in Eq. (16.116) is the complete current of the theory which is related to the exact fermion Green's function (not the free Green's function) for the interacting theory at coincident points. Therefore, if we can determine the complete fermion Green's function of the theory, which satisfies the complete dynamical equations of motion, satisfies the relevant symmetry properties as well as the anti-periodicity condition needed at finite temperature (for fermions), we can determine the complete effective action by integrating either of the equations in Eq. (16.116). Let us illustrate this with a simple example, namely, the $0+1$ dimensional fermion in an external electromagnetic field discussed in Eq. (16.91) (it can also be carried out in more complicated theories, at least formally).

At finite temperature, the theory defined on the contour in Fig. 16.1 has doubled fields. In particular, there are two external fields $A_\pm$ on the two real branches $C_\pm$ of the contour and the complete fermion Green's function $S_c(t, t')$ satisfies the dynamical equations

$$(i\partial_t - m - A_c(t))S_c(t, t') = \delta_c(t - t'),$$

$$S_c(t, t')(i\overleftarrow{\partial}_{t'} + m + A_c(t')) = -\delta_c(t - t'). \tag{16.117}$$

Here the subscript c denotes a function defined on the contour in Fig. 16.1 and the step function on the contour is defined as

$$\theta_c(t - t') = \begin{cases} \theta(t - t') & \text{if both } t, t' \in C_+, \\ \theta(t' - t) & \text{if both } t, t' \in C_-, \\ 1 & \text{if } t \in C_- \text{ and } t' \in C_+, \\ 0 & \text{if } t \in C_+ \text{and } t' \in C_-, \end{cases} \tag{16.118}$$

leading to the delta function on the contour in the standard manner as

$$\delta_c(t - t') = \partial_t \theta_c(t - t'). \tag{16.119}$$

Equation (16.117) can be exactly solved and the Green's function can determined to be

$$S_c(t, t') = -i(\theta_c(t - t')C - \theta_c(t' - t)D)$$

$$\times e^{-im(t-t') - i\int_{t'}^{t} dt''_c A_c(t''_c)}, \tag{16.120}$$

where C and D are two constants (to be determined) satisfying

$$C + D = 1, \tag{16.121}$$

for Eq. (16.117) to hold. Furthermore, the anti-periodicity of the fermion Green's function, namely,

$$S_c(-\infty, t') = -S_c(-\infty - i\beta, t'), \tag{16.122}$$

where β is defined in Eq. (14.6) leads to

$$D = Ce^{-\beta m - i \int_{-\infty}^{-\infty - i\beta} dt_c \, A_c(t_c)}$$

$$= Ce^{-\beta m - i a_c} = Ce^{-\beta m - i(a_+ - a_-)}. \qquad (16.123)$$

Here we have used the fact that the vertical branch of the contour decouples and have defined

$$a_c = \int_{-\infty C_+}^{-\infty C_-} dt_c \, A_c(t_c)$$

$$= \int_{-\infty}^{\infty} dt \, A_+(t) - \int_{-\infty}^{\infty} dt \, A_-(t) = (a_+ - a_-), \qquad (16.124)$$

where in the first line, the lower limit of integration corresponds to the left most point on C_+ while the upper limit corresponds to the left most point on C_- (see Fig. 16.1). (The relative negative sign in the second term arises because the sense of time flow on C_- is opposite to that on C_+.) Equations (16.121) and (16.123) determine

$$C = 1 - n_F \left(m + \frac{i a_c}{\beta} \right) = 1 - n_F \left(m + \frac{i}{\beta}(a_+ - a_-) \right),$$

$$D = 1 - C = n_F \left(m + \frac{i a_c}{\beta} \right) = n_F \left(m + \frac{i}{\beta}(a_+ - a_-) \right), \qquad (16.125)$$

where n_F denotes the Fermi-Dirac distribution function and this determines the complete fermion Green's function, for this theory, to be (see Eq. (16.120))

$$S_c(t, t') = -i \left(\theta_c(t - t') - n_F \left(m + \frac{i a_c}{\beta} \right) \right)$$

$$\times e^{-im(t - t') - i \int_{t'}^{t} dt_c'' \, A_c(t_c'')}. \qquad (16.126)$$

It is easy to check that this Green's function also satisfies the anti-periodicity condition

$$S_c(t, -\infty) = -S_c(t, -\infty - i\beta), \qquad (16.127)$$

as it should.

At the coincident point $t = t'$, the contour Green's function leads to (we use $\theta(0) = \frac{1}{2}$)

$$S_c(t, t) = -\frac{i}{2}\left(1 - 2n_F\left(m + \frac{ia_c}{\beta}\right)\right)$$

$$= -\frac{i}{2}\tanh\frac{\beta m + ia_c}{2}, \qquad (16.128)$$

so that Eq. (16.116) takes the form (there is no Dirac trace)

$$\frac{\delta\Gamma[a_c]}{\delta A_c(t)} = \frac{\partial\Gamma[a_c]}{\partial a_c}\frac{\delta a_c}{\delta A_c(t)} = \frac{\partial\Gamma[a_c]}{\partial a_c} = \frac{1}{2}\tanh\frac{\beta m + ia_c}{2}, \qquad (16.129)$$

where we have used the definition of a_c in Eq. (16.124) in the intermediate step. This can be easily integrated to give

$$\Gamma[a_c] = -i\ln\cosh\frac{\beta m + ia_c}{2}$$

$$= -i\ln\cosh\frac{\beta m + i(a_+ - a_-)}{2}. \qquad (16.130)$$

We can subtract out the value of the action at zero fields to obtain the normalized effective action as

$$\Gamma[a_c] = -i\ln\frac{\cosh\frac{\beta m + ia_c}{2}}{\cosh\frac{\beta m}{2}}$$

$$= -i\ln\left(\cos\frac{a_c}{2} + i\tanh\frac{\beta m}{2}\sin\frac{a_c}{2}\right)$$

$$= -i\ln\left(\cos\frac{a_+ - a_-}{2} + i\tanh\frac{\beta m}{2}\sin\frac{a_+ - a_-}{2}\right). \qquad (16.131)$$

If we now restrict the effective action to only the original fields (on the C_+ branch), this leads to

$$\Gamma[a_+, a_- = 0] = -i\ln\left(\cos\frac{a_+}{2} + i\tanh\frac{\beta m}{2}\sin\frac{a_+}{2}\right). \qquad (16.132)$$

This coincides with the effective action for the original fields calculated from perturbation theory as well as other methods. However, Eq. (16.131) describes the complete effective action at finite temperature which contains a lot more information. (Note that this effective action reduces to Eq. (16.101) at zero temperature, $\beta \to \infty$, as it should.

16.4 References

A. Das, *Finite Temperature Field Theory*, World Scientific Publishing (1997).

A. Das and G. V. Dunne, Phys. Rev. D **57**, 5023 (1998) .

A. Das and C. Farina *Proper time method for fermions*, arXiv:hep-th/9807152.

A. Das and J. Frenkel, Phys. Rev. D **80**, 125039 (2009).

A. Das and A. Karev, Phys. Rev. D **36**, 623 (1987).

G. V. Dunne, K. Lee and C. Lu, Phys. Rev. Lett. **78**, 3434 (1997).

V. A. Fock, Izv. Akad. Nauk. USSR (Phys.) **4-5**, 551 (1937).

T. Matsubara, Prog. Theor. Phys., **14**, 351 (1955).

J. Schwinger, J. Math. Physics **2** (1961) 407.

J. Schwinger, *Lecture Notes of Brandeis Summer Institute in Theoretical Physics* (1960).

J. Schwinger, Phys. Rev. **82**, 664 (1951); Lett. Math. Phys. **24**, 59 (1992).

CHAPTER 17

Zeta function regularization

17.1 Riemann zeta function

In this section, we will describe briefly how the Riemann zeta function developed from Euler's zeta function and study various of its properties.

17.1.1 Euler's zeta function

The zeta function was originally introduced and studied by Euler which he defined as

$$\zeta(s) = \sum_{n=1}^{\infty} a_n = \sum_{n=1}^{\infty} \frac{1}{n^s} = \frac{1}{1^s} + \frac{1}{2^s} + \frac{1}{3^s} + \cdots, \quad s > 1, \quad (17.1)$$

where s is a real parameter. Euler studied this function for real parameters s because complex analysis had not yet been developed. This series is convergent only for $s > 1$ as can be seen below. The ratio test for a series of the form $\sum_{n=1}^{\infty} a_n$ is inconclusive in this case since

$$\lim_{n \to \infty} \left| \frac{a_{n+1}}{a_n} \right| = \lim_{n \to \infty} \left| \left(\frac{n}{n+1} \right)^s \right| \to 1. \quad (17.2)$$

Note that for convergence of the series, this ratio should be (< 1). Therefore, one has to go to another test. There are many, but let us consider the Cauchy condensation test where, given the series in Eq. (17.1), we look at the series

$$\sum_{n=0}^{\infty} 2^n a_{2^n} = \sum_{n=0}^{\infty} 2^n \times \frac{1}{(2^n)^s} = \sum_{n=0}^{\infty} 2^{(1-s)n}. \quad (17.3)$$

This is a geometric series with ratio 2^{1-s} which is smaller than 1 for $s > 1$. Therefore, for $s > 1$, this series will be finitely convergent and $\zeta(s)$ is (absolutely) convergent for $s > 1$.

The reason Euler wanted to study this series is because he was interested in the prime numbers. (Prime numbers are integers greater than 1 which are divisible only by 1 and the number itself.) Euler showed that the zeta function can also be written as a product involving only prime numbers as

$$\zeta(s) = \prod_p \left(1 - \frac{1}{p^s}\right)^{-1}, \quad s > 1, \tag{17.4}$$

where p stands for a prime number and the product is over all the prime numbers. This can be seen as follows

$$\prod_p \left(1 - \frac{1}{p^s}\right)^{-1}$$

$$= \left(1 - \frac{1}{2^s}\right)^{-1} \left(1 - \frac{1}{3^s}\right)^{-1} \left(1 - \frac{1}{5^s}\right)^{-1} \cdots$$

$$= \left(1 + \frac{1}{2^s} + \frac{1}{2^{2s}} + \cdots\right) \left(1 + \frac{1}{3^s} + \frac{1}{3^{2s}} + \cdots\right)$$

$$\times \left(1 + \frac{1}{5^s} + \frac{1}{5^{2s}} + \cdots\right) \cdots$$

$$= 1 + \frac{1}{2^s} + \frac{1}{3^s} + \frac{1}{4^s} + \frac{1}{5^s} + \frac{1}{6^s} + \frac{1}{7^s} + \cdots$$

$$= \sum_{n=1}^{\infty} \frac{1}{n^s} = \zeta(s). \tag{17.5}$$

Basically, this uses the fact that any positive integer can be expressed as a product of prime numbers and the product on the right hand side of Eq. (17.4) generates the inverse of all possible products of primes raised to the power s. This is a fundamental relation in the sense that the zeta function, which is defined in terms of natural numbers, can also be written in terms of prime numbers. This connection opened up a new branch of mathematics known as analytic number theory.

Euler evaluated the exact values of $\zeta(2k)$ in a remarkable manner. For example, let us look at

$$\zeta(2) = \sum_{n=1}^{\infty} \frac{1}{n^2}. \tag{17.6}$$

To evaluate this, let us recall that any quadratic polynomial can be written as a product

$$ax^2 + bx + c = c\left(1 - \frac{x}{x_1}\right)\left(1 - \frac{x}{x_2}\right), \tag{17.7}$$

where x_1, x_2 are the two roots of the quadratic polynomial equation, $x_1 + x_2 = -\frac{b}{a}$, $x_1 x_2 = \frac{c}{a}$. This can be generalized to any polynomial in that any polynomial can be factored in terms of its roots (values where the function vanishes). Let us consider now the ratio $\frac{\sin x}{x}$ which can be written in terms of its roots as

$$\frac{\sin x}{x} = \left(1 - \frac{x^2}{\pi^2}\right)\left(1 - \frac{x^2}{(2\pi)^2}\right)\left(1 - \frac{x^2}{(3\pi)^2}\right)\cdots$$

$$\times \left(1 - \frac{x^2}{(n\pi)^2}\right)\cdots. \tag{17.8}$$

Here we have used the fact that any function can be factorized in terms of its roots and that the roots of $\frac{\sin x}{x}$ are given by $x = \pm n\pi, n = 1, 2, 3, \cdots$. The overall multiplicative constant is fixed from the normalization to be

$$\lim_{x \to 0} \frac{\sin x}{x} = 1. \tag{17.9}$$

Multiplying out all the factors in Eq. (17.8), we obtain

$$\frac{\sin x}{x} = 1 - x^2\left(\frac{1}{\pi^2} + \frac{1}{(2\pi)^2} + \frac{1}{(3\pi)^2} + \cdots\right) + O(x^4)$$

$$= 1 - \frac{x^2}{\pi^2}\sum_{n=1}^{\infty}\frac{1}{n^2} + O(x^4) = 1 - \frac{x^2}{\pi^2}\zeta(2) + O(x^4). \tag{17.10}$$

On the other hand, we can Taylor expand and write

$$\frac{\sin x}{x} = \frac{1}{x}\left(x - \frac{x^3}{3!} + \frac{x^5}{5!} - \frac{x^7}{7!} + \cdots\right)$$

$$= 1 - \frac{x^2}{3!} + \frac{x^4}{5!} - \frac{x^6}{7!} + \cdots . \tag{17.11}$$

Comparing the coefficients of the x^2 terms in Eqs. (17.10) and (17.11) we obtain

$$-\frac{1}{\pi^2}\zeta(2) = -\frac{1}{6}, \quad \text{or,} \quad \zeta(2) = \frac{\pi^2}{6}. \tag{17.12}$$

In fact, using this logic, Euler evaluated all the $\zeta(2k)$ functions in terms of Bernoulli's numbers as follows. Note that the series representation of $x\cot x$ leads to (actually, we have $x\cot x = \sum_{n=-\infty}^{\infty} \frac{x^2}{x^2 - n^2\pi^2}$)

$$x\cot x = 1 - 2\sum_{n=1}^{\infty} \frac{x^2}{n^2\pi^2 - x^2} = 1 - 2\sum_{n=1}^{\infty} \frac{x^2}{n^2\pi^2} \frac{1}{1 - \frac{x^2}{n^2\pi^2}}$$

$$= 1 - 2\sum_{n=1}^{\infty} \frac{x^2}{n^2\pi^2} \sum_{k=0}^{\infty} \left(\frac{x^2}{n^2\pi^2}\right)^k$$

$$= 1 - 2\sum_{k=1}^{\infty} \left(\sum_{n=1}^{\infty} \frac{1}{n^{2k}}\right) \left(\frac{x^2}{\pi^2}\right)^k$$

$$= 1 - 2\sum_{k=1}^{\infty} \zeta(2k) \left(\frac{x^2}{\pi^2}\right)^k . \tag{17.13}$$

On the other hand, Euler knew (from his friend Bernoulli) that we can write

$$x\cot x = 1 - \sum_{k=1}^{\infty} \frac{2^{2k}|B_{2k}|}{(2k)!}(x^2)^k, \tag{17.14}$$

where B_{2k} denote the even Bernoulli's numbers. Therefore, comparing the two expressions in Eqs. (17.13) and (17.14), we determine

$$\zeta(2k) = \frac{(2\pi)^{2k}}{2(2k)!} |B_{2k}| = \frac{(2\pi)^{2k}}{2(2k)!} (-1)^{k+1} B_{2k}. \tag{17.15}$$

Here we have used the property of the Bernoulli's numbers that B_{2k} is positive (negative) for k odd (even) for $k \geq 1$. These values are related to moments (Mellin transformation) of the Einstein integral (involving Bose-Einstein distribution functions). The values of the zeta function for a few of the odd integer values, $\zeta(2k+1)$, can also be calculated, but they are not very interesting. ($\zeta(3)$ is known to be irrational.)

Euler's zeta function diverges for $s = 1$ (remember $\ln(1-x) = -\sum_{n=1}^{\infty} \frac{x^n}{n}$)

$$\zeta(1) = 1 + \frac{1}{2} + \frac{1}{3} + \frac{1}{4} + \cdots = -\ln(1-1) \to \infty. \tag{17.16}$$

Even though the divergence appears to be logarithmic, we will see later that there is a simple pole singularity at $s = 1$ when viewed in a regularized manner. A simple example of how a naively logarithmic singularity may manifest as a pole when regularized is as follows

$$\int_0^1 \frac{dt}{t} = \ln t \Big|_0^1 = \ln 1 - \ln 0 = -\ln 0 \to \infty, \tag{17.17}$$

which is a logarithmic singularity. However, if we view the integral in a limiting (regularized) manner, we obtain

$$\lim_{\epsilon \to 0^+} \int_0^1 \frac{dt}{t^{1-\epsilon}} = \lim_{\epsilon \to 0^+} \int_0^1 dt\, t^{\epsilon-1} = \lim_{\epsilon \to 0^+} \frac{t^\epsilon}{\epsilon} \Big|_0^1 = \lim_{\epsilon \to 0^+} \frac{1}{\epsilon}, \tag{17.18}$$

which has a simple pole behavior.

Euler's zeta function also diverges for $s = 0$ (of course, the series is defined only for $s > 1$)

$$\zeta(0) = 1 + 1 + 1 + 1 + \cdots \to \infty. \tag{17.19}$$

In fact, it diverges for all negative integer values of s

$$\zeta(-1) = 1 + 2 + 3 + 4 + \cdots \to \infty,$$

$$\zeta(-2) = 1 + 2^2 + 3^2 + 4^2 + \cdots \to \infty,$$

$$\vdots$$

$$\zeta(-n) = 1 + 2^n + 3^n + 4^n + \cdots \to \infty. \tag{17.20}$$

17.1.2 Riemann's zeta function

During the time of Riemann, complex analysis had been developed fully and so he viewed Euler's zeta function as a holomorphic function in the parameter s and analytically continued it to the entire complex plane. He did this in three essential steps and this regularized away the divergences in Eqs. (17.19) and (17.20) except for the one at $s = 1$ in Eq. (17.16).

As the first step he defined the zeta function for complex values of s as (see Eq. (17.1))

$$\zeta(s) = \sum_{n=1}^{\infty} \frac{1}{n^s}, \quad s = \sigma + i\lambda, \quad \operatorname{Re} s > 1. \tag{17.21}$$

Here σ, λ are assumed to be real. This series can be given an integral representation

$$\zeta(s) = \sum_{n=1}^{\infty} \frac{1}{n^s} = \sum_{n=1}^{\infty} \frac{1}{\Gamma(s)} \int_0^{\infty} dx\, x^{s-1} e^{-nx}, \quad \operatorname{Re} s > 1$$

$$= \frac{1}{\Gamma(s)} \sum_{n=0}^{\infty} \int_0^{\infty} dx\, x^{s-1} e^{-(n+1)x}$$

$$= \frac{1}{\Gamma(s)} \int_0^{\infty} dx\, x^{s-1} e^{-x} \left(1 - e^{-x}\right)^{-1}$$

$$= \frac{1}{\Gamma(s)} \int_0^{\infty} dx\, \frac{x^{s-1}}{e^x - 1}, \quad \operatorname{Re} s > 1. \tag{17.22}$$

This integral representation is known as the Mellin transform of the Bose-Einstein distribution function ($M f(s) = \int_0^{\infty} dx\, x^{s-1} f(x)$). (We note parenthetically that the Gamma function is defined for $\operatorname{Re} s > 0$

which is compatible with our function defined for $\mathrm{Re}\,s > 1$. When $\mathrm{Re}\,s$ is not greater than 1, interchanging the sum and integral in the intermediate step in Eq. (17.22) is not permitted. Therefore, Riemann's zeta function coincides with Euler's only in the region where the series is convergent, namely, $\mathrm{Re}\,s > 1$.)

The next step is to define this function for $\mathrm{Re}\,s > 0$ except at $s = 1$ (namely, on the right half of the complex parameter space except for the point $s = 1$). To do this, we note that the zeta function in Eq. (17.21) can be written in terms of the alternating zeta function in the following way

$$\sum_{n=1}^{\infty} \frac{(-1)^n}{n^s} + \zeta(s) = \sum_{n=1}^{\infty} ((-1)^n + 1)\frac{1}{n^s} = 2\sum_{n=1}^{\infty} \frac{1}{(2n)^s}$$

$$= 2^{1-s}\sum_{n=1}^{\infty} \frac{1}{n^s} = 2^{1-s}\zeta(s), \qquad (17.23)$$

so that we can write

$$\zeta(s) = \frac{1}{1 - 2^{1-s}}\sum_{n=1}^{\infty} \frac{(-1)^{n+1}}{n^s} = \frac{1}{1 - 2^{1-s}}\eta(s), \qquad (17.24)$$

where $\eta(s)$ is known as the alternating zeta function. Here the singularity at $s = 1$ has been explicitly factored out and the pole structure at $s = 1$ is already manifest. Alternating series, in general, have much better convergence properties and it can be checked that the series in Eq. (17.24) is defined for $\mathrm{Re}\,s > 0, s \neq 1$.

This has the integral representation in terms of the Mellin transform of the Fermi-Dirac distribution function. In fact, following the same steps as in Eq. (17.22), we obtain

$$\zeta(s) = \frac{1}{1 - 2^{1-s}}\sum_{n=1}^{\infty} \frac{(-1)^{n+1}}{n^s}$$

$$= \frac{1}{(1 - 2^{1-s})}\sum_{n=1}^{\infty} (-1)^{n+1}\frac{1}{\Gamma(s)}\int_0^{\infty} \mathrm{d}x\, x^{s-1}e^{-nx}$$

$$= \frac{1}{(1 - 2^{1-s})\Gamma(s)}\int_0^{\infty} \mathrm{d}x\, x^{s-1}\sum_{n=0}^{\infty}(-1)^n e^{-(n+1)x}$$

$$= \frac{1}{(1 - 2^{1-s})\Gamma(s)} \int_0^\infty dx\, x^{s-1} e^{-x} \left(1 + e^{-x}\right)^{-1}$$

$$= \frac{1}{(1 - 2^{1-s})\Gamma(s)} \int_0^\infty dx\, \frac{x^{s-1}}{e^x + 1}. \tag{17.25}$$

This gives an integral representation for the Riemann zeta function on the right half of the complex plane except at $s = 1$.

The last step involves analytically continuing the function to the left half of the complex plane. To this end, Riemann derived two equivalent functional relations of the forms

$$\zeta(1 - s) = 2(2\pi)^{-s} \cos\frac{\pi s}{2}\, \Gamma(s)\, \zeta(s), \quad \mathrm{Re}\, s > 0,$$

$$\zeta(s) = 2(2\pi)^{s-1} \sin\frac{\pi s}{2}\, \Gamma(1 - s)\, \zeta(1 - s), \quad \mathrm{Re}\, s < 1. \tag{17.26}$$

The two relations are obtained from each other by letting $s \to 1 - s$ and they allow us to extend the zeta function to the left half of the complex plane.

The zeta function is now defined on the entire complex plane except at $s = 1$. (It coincides with the series representation only when the series does not diverge, otherwise it is a different function which is holomorphic everywhere in the complex parameter space except for a simple pole at $s = 1$ which we show next.) Let us evaluate $\zeta(s)$ from its integral representation in Eq. (17.22),

$$\zeta(s) = \frac{1}{\Gamma(s)} \int_0^\infty dx\, \frac{x^{s-1}}{e^x - 1} = \frac{1}{\Gamma(s)} \int_0^\infty dx\, \frac{x^{s-1} e^{-x}}{1 - e^{-x}}$$

$$= \frac{1}{\Gamma(s)} \int_0^\infty dx\, x^{s-2} e^{-x} \left(1 + \frac{x}{2} + \frac{x^2}{12} + \cdots\right)$$

$$= \frac{1}{\Gamma(s)} \int_0^\infty dx\, \left(x^{s-2} + \frac{x^{s-1}}{2} + \frac{x^s}{12} + \cdots\right) e^{-x}$$

$$= \frac{1}{\Gamma(s)} \left(\Gamma(s - 1) + \frac{1}{2}\Gamma(s) + \frac{1}{12}\Gamma(s + 1) + \cdots\right)$$

$$= \frac{1}{s - 1} + \frac{1}{2} + \frac{s}{12} + \frac{1}{\Gamma(s)} \sum_{n=2}^\infty a_n \Gamma(s + n). \tag{17.27}$$

This shows that the zeta function has only a simple pole at $s = 1$ with residue 1. In fact, the series of terms in Eq. (17.27) can be summed exactly as $s \to 1$ and leads to

$$\zeta(s) \xrightarrow{s \to 1} \frac{1}{s-1} + \gamma + O(s-1), \tag{17.28}$$

where γ denotes Euler's constant. This relation can also be obtained from the (second) functional relation in Eq. (17.26) which we don't go into.

We also note from Eq. (17.27) that

$$\zeta(0) = -1 + \frac{1}{2} = -\frac{1}{2}, \quad \left(\zeta'(0) = -\frac{1}{2}\ln 2\pi\right). \tag{17.29}$$

Here we have used the fact that $\Gamma(0) \to \infty$ so that the remainder of the terms in Eq. (17.27) vanish. Therefore, the singular behavior of $\zeta(0)$ (see Eq. (17.19)) has been regularized and it is well behaved there. Note also from Eq. (17.27) that

$$\zeta(-1) = -\frac{1}{2} + \frac{1}{2} - \frac{1}{12} = -\frac{1}{12}, \tag{17.30}$$

which is useful in the studies of string theory, Casimir energy and other physical problems. We note from the second functional relation in Eq. (17.26) that, for $s = -2k$ with k a positive integer, we have

$$\zeta(-2k) = 2(2\pi)^{-2k-1}\sin\frac{\pi}{2}(-2k)\,\Gamma(1+2k)\zeta(1+2k)$$

$$= -2(2\pi)^{-2k-1}\sin k\pi\,\Gamma(1+2k)\zeta(1+2k) = 0, \tag{17.31}$$

where we have used the fact that $\Gamma(1+2k)$ and $\zeta(1+2k)$ are both finite for k positive. Similarly, for $s = 1-2k$ with k a positive integer, the second functional relation leads to

$$\zeta(1-2k) = 2(2\pi)^{-2k}\sin\frac{\pi}{2}(1-2k)\Gamma(2k)\zeta(2k)$$

$$= 2(2\pi)^{-2k}(-1)^k(2k-1)!\zeta(2k) = -\frac{B_{2k}}{2k}, \tag{17.32}$$

where we haves used Eq. (17.15). This analysis shows that all the divergences of the zeta function have been regularized except for the one at $s = 1$.

17.2 Zeta function regularization

We have seen that the Riemann zeta function regularizes all the singularities of the Euler zeta function, except for the one at $s = 1$, through analytic continuation. In quantum field theory, we know that there are divergences and the idea is to explore if the techniques of the zeta function can be applied there as well providing a regularization for the singularities. If this can be done, it will correspond to a gauge invariant regularization.

To proceed, let us recall from Eq. (17.22) that we can write (we have changed the integration variable from x to τ for reasons that will be clear shortly)

$$\zeta(s) = \frac{1}{\Gamma(s)} \int_0^\infty d\tau \, \frac{\tau^{s-1}}{e^\tau - 1} = \frac{1}{\Gamma(s)} \int_0^\infty d\tau \, \tau^{s-1} \sum_{n=1}^\infty e^{-n\tau}, \quad (17.33)$$

so that if we define a diagonal matrix, $\bar{A}$, of the form

$$\bar{A} = \begin{pmatrix} 1 & & & \\ & 2 & & \\ & & 3 & \\ & & & \ddots \end{pmatrix}, \quad (17.34)$$

then, we can write Eq. (17.33) also as

$$\zeta(s) = \frac{1}{\Gamma(s)} \int_0^\infty d\tau \, \tau^{s-1} \operatorname{Tr} e^{-\bar{A}\tau}. \quad (17.35)$$

This suggests that we can define a zeta function associated with a $N \times N$ Hermitian matrix A with real, positive eigenvalues $\lambda_1, \lambda_2, \cdots, \lambda_N$ as

$$\zeta_A(s) = \sum_{n=1}^N \frac{1}{\lambda_n^s} = \frac{1}{\Gamma(s)} \int_0^\infty d\tau \, \tau^{s-1} \operatorname{Tr} e^{-A\tau}. \quad (17.36)$$

Here the eigenvalues are assumed to be real and positive, but are not necessarily nondegenerate. Looking back at Eq. (16.43) for matrices, we recognize that we can identify from Eq. (17.36) that the zeta function for such a matrix can also be written as

$$\zeta_A(s) = \text{Tr}\, A^{-s} = \frac{1}{\Gamma(s)} \int_0^\infty d\tau\, \tau^{s-1}\, \text{Tr}\, e^{-A\tau}. \tag{17.37}$$

Either from the definition in Eq. (17.36) in terms of the series or from Eq. (17.37) we can derive the following properties (for example, using Eq. (17.37))

$$\zeta_{\frac{A}{\lambda}}(s) = \text{Tr}\left(\frac{A}{\lambda}\right)^{-s} = \lambda^s \,\text{Tr}\, A^{-s} = \lambda^s \zeta_A(s),$$

$$\zeta_A(0) = \text{Tr}\, \mathbb{1} = N = \text{dimension of the matrix},$$

$$\zeta_A'(0) = \left.\frac{d\zeta_A(s)}{ds}\right|_{s=0} = \text{Tr}\, \frac{d}{ds} e^{-s \ln A}\bigg|_{s=0}$$

$$= -\text{Tr}\,\ln A = -\ln \det A, \tag{17.38}$$

where we have used the definition of $\zeta_A(s)$ in Eq. (17.37) as well as the relation $\det A = e^{\text{Tr}\,\ln A}$ in the last step. Although we have derived these properties for finite dimensional matrices, they hold equally well in the infinite dimensional case and are useful in various calculations.

In quantum mechanics as well as in quantum field theory, square matrices are associated with operators and, in this way, one can also define zeta functions associated with infinite dimensional operators. This leads naturally to what is commonly known as the zeta function regularization in quantum field theory. For example, for an infinite dimensional self-adjoint operator $\hat{O}$ with (real) positive eigenvalues, we can define the zeta function as

$$\zeta_{\hat{O}}(s) = \text{Tr}\,(\hat{O})^{-s} = \frac{1}{\Gamma(s)} \int_0^\infty d\tau\, \tau^{s-1}\, \text{Tr}\, e^{-\hat{O}\tau}, \tag{17.39}$$

where the "Trace" can be taken in any complete basis, as we have discussed earlier. (For continuum operators, trace involves an integral as we have already discussed in Eq. (16.48) in addition to a sum over any matrix indices.) All the properties derived in Eq. (17.38) carry over to this infinite dimensional case as well. In particular, we note that, in an infinite dimensional quantum field theory, $\zeta(0)$ is divergent and if it does not depend on the dynamics, it can be neglected in various calculations. As is clear from Eq. (17.38), zeta function can easily calculate $\det \hat{O}$ or $\mathrm{Tr}\ln \hat{O}$ of any operator $\hat{O}$. This is what is involved in the calculation of a partition function at finite temperature or the effective action in a theory interacting with a background field. Let us next show an example of how the partition function of the simple bosonic harmonic oscillator can be calculated using this method.

17.2.1 Partition function for the bosonic oscillator

Let us start with the simple example of the bosonic harmonic oscillator which is already discussed in Section **16.2.1**. The partition function is already defined in Eq. (16.42) to be

$$Z(\beta) = \left[\det\left(-\frac{\mathrm{d}^2}{\mathrm{d}t^2} + \omega^2\right)\right]^{-\frac{1}{2}}, \tag{17.40}$$

where, as explained in the last chapter, t is the imaginary (Euclidean) time. Therefore, the relevant operator, in this case, is $\hat{O} = -\frac{\mathrm{d}^2}{\mathrm{d}t^2} + \omega^2$. To calculate the zeta function (determinant) associated with this operator, we need the eigenvalues of the operator $\hat{O}$ on the space of periodic eigenfunctions (see discussion following Eq. (16.42)), namely, we are looking for solutions to

$$\left(-\frac{\mathrm{d}^2}{\mathrm{d}t^2} + \omega^2\right)\psi_n(t) = \lambda_n \psi_n(t), \quad \psi_n(t+\beta) = \psi_n(t). \tag{17.41}$$

Clearly, there are two general solutions to this second order equation

$$\psi_n(t) \sim \begin{cases} \sin\frac{2\pi nt}{\beta} \\ \cos\frac{2\pi nt}{\beta} \end{cases}, \quad n = 0, \pm 1, \pm 2, \cdots, \tag{17.42}$$

which satisfy the periodicity condition and lead to the eigenvalues

$$\lambda_n = \left(\frac{2\pi n}{\beta}\right)^2 + \omega^2 = \left(\frac{2\pi}{\beta}\right)^2 (n^2 + \nu^2), \quad \nu = \frac{\beta\omega}{2\pi}, \tag{17.43}$$

where $n = 0, \pm 1, \pm 2, \cdots$. As a result, we can write the zeta function for the harmonic oscillator to be (see Eq. (17.36))

$$
\begin{aligned}
\zeta_{\text{HO}}(s) &= \sum_{n=-\infty}^{\infty} \frac{1}{\lambda_n^s} = \left(\frac{\beta}{2\pi}\right)^{2s} \sum_{n=-\infty}^{\infty} \frac{1}{(n^2 + \nu^2)^s} \\
&= \frac{1}{\omega^{2s}} + 2 \left(\frac{\beta}{2\pi}\right)^{2s} \sum_{n=1}^{\infty} \frac{1}{(n^2 + \nu^2)^s} \\
&= \frac{1}{\omega^{2s}} + 2 \left(\frac{\beta}{2\pi}\right)^{2s} E_1^{\nu^2}(s; 1),
\end{aligned}
\tag{17.44}
$$

where the Epstein function $E_1^{\nu^2}(s; 1)$ is given by

$$
\begin{aligned}
E_1^{\nu^2}(s; 1) &= -\frac{\nu^{-2s}}{2} + \frac{\sqrt{\pi}\,\Gamma(s - \frac{1}{2})}{2\Gamma(s)\nu^{2s-1}} \\
&\quad + \frac{2\sqrt{\pi}}{\Gamma(s)} \sum_{n=1}^{\infty} \left(\frac{n\pi}{\nu}\right)^{s-\frac{1}{2}} K_{s-\frac{1}{2}}(2\pi n\nu),
\end{aligned}
\tag{17.45}
$$

with $K_\nu(z)$ denoting the modified Bessel function. Equation (17.44), together with Eq. (17.45), can now be written as

$$
\zeta_{\text{HO}}(s) = \frac{1}{\Gamma(s)} F(s),
\tag{17.46}
$$

where

$$
\begin{aligned}
F(s) = \sqrt{\pi} \left(\frac{\beta}{2\pi}\right)^{2s} \Bigg[&\nu^{1-2s} \Gamma\left(s - \frac{1}{2}\right) \\
&+ 4 \sum_{n=1}^{\infty} \left(\frac{n\pi}{\nu}\right)^{s-\frac{1}{2}} K_{s-\frac{1}{2}}(2\pi n\nu) \Bigg].
\end{aligned}
\tag{17.47}
$$

We note that $F(s)$ is an analytic function near the origin $(s \to 0)$. We also recall that as $s \to 0$,

$$
\Gamma(s) \to \frac{1}{s}, \quad \left(\frac{1}{\Gamma(s)}\right)' \to 1,
\tag{17.48}
$$

so that

$$\frac{\partial \zeta_{\mathrm{HO}}(s)}{\partial s}\bigg|_{s=0} = \left[\left(\frac{1}{\Gamma(s)}\right)' F(s) + \frac{1}{\Gamma(s)} F'(s)\right]_{s=0} = F(0). \quad (17.49)$$

$F(0)$ can be calculated from Eq. (17.47) and with a little bit of algebra, as in Section **16.2.1**, leads to

$$F(0) = -2\ln\left(2\sinh \pi\nu\right) = 2\ln\left(2\sinh \pi\nu\right)^{-1}$$

$$= 2\ln\left(2\sinh \frac{\beta\omega}{2}\right)^{-1} = \frac{\partial \zeta_{\mathrm{HO}}(s)}{\partial s}\bigg|_{s=0}, \quad (17.50)$$

where we have used the definition $K_{-\frac{1}{2}}(z) = \sqrt{\frac{\pi}{2z}}e^{-z}$ as well as relations derived earlier such as Eq. (16.65) (and made the identification in Eq. (17.49) in the last step). Using Eqs. (17.38) and (17.40), we now obtain

$$Z(\beta) = \left[\det\left(-\frac{d^2}{dt^2} + \omega^2\right)\right]^{-\frac{1}{2}} = \exp\left[\frac{1}{2}\frac{\partial \zeta_{\mathrm{HO}}(s)}{\partial s}\right]_{s=0}$$

$$= \frac{1}{2\sinh \frac{\beta\omega}{2}}, \quad (17.51)$$

which was already derived in Eq. (16.67) in the proper time formalism. This illustrates how partition functions can be obtained with zeta function regularization.

17.3 Heat kernel method

In the last chapter, we calculated the Green's function and the effective action for simple theories through the proper time method. This method, however, becomes complicated when there are nontrivial backgrounds present. For example, a quantum field theory may be defined on a curved manifold (namely, interacting with a background gravitational field) with or without boundaries; if it is charged, it may be interacting with a background electromagnetic field or any other kind of external sources. In these cases, the solutions of the proper time evolution equations may be difficult to obtain in a useful form and the heat kernel method comes in handy. It leads to the one loop Green's function as well as the one loop effective action for the theory interacting with a background in a standard manner which we will describe in this section.

Suppose, we have a quantum mechanical operator $\hat{O}$ in the coordinate representation. Then the heat kernel associated with this operator is defined as (see also Eq. (17.39))

$$K_{\hat{O}}(x, x'; \tau) = \langle x | e^{-\tau \hat{O}} | x' \rangle = \langle x, \tau | x', 0 \rangle, \tag{17.52}$$

where τ represents the proper time introduced earlier, with the evolution generated by the Hamiltonian $\hat{O}$. The name "heat kernel" arises because this quantity satisfies the evolution equation

$$\partial_\tau K_{\hat{O}}(x, x'; \tau) = -O(x) K_{\hat{O}}(x, x'; \tau), \tag{17.53}$$

with the initial condition (see Eq. (17.52))

$$K_{\hat{O}}(x, x'; 0) = \delta(x - x'). \tag{17.54}$$

$O(x)$, in Eq. (17.53), represents the coordinate representation of $\hat{O}$ and this equation resembles the heat diffusion equation. Let us recall from Eq. (17.39) that, in the coordinate basis, we can write

$$\begin{aligned}
\zeta_{\hat{O}}(s) &= \frac{1}{\Gamma(s)} \int_0^\infty \mathrm{d}\tau \, \tau^{s-1} \, \mathrm{Tr} \, e^{-\tau \hat{O}} \\
&= \int \mathrm{d}^D x \, |g(x)|^{\frac{1}{2}} \frac{1}{\Gamma(s)} \int_0^\infty \mathrm{d}\tau \, \tau^{s-1} \langle x | e^{-\tau \hat{O}} | x \rangle \\
&= \int \mathrm{d}^D x \, |g(x)|^{\frac{1}{2}} \frac{1}{\Gamma(s)} \int_0^\infty \mathrm{d}\tau \, \tau^{s-1} \, K_{\hat{O}}(x, x; \tau) \\
&= \int \mathrm{d}^D x \, |g(x)|^{\frac{1}{2}} \zeta_{\hat{O}}(s, x),
\end{aligned} \tag{17.55}$$

where D denotes the number of space-time dimensions and we have used the definition of the heat kernel in Eq. (17.52). In addition, we have allowed for the space to be curved with a metric $g_{\mu\nu}(x)$ which leads to a factor of $|g(x)|^{\frac{1}{2}}$ in the integration measure and makes the volume element invariant under diffeomorphisms. (Depending on the number of dimensions $g(x) = \det g_{\mu\nu}(x)$ can be positive or negative in a space with Minkowski signatures which is why we have the magnitude of $|g(x)|$. In flat space-time, $|g(x)| = 1$.) We have also defined a generalized zeta function as

$$\zeta_{\hat{O}}(s, x) = \frac{1}{\Gamma(s)} \int_0^\infty d\tau\, \tau^{s-1} \langle x | e^{-\tau \hat{O}} | x \rangle$$

$$= \frac{1}{\Gamma(s)} \int_0^\infty d\tau\, \tau^{s-1} K_{\hat{O}}(x, x; \tau). \tag{17.56}$$

Furthermore, introducing a complete set of basis states of the operator $\hat{O}$, namely,

$$\hat{O} | \psi_n \rangle = \lambda_n | \psi_n \rangle, \tag{17.57}$$

we can write Eq. (17.56) as

$$\zeta_{\hat{O}}(s, x) = \frac{1}{\Gamma(s)} \int_0^\infty d\tau\, \tau^{s-1} \sum_n \langle x | \psi_n \rangle \langle \psi_n | e^{-\tau \hat{O}} | x \rangle$$

$$= \frac{1}{\Gamma(s)} \int_0^\infty d\tau\, \tau^{s-1} \sum_n e^{-\tau \lambda_n} \langle x | \psi_n \rangle \langle \psi_n | x \rangle$$

$$= \sum_n \frac{\psi_n(x) \psi_n^*(x)}{\lambda_n^s}, \tag{17.58}$$

which is also known as the spectral resolution of the generalized zeta function.

Let us consider a real, massive scalar field interacting with a gravitational background field (in arbitrary dimensions), we can write the action as

$$S = \frac{1}{2} \int d^D x\, |g(x)|^{\frac{1}{2}} \left(g^{\mu\nu} \partial_\mu \phi \partial_\nu \phi - m^2 \phi^2 \right)$$

$$= -\frac{1}{2} \int d^D x |g(x)|^{\frac{1}{2}} \phi \left(\frac{1}{|g|^{\frac{1}{2}}} \partial_\mu (|g|^{\frac{1}{2}} g^{\mu\nu} \partial_\nu) + m^2 \right) \phi, \tag{17.59}$$

where we have integrated the first term by parts and have neglected the surface term. Here $g^{\mu\nu}(x)$ denotes the contravariant metric tensor of the manifold (and $g(x) = \det g_{\mu\nu}(x)$). Furthermore, if we identify

$$O(x) = -\left(\frac{1}{|g(x)|^{\frac{1}{2}}}\partial_\mu(|g(x)|^{\frac{1}{2}}g^{\mu\nu}(x)\partial_\nu) + m^2\right)$$

$$= -(D_\mu D^\mu + m^2), \tag{17.60}$$

where D_μ denotes the gravitational covariant derivative, then this is simply the inverse of the Green's function (see also Eqs. (16.4) and (16.5)) in the presence of gravitational background fields, namely,

$$\hat{G} = \frac{1}{\hat{O}} = \int_0^\infty d\tau\, e^{-\tau\hat{O}}. \tag{17.61}$$

Therefore, it follows that the Green's function can be identified with (recall Eqs. (16.4), (16.5) and (17.52))

$$G(x - x') = \langle x|\hat{G}|x'\rangle = \int_0^\infty d\tau\, K_{\hat{O}}(x, x'; \tau), \tag{17.62}$$

and satisfies the equation

$$O(x)G(x - x') = \frac{1}{|g|^{\frac{1}{2}}}\delta^D(x - x'). \tag{17.63}$$

The factor of $|g|^{-\frac{1}{2}}$ on the right hand side (which can be written in a more symmetric form as well) arises from the fact that, in a curved manifold,

$$\langle x|x'\rangle = \frac{1}{|g|^{\frac{1}{2}}}\delta^D(x - x'), \quad \int d^D x\, |g|^{\frac{1}{2}}\, |x\rangle\langle x| = \mathbb{1}, \tag{17.64}$$

so that we naturally have

$$\langle x|x' \rangle = \langle x|\mathbb{1}|x' \rangle = \int d^D y \, |g|^{\frac{1}{2}} \, \langle x|y \rangle \langle y|x' \rangle$$

$$= \int d^D y \, |g|^{\frac{1}{2}} \, \frac{1}{|g|^{\frac{1}{2}}} \, \delta^D(x-y) \, \frac{1}{|g|^{\frac{1}{2}}} \, \delta^D(y-x')$$

$$= \frac{1}{|g|^{\frac{1}{2}}} \, \delta^D(x-x'). \tag{17.65}$$

Here we have suppressed the coordinate dependencies of the factors $|g|^{\frac{1}{2}}$ for simplicity which are quite easy to figure out. (We parenthetically note here that the operator $O(x)$ in Eq. (17.60) is self-adjoint only with an integration measure $\int d^D x \, |g|^{\frac{1}{2}}$.) Furthermore, we note that, since the action in Eq. (17.59) is quadratic in the field variables $\phi(x)$, the generating functional is given by

$$Z[g_{\mu\nu}] = e^{iW[g_{\mu\nu}]} = N[\det \hat{O}]^{-\frac{1}{2}}, \tag{17.66}$$

so that, up to a constant, the effective action is given by (see, for example, Eq. (16.84))

$$W[g_{\mu\nu}] = \frac{i}{2} \ln \det \hat{O} = \frac{i}{2} \operatorname{Tr} \ln \hat{O} = -\frac{i}{2} \frac{d\zeta_{\hat{O}}(s)}{ds} \Big|_{s=0}$$

$$= -\frac{i}{2} \int_0^\infty \frac{d\tau}{\tau} \operatorname{Tr} e^{-\tau \hat{O}} = -\frac{i}{2} \int d^D x \, |g(x)|^{\frac{1}{2}} \frac{d\zeta_{\hat{O}}(s,x)}{ds} \Big|_{s=0}$$

$$= -\frac{i}{2} \int d^D x \, |g(x)|^{\frac{1}{2}} \int_0^\infty \frac{d\tau}{\tau} K_{\hat{O}}(x,x;\tau). \tag{17.67}$$

Here we have used Eqs. (16.47) and (17.52) (see also Eqs. (17.38), (17.49), (17.55) and (17.56)). From Eqs. (17.62) and (17.67) we see that we can determine the one loop Green's function of the theory as well as the one loop effective action if we know the heat kernel $K_{\hat{O}}(x,x';\tau)$ of the theory. Some comments are in order here. We have used the Minkowski space form of the generating functional in Eq. (17.66). However, if the positivity requirement for the operator $\hat{O}$ needs to analytically continue it to the Euclidean space, that can also be done without any problem.

We will next illustrate these ideas with some simple examples which will also make some aspects of this formulation more clear.

17.3.1 Bosonic propagator in flat space-time

We have already calculated the bosonic propagator in flat space-time using the proper time method in Section **16.1**. Here we will rederive the results using the heat kernel in two different ways to stress some of the aspects of this method.

In flat space-time, $g^{\mu\nu} = \eta^{\mu\nu}$, $|g| = 1$ so that the operator $O(x)$ in Eq. (17.60) takes the simple form $O(x) = -(\Box + m^2)$. This operator is not positive definite and there are two ways we can proceed. We can either analytically continue the operator in Eq. (17.52) (and Eq. (17.61)) to the Euclidean space as (see, for example, Eq. (4.9) or Eq. (8.5) or Eq. (16.41))

$$O(x) \rightarrow O_E(x_E) = (-\Box_E + m^2), \tag{17.68}$$

and at the end of the calculations rotate back to Minkowski space. Or the alternative is to let

$$\tau \rightarrow \bar{\tau} = i\tau, \quad O(x) \rightarrow (O(x) + i\epsilon), \tag{17.69}$$

in Eq. (17.52) (and Eq. (17.61)), where it is understood that $\epsilon \rightarrow 0^+$ at the end as in Eq. (16.5). Here, we will rotate the operator to the Euclidean space as in Eq. (17.68) and rotate back the results to Minkowski space at the end. (Afterwards, we will evaluate the heat kernel also doing the analytic continuation as described in Eq. (17.69).) The operator $O_E(x_E)$ is a positive operator and the normalized eigenfunctions of this operator are the plane waves, namely,

$$O_E(x_E)\psi_{k_E}(x_E) = \lambda_{k_E}\psi_{k_E}(x_E), \tag{17.70}$$

with

$$\psi_{k_E}(x_E) = \langle x_E | k_E \rangle = \frac{e^{-ik_E \cdot x_E}}{(2\pi)^{\frac{D}{2}}}, \quad \lambda_{k_E} = (k_E^2 + m^2) > 0. \tag{17.71}$$

Following Eq. (17.52), we now obtain

$$K_{\hat{O}_E}(x_E, x_E'; \tau) = \int d^D k_E \, \langle x_E | k_E \rangle \langle k_E | e^{-\tau O_E(x_E)} | x_E' \rangle$$

$$= \int d^D k_E \, \psi_{k_E}(x_E)\psi_{k_E}^*(x_E')e^{-\tau \lambda_{k_E}}, \tag{17.72}$$

and this leads to (see Eq. (17.62))

$$G_E(x_E - x'_E) = \int_0^\infty d\tau \int d^D k_E \, \psi_{k_E}(x_E) \psi^*_{k_E}(x'_E) e^{-\tau \lambda_{k_E}}$$

$$= \int d^D k_E \, \frac{\psi_{k_E}(x_E) \psi^*_{k_E}(x'_E)}{\lambda_{k_E}}$$

$$= \int d^D k_E \, \frac{\psi_{k_E}(x_E) \psi^*_{k_E}(x'_E)}{k_E^2 + m^2}, \tag{17.73}$$

which describes the spectral resolution of the heat kernel (see also Eq. (17.58)). It is straightforward to check from the spectral decomposition in Eq. (17.74) (using Eq. (17.70) and the completeness of the eigenfunctions) that

$$O_E(x_E) G_E(x_E - x'_E) = \delta^D(x_E - x'_E). \tag{17.74}$$

The heat kernel, in the present simple case, can be exactly evaluated (substituting the eigenfunctions $\psi_{k_E}(x)$ and the eigenvalues λ_{k_E} into Eq. (17.71) in Eq. (17.72))

$$K_{\hat{O}_E}(x_E, x'_E; \tau) = \int \frac{d^D k_E}{(2\pi)^D} e^{-i k_E \cdot (x_E - x'_E)} e^{-\tau(k_E^2 + m^2)}$$

$$= \frac{e^{-\tau m^2 - \frac{(x_E - x'_E)^2}{4\tau}}}{(2\pi)^D} \int d^D \tilde{k}_E \, e^{-\tau \tilde{k}_E^2}$$

$$= \frac{e^{-\tau m^2 - \frac{(x_E - x'_E)^2}{4\tau}}}{(2\pi)^D} \left(\frac{\pi}{\tau}\right)^{\frac{D}{2}}$$

$$= \frac{e^{-\tau m^2 - \frac{(x_E - x'_E)^2}{4\tau}}}{(4\pi\tau)^{\frac{D}{2}}}, \tag{17.75}$$

which can be compared with Eq. (16.34) with appropriate analytic continuation. It follows from Eq. (16.25) that

$$K(x_E, x'_E; \tau) \xrightarrow{\tau \to 0} \delta^D(x_E - x'_E), \tag{17.76}$$

as it should (see Eq. (17.54)). Furthermore, it can be checked in a straightforward manner now that Eq. (17.53) holds, namely,

$$\partial_\tau K_{\hat{O}_E}(x_E - x'_E; \tau) = -\hat{O}_E(x) K_{\hat{O}_E}(x_E - x'_E; \tau). \tag{17.77}$$

The Green's function can also be exactly evaluated in this simple case as (see Eq. (17.62))

$$G_E(x_E - x'_E) = \int_0^\infty d\tau \, K_{\hat{O}_E}(x - x'; \tau)$$

$$= \int_0^\infty d\tau \, \frac{e^{-\tau m^2 - \frac{(x_E - x'_E)^2}{4\tau}}}{(4\pi\tau)^{\frac{D}{2}}}, \tag{17.78}$$

which can be evaluated in terms of (modified) Bessel functions (see, for example, Eqs. (16.35)–(16.36)).

We will now evaluate the heat kernel by analytically continuing the τ parameter as discussed in Eq. (17.69) for later use. Under the analytic continuation, the heat kernel associated with $O(x) = -(\Box + m^2)$ is defined as (we are renaming $\bar{\tau}$ in Eq. (17.69) as τ for simplicity)

$$K_{\hat{O}}(x, x'; \tau) = \langle x | e^{i\tau\hat{O}} | x' \rangle, \tag{17.79}$$

where the $i\epsilon$ prescription is understood and this can be compared with Eq. (16.18). This heat kernel satisfies the equation

$$\partial_\tau K_{\hat{O}}(x, x'; \tau) = i\langle x | \hat{O} e^{i\tau\hat{O}} | x' \rangle = iO(x) K_{\hat{O}}(x, x'; \tau), \tag{17.80}$$

which can be compared with Eq. (16.19). The constant term in the operator $O(x) = -(\Box + m^2)$ can be factored out by redefining

$$K_{\hat{O}}(x, x'; \tau) = e^{-im^2\tau} \tilde{K}_{-\Box}(x, x'; \tau), \tag{17.81}$$

which leads to the equation (see Eq. (17.80))

$$\partial_\tau \tilde{K}_{-\Box}(x, x'; \tau) = -i\Box \tilde{K}_{-\Box}(x, x'; \tau)$$

$$= -i\partial_\mu \partial^\mu \tilde{K}_{-\Box}(x, x'; \tau). \tag{17.82}$$

Equation (17.82) simply corresponds to the quantum mechanical free particle equation in D dimensional (Minkowski) space so that we can write the wave packet solution as

$$K_{\hat{O}}(x, x'; \tau) = \langle x, \tau | x', 0 \rangle = e^{-im^2\tau} \tilde{K}_{-\Box}(x, x'; \tau)$$

$$= \frac{i}{(4\pi i \tau)^{\frac{D}{2}}} e^{-im^2\tau - \frac{i(x-x')^2}{4\tau}}. \tag{17.83}$$

Here the constant $i \left(\frac{-i}{4\pi} \right)^{\frac{D}{2}}$ has been fixed from the initial condition in Eq. (17.54) (see also Eq. (16.32)) and Eq. (17.83) can be compared with Eq. (16.34). Furthermore, this leads to the Green's function as (see Eqs. (17.61)–(17.62) with appropriate analytic continuation)

$$G(x - x') = -i \int_0^\infty d\tau \, K_{\hat{O}}(x, x'; \tau)$$

$$= \frac{1}{(4\pi i)^{\frac{D}{2}}} \int_0^\infty \frac{d\tau}{\tau^{\frac{D}{2}}} e^{-im^2\tau - \frac{i(x-x')^2}{4\tau}}, \tag{17.84}$$

which coincides with Eq. (16.35) (remember Eq. (17.81) as well as the $i\epsilon$ prescription that is understood). We will see in the next section how such a form becomes useful in the asymptotic expansion of the heat kernel.

To conclude this discussion, let us define a parameter related to the geodesic distance between the two coordinates (remember, we are in flat space-time, otherwise we have to consider the geodetic interval or the world function) as

$$\sigma(x, x') = \frac{1}{2}(x - x')^2, \quad \sigma(x, x')\big|_{x'=x} = 0,$$

$$\frac{\partial \sigma(x, x')}{\partial x^\mu} = (x - x')_\mu, \quad \frac{\partial^2 \sigma(x, x')}{\partial x^\mu \partial x^\nu} = \eta_{\mu\nu}, \quad \eta^{\mu\nu} \frac{\partial^2 \sigma(x, x')}{\partial x^\mu \partial x^\nu} = D,$$

$$\eta^{\mu\nu} \frac{\partial \sigma(x, x')}{\partial x^\mu} \frac{\partial \sigma(x, x')}{\partial x^\nu} = 2\sigma(x, x'). \tag{17.85}$$

In terms of this new variable, the operator $O(x)$ takes the form

$$O(x) = -(\Box + m^2) = -\left(2\sigma\frac{\partial^2}{\partial\sigma^2} + D\frac{\partial}{\partial\sigma} + m^2\right), \qquad (17.86)$$

and the solutions in Eqs. (17.83) and (17.84) can be rewritten in terms of the variable $\sigma(x, x')$ as

$$K_{\hat{O}}(x, x'; \tau) = \frac{i}{(4\pi i\tau)^{\frac{D}{2}}} e^{-im^2\tau - \frac{i\sigma(x,x')}{2\tau}},$$

$$G(x - x') = \frac{1}{(4\pi i)^{\frac{D}{2}}} \int_0^\infty \frac{d\tau}{\tau^{\frac{D}{2}}} e^{-im^2\tau - \frac{i\sigma(x,x')}{2\tau}}. \qquad (17.87)$$

17.4 Expansion of the heat kernel

In the last section, we studied the heat kernel as well as the Green's function for a free scalar field theory in flat space-time which can be worked out explicitly (in closed form). However, if a theory is in a curved background or in the background of interacting external sources, we may no longer be able to determine the heat kernel exactly. In this case, we can write a general solution of the heat kernel as a power series expansion in τ as (in D dimensions)

$$K_{\hat{O}}(x, x'; \tau) = \frac{i\Delta^{\frac{1}{2}}(x, x')}{(4\pi i\tau)^{\frac{D}{2}}} e^{-im^2\tau - \frac{i\sigma(x,x')}{2\tau}} F(x, x'; \tau),$$

$$F(x, x'; \tau) = \sum_{n=0}^\infty a_n(x, x')(i\tau)^n, \qquad (17.88)$$

where

$$\Delta(x, x') = |g(x)|^{-\frac{1}{2}} |\det \partial_\mu^x \partial_\nu^{x'} \sigma(x, x')| |g(x')|^{-\frac{1}{2}}, \qquad (17.89)$$

is known as the biscalar Van Vleck-Morette determinant. $\sigma(x, x')$ here is one half of the length of the proper interval (geodesic length) squared along the geodesic between the points x and x', namely,

$$\sigma(x, x') = \frac{1}{2}\ell^2(x, x').$$ (17.90)

(Normally, the proper interval is denoted by τ or s both of which we are using in other contexts and, therefore, we denote the invariant length as ℓ. Furthermore, we assume that there is only one geodesic between the two points x, x'. If there are more geodesics, then, the contributions from all of them need to be summed.) The important thing to note from Eq. (17.88) is that the coefficients of expansion $a_n(x, x')$ depend only on the coordinates and not on τ. They also do not depend on the dimensionality of space-time. Furthermore, the initial condition on the heat kernel, Eq. (17.54), leads to

$$a_0(x, x') = 1,$$ (17.91)

and we can also derive the covariantized relations (see Eq. (17.85))

$$g^{\mu\nu}\frac{\partial\sigma}{\partial x^\mu}\frac{\partial\sigma}{\partial x^\nu} = (D_\mu\sigma)(D^\mu\sigma) = 2\sigma,$$

$$\Delta^{-1}D_\mu(\Delta D^\mu\sigma) = D.$$ (17.92)

Since the coordinate and τ dependencies are explicitly separated out in the expansion Eq. (17.88), we can now go back to the heat equation (see the Minkowski space equation, Eq. (17.80)) and the two sides of the equation lead to

$$\partial_\tau K_{\hat{O}} = \sum_{n=0}^{\infty} \frac{i\Delta^{\frac{1}{2}}e^{-im^2\tau - \frac{i\sigma}{2\tau}}}{(4\pi i\tau)^{\frac{D}{2}}}$$

$$\times \left[\left(\frac{i\sigma}{2\tau^2} - \frac{D}{2\tau} - im^2\right)a_n + i(n+1)a_{n+1}\right](i\tau)^n,$$

$$iO(x)K_{\hat{O}} = i\sum_{n=0}^{\infty} O(x)\left(\frac{i\Delta^{\frac{1}{2}}e^{-im^2\tau - \frac{i\sigma}{2\tau}}a_n}{(4\pi i\tau)^{\frac{D}{2}}}\right)(i\tau)^n.$$ (17.93)

Here we have suppressed the coordinate dependence in various terms for simplicity. The right hand side of the second relation in Eq. (17.93) depends explicitly on the exact form of $O(x)$ and once a form is specified (by the theory), equating the two relations in Eq. (17.92) (see Eq. (17.80)) would generate a recursion relation between the coefficients $a_n(x, x')$ (depending on the form of $O(x)$) which can be solved. Let us illustrate this with a general form of the operator $O(x)$.

Consider an operator of the form (compare with Eq. (17.60))

$$O(x) = -(\Box + m^2 + V) = -(D_\mu D^\mu + m^2 + V), \qquad (17.94)$$

where D_μ denotes the gravitational covariant derivative and V represents a potential term (which does not involve derivatives). Such an operator may arise, say from the Lagrangian of a charged particle (bosonic/fermionic) in a gravitational background interacting with an external electromagnetic field. In this case, the second relation in Eq. (17.93) can be worked out to give

$$iO(x)K_{\hat{O}} = \sum_{n=0}^{\infty} \frac{i\Delta^{\frac{1}{2}} e^{-im^2\tau - \frac{i\sigma}{2\tau}}}{(4\pi i\tau)^{\frac{D}{2}}}$$

$$\times \left[\left(\frac{iD_\mu\sigma D^\mu\sigma}{4\tau^2} - \frac{\Delta^{-1}D_\mu(\Delta D^\mu\sigma)}{2\tau} - i(m^2 + V) \right) a_n \right.$$

$$\left. - i\Delta^{-\frac{1}{2}}D_\mu D^\mu(\Delta^{\frac{1}{2}}a_n) - i(D^\mu\sigma)(D_\mu a_{n+1}) \right] (i\tau)^n. \quad (17.95)$$

It is now straightforward to compare the first relation in Eq. (17.93) with Eq. (17.95). The $-im^2$ terms cancel on both sides of Eq. (17.80). The $\frac{1}{\tau^2}$ terms cancel because of the first identity in Eq. (17.92). The $\frac{1}{\tau}$ terms cancel because of the second identity in Eq. (17.92). Comparing the coefficients of $(i\tau)^n$ terms on both sides of Eq. (17.80), we obtain the recursion relation (recall from Eq. (17.91) that $a_0(x, x') = 1$)

$$(n + 1)a_{n+1} + (D^\mu\sigma)(D_\mu a_{n+1})$$

$$= -Va_n - \Delta^{-\frac{1}{2}}D_\mu D^\mu(\Delta^{\frac{1}{2}}a_n), \quad n = 0, 1, 2, \cdots . \quad (17.96)$$

We emphasize here that, while $V = V(x)$, other variables depend on two coordinates, namely, $\sigma = \sigma(x, x')$, $\Delta = \Delta(x, x')$, $a_n = a_n(x, x')$

and we have suppressed these coordinate dependencies for simplicity. Once, we have determined the coefficients $a_n(x, x')$, we can substitute them into Eq. (17.88) to determine the heat kernel $K_{\hat{o}}(x, x'; \tau)$ which can give, for example, the Green's function of the theory.

On the other hand, if we are interested in the effective action of the theory, that depends only on the trace of the heat kernel and, therefore, on the heat kernel in the coincidence limit $K_{\hat{o}}(x, x; \tau)$ (see, for example, Eq. (17.67)). The recursion relation in Eq. (17.96) becomes much simpler in the coincidence limit. We note that when $x' \to x$, we can write (see Eq. (17.90))

$$\sigma(x, x') = \frac{1}{2} d\ell^2 = \frac{1}{2} g_{\mu\nu}(x) dx^\mu dx^\nu, \tag{17.97}$$

where $dx^\mu = (x' - x)^\mu$ denotes the infinitesimal interval as $x' \to x$. Therefore, it follows immediately that in the coincidence limit ($x' = x$ corresponds to $dx^\mu = 0$)

$$\sigma(x, x')\big|_{x'=x} = 0, \ \ D_\mu \sigma(x, x')\big|_{x'=x} = g_{\mu\nu}(x) dx^\nu\big|_{dx=0} = 0,$$

$$\frac{\partial \sigma(x, x')}{\partial x^\mu \partial x^\nu}\bigg|_{x'=x} = g_{\mu\nu}(x), \ \ \Delta(x, x')\big|_{x'=x} = 1, \tag{17.98}$$

and so on. In fact, taking successive derivatives of the two relations in Eq. (17.92) one can derive many more such useful relations in the coincident limit. In particular, we have

$$D_\mu(\Delta^{\frac{1}{2}}(x, x'))\big|_{x'=x} = 0, \ \ D_\mu D_\nu(\Delta^{\frac{1}{2}}(x, x'))\big|_{x'=x} = \frac{1}{6} R_{\mu\nu}, \tag{17.99}$$

where $R_{\mu\nu}$ denotes the Riemann curvature tensor of the manifold. As a result, the recursion relation in Eq. (17.96) simplifies in this limit to

$$(n + 1)a_{n+1}(x) = -V a_n(x)$$

$$- D_\mu D^\mu(\Delta^{\frac{1}{2}}(x, x') a_n(x, x'))\big|_{x'=x}, \tag{17.100}$$

where $n = 0, 1, 2, \cdots$ and we have identified $a_n(x, x')|_{x'=x} = a_n(x)$. Recalling that $a_0(x) = 1$ (see Eq. (17.91)), it follows now that

$$a_0(x) = 1, \quad a_1(x) = -\frac{1}{6}g^{\mu\nu}R_{\mu\nu} - V = -\frac{1}{6}R - V, \quad \cdots, \quad (17.101)$$

where R stands for the Ricci scalar curvature. Therefore, in the coincident limit, we can write

$$K_{\hat{O}}(x, x; \tau) = \frac{ie^{-im^2\tau}}{(4\pi i\tau)^{\frac{D}{2}}} \sum_{n=0}^{\infty} a_n(x)(i\tau)^n, \qquad (17.102)$$

where the coefficients $a_n(x)$ are determined from the recursion relation Eq. (17.100).

In evaluating the effective action, Eq. (17.67), the basic quantity of interest is

$$-i \int_0^{\infty} \frac{d\tau}{\tau} K_{\hat{O}}(x, x; \tau) = (4\pi)^{-\frac{D}{2}} \sum_{n=0}^{\infty} (i)^{n-\frac{D}{2}} a_n(x)$$

$$\times \int_0^{\infty} d\tau \, (\tau)^{n-\frac{D}{2}-1} e^{-im^2\tau}. \quad (17.103)$$

Recalling that there is an imaginary part in the mass term (namely, $m^2 \to m^2 - i\epsilon$), the integral over τ in Eq. (17.103) can be done trivially and gives

$$\int_0^{\infty} d\tau \, (\tau)^{n-\frac{D}{2}-1} e^{-im^2\tau} = \frac{1}{(im^2)^{n-\frac{D}{2}}} \Gamma\left(n - \frac{D}{2}\right), \qquad (17.104)$$

so that we have (see Eqs. (17.103) and (17.104))

$$-i \int_0^{\infty} \frac{d\tau}{\tau} K_{\hat{O}}(x, x; \tau) = \sum_{n=0}^{\infty} \frac{\Gamma(n - \frac{D}{2})}{(m^2)^{n-\frac{D}{2}}} \frac{a_n(x)}{(4\pi)^{\frac{D}{2}}}. \qquad (17.105)$$

As a result, the (subtracted or normalized) effective action for the theory (see Eqs. (17.67) and (16.85)) is obtained to be (the field independent $a_0(x)$ term cancels out)

$$W[g_{\mu\nu}] = \frac{1}{2} \int \mathrm{d}^D x \, |g(x)|^{\frac{1}{2}} \sum_{n=1}^{\infty} \frac{\Gamma(n - \frac{D}{2})}{(m^2)^{n - \frac{D}{2}}} \frac{a_n(x)}{(4\pi)^{\frac{D}{2}}}. \tag{17.106}$$

There are several things to note from this result. First, we have continued to denote the normalized effective action as W and not Γ to avoid confusion with the gamma functions. We note that for odd D dimensions, the gamma function is well behaved and, therefore, there are no divergences in the effective action in odd space-time dimensions. For even space-time dimensions (D even), on the other hand, the gamma function diverges for $n - \frac{D}{2} \leq 0$. Consequently, there are divergent terms in the effective action which can be thought of as renormalization parts which need to be removed by adding counter terms to the starting theory. For a given D, such divergent terms are finite in number and, consequently, the number of counter terms needed is also finite. Using the gamma function recursion relation $\Gamma(s + 1) = s\Gamma(s)$ repeatedly, it can be shown that

$$\Gamma(s) = \frac{1}{s(s+1)(s+2)\cdots(s+k-1)} \Gamma(s+k), \tag{17.107}$$

which leads to

$$\Gamma(-k) = \frac{(-1)^k}{k!} \left. \frac{1}{s+k} \right|_{s \to -k}, \tag{17.108}$$

and shows that all the divergent terms in Eq. (17.106) have simple pole structures. For $n - \frac{D}{2} \geq 1$, the terms in the effective action in Eq. (17.106) are finite and give the quantum corrections to the theory. Symbolically, we can see that the zeta function regularization naturally divides the effective action into a divergent renormalization part and a finite part which represents the genuine quantum corrections (we remind the readers that the normalized effective action which vanishes when external backgrounds are set to zero is denoted by $\Gamma[g_{\mu\nu}, \cdots]$, see Eq. (16.85))

$$W[g_{\mu\nu}] = W_{\mathrm{div}}[g_{\mu\nu}] + W_{\mathrm{finite}}[g_{\mu\nu}]. \tag{17.109}$$

We note from Eq. (17.106) that the effective action, in this massive case, is an expansion in inverse powers of the mass parameter. Therefore, it does not allow us to take the massless limit directly. If the theory under consideration involves massless particles, the calculation of the effective action proceeds as follows. In evaluating Eq. (17.104) (with $m = 0$), we note that since the operator $\hat{O}$ in the exponent has an imaginary part (remember $m^2 - i\epsilon$), large values of τ will be damped out and the singularities as well as the dominant contributions will come only from small values of τ. Therefore, we can divide the τ integral into two parts at some finite small value τ_0 and write

$$\int_0^\infty d\tau\, \tau^{s+n-\frac{D}{2}} = \int_0^{\tau_0} d\tau\, \tau^{s+n-\frac{D}{2}} + \int_{\tau_0}^\infty d\tau\, \tau^{s+n-\frac{D}{2}}. \quad (17.110)$$

As a result, for $\operatorname{Re} s > \frac{D}{2}$, we can write (see Eqs. (17.56) and (17.102) with $m = 0$)

$$\zeta_{\hat{O}}(s, x) = \sum_{n=0}^\infty \frac{(i)^{n+1-\frac{D}{2}}}{(4\pi)^{\frac{D}{2}}} a_n(x) \frac{1}{\Gamma(s)} \int_0^{\tau_0} d\tau\, \tau^{s+n-\frac{D}{2}-1}$$

$$= \sum_{n=0}^\infty \frac{(i)^{n+1-\frac{D}{2}}}{(4\pi)^{\frac{D}{2}}} a_n(x) \frac{1}{\Gamma(s)} \frac{\tau_0^{s+n-\frac{D}{2}}}{s + n - \frac{D}{2}}, \quad (17.111)$$

where the integral vanishes at the lower limit for $\operatorname{Re} s > \frac{D}{2}$ (and the second integral is assumed to be damped). There is only a pole singularity in Eq. (17.111) and we can now analytically continue this result to $\operatorname{Re} s < \frac{D}{2}$ as well. Furthermore, using Eq. (17.108), we note that in even dimensions (D even)

$$\zeta_{\hat{O}}(-m, x) = \lim_{s \to -m} \sum_{n=0}^\infty \frac{(i)^{n+1-\frac{D}{2}}}{(4\pi)^{\frac{D}{2}}} (-1)^m\, m!$$

$$\times (s + m) \frac{\tau_0^{s+n-\frac{D}{2}}}{s + n - \frac{D}{2}} a_n(x)$$

$$= \frac{i(-i)^m\, m!}{(4\pi)^{\frac{D}{2}}} a_{m+\frac{D}{2}}(x). \quad (17.112)$$

As is clear from the second line of Eq. (17.112), $\zeta_{\hat{o}}(-m, x)$ vanishes in odd dimensions (D odd) since $(s + m) \xrightarrow{s \to -m} 0$ and there is no pole term to cancel this. Equation (17.112) leads, in particular to

$$\zeta_{\hat{o}}(0, x) = \frac{i}{(4\pi)^{\frac{D}{2}}} \, a_{\frac{D}{2}}(x), \tag{17.113}$$

which, as we will see in the next section, is very useful in the calculation of anomalies. This also leads to (see Eq. (17.55))

$$\zeta_{\hat{o}}(0) = \int \mathrm{d}^D x \, |g(x)|^{\frac{1}{2}} \, \zeta_{\hat{o}}(0, x)$$

$$= \frac{i}{(4\pi)^{\frac{D}{2}}} \int \mathrm{d}^D x \, |g(x)|^{\frac{1}{2}} \, a_{\frac{D}{2}}(x). \tag{17.114}$$

We note here that we can also calculate $\zeta'(0)$ from Eq. (17.112) which, as we have seen, leads to the effective action, but the calculation is not as simple.

17.5 Schwinger model in curved space-time

As an application of the zeta function regularization, let us study the Schwinger model in a curved space-time background. As we have discussed in great detail in Section **13.2**, Schwinger model describes massless QED (quantum electrodynamics) in $1 + 1$ dimensions. In flat space-time, the Lagrangian density is given by (see Eq. (13.43))

$$\mathcal{L} = -\frac{1}{4} F_{\mu\nu} F^{\mu\nu} + i\bar{\psi}\eta^{\mu\nu}\gamma_\nu(\partial_\mu + ieA_\mu)\psi, \quad \mu, \nu = 0, 1, \tag{17.115}$$

leading to the action

$$S = \int \mathrm{d}^2 x \, \mathcal{L}. \tag{17.116}$$

The properties of the Dirac gamma matrices are described in Eqs. (13.48)–(13.51) and a representation for these 2×2 matrices is given by the Pauli matrices in Eq. (13.52). Various identities associated with these two dimensional matrices are discussed in Eqs. (13.57)–(13.62).

Let us next look at only the fermion part of the theory. In a curved space-time manifold, of course, the metric $\eta^{\mu\nu} \to g^{\mu\nu}(x)$, but since the fermions are defined only in the tangent space (to the manifold), the part of the Lagrangian density involving the fermions is given by (e denotes the electric charge which has the dimension of mass, see Eq. (13.47))

$$\mathcal{L}_f = i\bar{\psi}e^{\mu a}(x)\gamma_a \mathcal{D}_\mu \psi = i\bar{\psi}e^{\mu a}(x)\gamma_a(D_\mu + ieA_\mu)\psi, \qquad (17.117)$$

with the action given by

$$S_f = \int \mathrm{d}^2 x \, |g(x)|^{\frac{1}{2}} \mathcal{L}_f. \qquad (17.118)$$

We note that, in Eq. (17.117), $\mu = 0, 1$ and $a = 0, 1$. Here the indices $a, b, \cdots$ are the tangent space (world) indices while $\mu, \nu, \cdots$ denote the indices of the Riemannian manifold. $e^{\mu a}(x)$ denotes the transformation matrix between the world vectors and the vectors on the Riemannian manifold and, in two dimensions, is known as the zweibein (in four dimensions it is known as the vierbein or is also known as the tetrad). The Riemannian indices are raised or lowered by the gravitational metric while the world indices are raised and lowered by the (flat space) Minkowski metric. The zweibeins satisfy various relations of the forms (repeated indices are summed)

$$e^{\mu a}e_a^\nu = g^{\mu\nu}, \;\; e^{\mu a}e_\mu^b = \eta^{ab}, \;\; e^{\mu a}e_{\nu a} = \delta_\nu^\mu, \;\; e^{\mu a}e_{\mu b} = \delta_b^a. \qquad (17.119)$$

D_μ, in Eq. (17.117), denotes the gravitational covariant derivative and acting on spinors, it has the form

$$D_\mu \psi = \left(\partial_\mu + \frac{1}{2}\,\omega_{\mu ab}\sigma^{ab} \right) \psi, \qquad (17.120)$$

where $\omega_{\mu ab}$ is known as the spin connection and is defined in terms of the zweibein as

$$\omega_{\mu ab} = \frac{1}{2}[e_a^\nu(\partial_\mu e_{\nu b} - \partial_\nu e_{\mu b}) + e_a^\rho e_b^\sigma(\partial_\sigma e_{\rho c})e_\mu^c - (a \leftrightarrow b)], \quad (17.121)$$

and (see also Eqs. (13.58) and (13.59))

$$\sigma_{ab} = \frac{1}{2}[\gamma_a, \gamma_b] = \epsilon_{ab}\gamma_5. \tag{17.122}$$

Although the Dirac matrices in flat space-time are constant matrices, in curved space-time they become space-time dependent, $\gamma^\mu(x) = e_a^\mu(x)\gamma^a$ (nonetheless, γ_5 does not change). As a result, the gamma matrix decomposition, Eq. (13.60), takes the form (recall that the contravariant Levi-Civita tensor transforms under a general coordinate transformation as a tensor density of weight -1 while the covariant (Levi-Civita) tensor transforms as a tensor density of weight $+1$)

$$\gamma^\mu\gamma^\nu = g^{\mu\nu} + \frac{1}{|g(x)|^{\frac{1}{2}}}\,\epsilon^{\mu\nu}\gamma_5, \tag{17.123}$$

and we have

$$\not{D}^2 = \gamma^\mu\gamma^\nu \mathcal{D}_\mu \mathcal{D}_\nu = \mathcal{D}_\mu \mathcal{D}^\mu + \frac{1}{2|g(x)|^{\frac{1}{2}}}\,\gamma_5\,\epsilon^{\mu\nu}\,[\mathcal{D}_\mu, \mathcal{D}_\nu]$$

$$= \mathcal{D}_\mu \mathcal{D}^\mu + \frac{ie}{2|g(x)|^{\frac{1}{2}}}\,\gamma_5\epsilon^{\mu\nu}\,F_{\mu\nu}. \tag{17.124}$$

The fermion part of the generating functional can be written as

$$Z_f[A] = e^{iW_f[A]} = \int \mathcal{D}\bar{\psi}\mathcal{D}\psi\, e^{iS_f}, \tag{17.125}$$

with S_f given in Eq. (17.118). If we now make an infinitesimal local chiral transformation (see Eq. (13.125)) of the fermion fields

$$\psi(x) \rightarrow \psi'(x) = \psi(x) - ie\epsilon(x)\gamma_5\psi(x),$$
$$\bar{\psi}(x) \rightarrow \bar{\psi}'(x) = \bar{\psi}(x) - ie\epsilon(x)\bar{\psi}(x)\gamma_5, \tag{17.126}$$

where $\epsilon(x)$ is the infinitesimal local parameter of transformation, the fermion Lagrangian density changes as

$$\mathcal{L}_f \to \mathcal{L}_f - e(\partial_\mu \epsilon(x)) \bar{\psi} \gamma_5 \gamma^\mu \psi, \tag{17.127}$$

leading to

$$S_f \to S_f - \int d^2x \, |g(x)|^{\frac{1}{2}} (\partial_\mu \epsilon(x)) J_5^\mu(x)$$

$$= S_f + \int d^2x \, |g(x)|^{\frac{1}{2}} \, \epsilon(x) \, D_\mu J_5^\mu(x), \tag{17.128}$$

where we note that (see Eqs. (13.63) and (13.64) in a curved background)

$$J^\mu(x) = e\bar{\psi}(x)\gamma^\mu \psi(x), \quad J_5^\mu(x) = e\bar{\psi}(x)\gamma_5\gamma^\mu \psi(x),$$

$$J_5^\mu(x) = \frac{1}{|g(x)|^{\frac{1}{2}}} \, \epsilon^{\mu\nu} J_\nu(x). \tag{17.129}$$

Furthermore, we have integrated by parts in the second line of Eq. (17.128) discarding a surface term and have used the fact that

$$D_\mu J_5^\mu(x) = \frac{1}{|g(x)|^{\frac{1}{2}}} \partial_\mu (|g(x)|^{\frac{1}{2}} J_5^\mu(x)). \tag{17.130}$$

The generating functional in Eq. (17.125) does not change under a field redefinition (since the path integral is an integral over all field configurations). Therefore, if there were no other contributions in the path integral coming from the field redefinition in Eq. (17.126), we would conclude that

$$\delta Z_f[A] = \int \mathcal{D}\bar{\psi}\mathcal{D}\psi \, (i\delta S_f) \, e^{iS_f} = 0,$$

$$\text{or,} \quad D_\mu \langle J_5^\mu(x) \rangle = -\frac{1}{|g(x)|^{\frac{1}{2}}} \epsilon^{\mu\nu} D_\mu \frac{\delta W_f[A]}{\delta A^\nu(x)} - 0, \tag{17.131}$$

where we have used Eq. (17.129). This simply expresses the covariantized form of the chiral curent conservation for a massless fermion (see Eq. (13.8)).

On the other hand, we know (see Chapter **13**) that the path integral measure is not invariant under a chiral transformation, Eq. (17.126), of the fermion fields. In Chapter **13**, we had calculated the change in the measure in the momentum space in a regularized manner. Here we will calculate the same using zeta function regularization. (The momentum space calculation can also be done here by using Riemann normal coordinates.) Let us consider the eigenvalue equation

$$i\not{D}\phi_n = \lambda_n\phi_n, \qquad (i\not{D})^2\phi_n = -\not{D}^2\phi_n = \lambda_n^2\phi_n, \qquad (17.132)$$

so that the zeta function associated with this operator can be defined as (see the spectral resolution of the zeta function in Eq. (17.58))

$$\zeta_{-\not{D}^2}(s, x) = \sum_n \frac{\phi_n(x)\phi_n^\dagger(x)}{(\lambda_n^2)^s}. \qquad (17.133)$$

We can expand the fermion fields (see, for example, Eqs. (13.17)–(13.25) as well as Eq. (17.126)) in these basis functions as

$$\psi(x) = \sum_n a_n\phi_n(x), \qquad \psi'(x) = \sum_{n,m} c_{nm}a_m\phi_n(x),$$

$$\bar{\psi}(x) = \sum_n b_n\phi_n^\dagger(x), \qquad \bar{\psi}'(x) = \sum_{n,m} c_{nm}b_m\phi_n^\dagger(x), \qquad (17.134)$$

where (see Eq. (13.22) with $\lambda(x) = e\epsilon(x)$)

$$c_{nm} = \delta_{nm} - ie \int d^2x \, |g(x)|^{\frac{1}{2}} \, \epsilon(x)\phi_n^\dagger(x)\gamma_5\phi_m(x). \qquad (17.135)$$

As a result, the path integral measures can be written as

$$\mathcal{D}\psi = \prod_n \mathrm{d}a_n, \qquad \mathcal{D}\bar{\psi} = \prod_n \mathrm{d}b_n,$$

$$\mathcal{D}\psi' = (\det c_{nm})^{-1} \prod_n \mathrm{d}a_n = (\det c_{nm})^{-1}\mathcal{D}\psi,$$

$$\mathcal{D}\bar{\psi}' = (\det c_{nm})^{-1} \prod_n \mathrm{d}b_n = (\det c_{nm})^{-1}\mathcal{D}\bar{\psi}, \qquad (17.136)$$

where the inverse of the determinant arises because we are dealing with fermion fields (see Eqs. (5.30) and (5.32)).

We see that the basic element in the study of the change in the path integral measure is (see Eq. (17.135))

$$\det c_{nm} = \det\left(\delta_{nm} - ie\int \mathrm{d}^2x\,|g(x)|^{\frac{1}{2}}\,\epsilon(x)\,\phi_n^\dagger(x)\gamma_5\phi_m(x)\right)$$

$$= \exp \operatorname{Tr} \ln\left(\delta_{nm} - ie\int \mathrm{d}^2x\,|g(x)|^{\frac{1}{2}}\,\epsilon(x)\,\phi_n^\dagger(x)\gamma_5\phi_m(x)\right)$$

$$= \exp\left(-ie\int \mathrm{d}^2x\,|g(x)|^{\frac{1}{2}}\,\epsilon(x)\,\operatorname{tr}(\gamma_5\phi_n(x)\phi_n^\dagger(x))\right)$$

$$= \exp\left(-ie\int \mathrm{d}^2x\,|g(x)|^{\frac{1}{2}}\,\epsilon(x)\,\operatorname{tr}(\gamma_5\zeta_{-\slashed{\mathcal{D}}^2}(0,x))\right). \qquad (17.137)$$

Here we have used Eqs. (17.133) as well as (17.135) and "tr" stands for trace over Dirac (spinor) indices. As a result, the change in the path integral measure takes the form

$$\delta\left(\mathcal{D}\bar{\psi}\mathcal{D}\psi\right) = \mathcal{D}\bar{\psi}'\mathcal{D}\psi' - \mathcal{D}\bar{\psi}\mathcal{D}\psi = \mathcal{D}\bar{\psi}\mathcal{D}\psi((\det c_{nm})^{-2} - 1)$$

$$= \mathcal{D}\bar{\psi}\mathcal{D}\psi\left(2ie\int \mathrm{d}^2x\,|g(x)|^{\frac{1}{2}}\,\epsilon(x)\,\operatorname{tr}(\gamma_5\zeta_{-\slashed{\mathcal{D}}^2}(0,x))\right)$$

$$= \mathcal{D}\bar{\psi}\mathcal{D}\psi\left(2ie\int \mathrm{d}^2x\,|g(x)|^{\frac{1}{2}}\,\epsilon(x)\,\operatorname{tr}\left(\gamma_5\frac{ia_1(x)}{4\pi}\right)\right)$$

$$= \mathcal{D}\bar{\psi}\mathcal{D}\psi\left(-\frac{e}{2\pi}\int \mathrm{d}^2x\,|g(x)|^{\frac{1}{2}}\,\epsilon(x)\,\operatorname{tr}(\gamma_5 a_1(x))\right), \qquad (17.138)$$

where we have used Eq. (17.113) for $D = 2$. Adding this contribution from the change in the path integral measure to Eq. (17.131), we obtain

$$\delta Z_f[A] = \int \mathcal{D}\bar{\psi}\mathcal{D}\psi \left(-\frac{e}{2\pi} \int d^2x \, |g(x)|^{\frac{1}{2}} \, \epsilon(x) \mathrm{tr}(\gamma_5 a_1(x)) \right.$$

$$\left. + i\delta S_f \right) e^{iS_f} = 0,$$

$$\text{or,} \quad D_\mu \langle J_5^\mu(x) \rangle - \frac{e}{2i\pi} \mathrm{tr}(\gamma_5 a_1(x)) = 0. \tag{17.139}$$

Here we have used the form of δS_f from Eq. (17.128). Compared with Eq. (17.131), we note that the change in the path integral measure leads to an anomaly in the conservation of the chiral current.

The coefficient $a_1(x)$ in the heat kernel expansion (for the operator $-\slashed{D}^2$) is already known (in fact, see Eqs. (17.101) and (17.124)) and has the form

$$a_1(x) = -\frac{1}{6}R + \cdots - \frac{ie}{2|g(x)|^{\frac{1}{2}}} \gamma_5 \epsilon^{\mu\nu} F_{\mu\nu}, \tag{17.140}$$

where $\cdots$ above denote terms without any γ_5. Since $\mathrm{tr}\,\gamma_5 = 0$ and $\mathrm{tr}\,\gamma_5^2 = \mathrm{tr}\,\mathbb{1} = 2$ (see Eqs. (13.51) and (13.52)), substituting the form of $a_1(x)$ from Eq. (17.140) into Eq. (17.139) we obtain

$$D_\mu \langle J_5^\mu(x) \rangle = \frac{e}{2i\pi} \times 2 \times \frac{(-ie)}{2|g(x)|^{\frac{1}{2}}} \epsilon^{\mu\nu} F_{\mu\nu}$$

$$= -\frac{e^2}{2\pi} \frac{1}{|g(x)|^{\frac{1}{2}}} \epsilon^{\mu\nu} F_{\mu\nu},$$

$$\text{or,} \quad \epsilon^{\mu\nu} D_\mu \frac{\delta W_f[A]}{\delta A^\nu(x)} = \frac{e^2}{2\pi} \epsilon^{\mu\nu} F_{\mu\nu}, \tag{17.141}$$

which can be integrated to determine the effective action for the photon (coming from the fermion integral) to be (we are suppressing coordinate dependencies for simplicity)

$$W_f[A] = \frac{e^2}{8\pi} \int d^2x \, |g|^{\frac{1}{2}} \frac{1}{|g|^{\frac{1}{2}}} \epsilon^{\lambda\rho} F_{\lambda\rho} \frac{1}{D^2} \frac{1}{|g|^{\frac{1}{2}}} \epsilon^{\sigma\tau} F_{\sigma\tau}. \tag{17.142}$$

Here, in the denominator, $D^2 = D_\mu D^\mu$ and the form of this effective action can be easily checked as follows. We note from Eq. (17.142) that

$$\frac{\delta W_f[A]}{\delta A^\nu(x)} = -\frac{e^2}{2\pi} \frac{1}{|g|^{\frac{1}{2}}} \epsilon^{\lambda\rho} g_{\rho\nu} D_\lambda \frac{1}{D^2} \frac{1}{|g|^{\frac{1}{2}}} \epsilon^{\sigma\tau} F_{\sigma\tau}, \qquad (17.143)$$

which leads to

$$\epsilon^{\mu\nu} D_\mu \frac{\delta W_f[A]}{\delta A^\nu(x)}$$

$$= -\frac{e^2}{2\pi} \frac{1}{|g|^{\frac{1}{2}}} \epsilon^{\mu\nu} \epsilon^{\lambda\rho} g_{\rho\nu} D_\mu D_\lambda \frac{1}{D^2} \frac{1}{|g|^{\frac{1}{2}}} \epsilon^{\sigma\tau} F_{\sigma\tau}. \qquad (17.144)$$

If we use the Levi-Civita tensor decomposition (in curved space)

$$\frac{1}{|g|} \epsilon^{\mu\nu} \epsilon^{\lambda\rho} = -g^{\mu\lambda} g^{\nu\rho} + g^{\mu\rho} g^{\nu\lambda}, \qquad (17.145)$$

Eq. (17.144) yields

$$\epsilon^{\mu\nu} D_\mu \frac{\delta W_f[A]}{\delta A^\nu(x)} = \frac{e^2}{2\pi} |g|^{\frac{1}{2}} D^2 \frac{1}{D^2} \frac{1}{|g|^{\frac{1}{2}}} \epsilon^{\sigma\tau} F_{\sigma\tau}$$

$$= \frac{e^2}{2\pi} \epsilon^{\mu\nu} F_{\mu\nu}, \qquad (17.146)$$

which coincides with Eq. (17.141). The effective action in Eq. (17.142) is recognized to be the covariantized form of Eq. (13.107) (without the photon kinetic energy term) and shows that the photon has become massive with a mass $m_{\mathrm{ph}}^2 = \frac{e^2}{\pi}$ as was also seen in Eq. (13.104) in the flat space-time case. This is better seen in the gauge $D_\mu A^\mu(x) = 0$ where the photon field can be written as

$$A_\mu = |g(x)|^{\frac{1}{2}} \epsilon_{\mu\nu} D^\nu \eta(x), \qquad A^\mu = \frac{1}{|g(x)|^{\frac{1}{2}}} \epsilon^{\mu\nu} D_\nu \eta(x). \qquad (17.147)$$

In this gauge, it follows that

$$\frac{1}{|g(x)|^{\frac{1}{2}}} \epsilon^{\mu\nu} D_\mu A_\nu(x) = \frac{1}{|g(x)|^{\frac{1}{2}}} \epsilon^{\mu\nu} |g(x)|^{\frac{1}{2}} \epsilon_{\nu\lambda} D_\mu D^\lambda \eta(x)$$

$$= D_\mu D^\mu \eta(x) = D^2 \eta(x), \qquad (17.148)$$

so that we can write Eq. (17.142) as

$$W_f[A] = \frac{e^2}{2\pi} \int d^2x \, |g(x)|^{\frac{1}{2}} \frac{1}{|g(x)|^{\frac{1}{2}}} \epsilon^{\mu\nu} D_\mu A_\nu \frac{1}{D^2} \frac{1}{|g(x)|^{\frac{1}{2}}} \epsilon^{\lambda\rho} D_\lambda A_\rho$$

$$= \frac{e^2}{2\pi} \int d^2x \, |g(x)|^{\frac{1}{2}} \frac{1}{|g(x)|^{\frac{1}{2}}} \epsilon^{\mu\nu} D_\mu A_\nu \frac{1}{D^2} D^2 \eta$$

$$= \frac{e^2}{2\pi} \int d^2x \, |g(x)|^{\frac{1}{2}} A_\nu(x) \frac{1}{|g(x)|^{\frac{1}{2}}} \epsilon^{\nu\mu} D_\mu \eta(x)$$

$$= \frac{e^2}{2\pi} \int d^2x \, |g(x)|^{\frac{1}{2}} A_\mu(x) A^\mu(x), \qquad (17.149)$$

where we have used Eq. (17.147) in the last step. Adding the photon kinetic energy term to this, the total effective action for the photon at one loop can be written as

$$W_{\text{tot}}[A] = \int d^2x \, |g(x)|^{\frac{1}{2}} \left(-\frac{1}{4} F_{\mu\nu} F^{\mu\nu} + \frac{m_{\text{ph}}^2}{2} A_\mu A^\mu \right), \qquad (17.150)$$

where $m_{\text{ph}}^2 = \frac{e^2}{\pi}$. The photon continues to be massive in the presence of gravitation which does not change the quantitative behavior of the theory including the chiral anomaly (other than covariantizing the flat space-time results).

We conclude this discussion by noting that all of this analysis of the effective action for the photon can also be done by making a finite chiral transformation which will decouple the photon field from the fermions as done in Chapter **13**. However, we have done it differently here (by integrating the anomaly equation) only to bring out alternative ways of deriving the effective action as also discussed in Section **16.3.2** where we have calculated the effective action at finite temperature.

17.6 References

J. Barcelos-Neto and A. Das, Phys. Rev. D **33**, 2262 (1986).

S. M. Christensen, Phys. Rev. D **14**, 2490 (1976).

A. Das and J. Frenkel, Phys. Rev. D **80**, 125039 (2009).

A. Das and P. Kalauni, Phys. Rev. D **92**, 104037 (2015).

B. S. DeWitt, *Dynamical Theory of Groups and Fields* (Blackie, London, 1965).

B. S. DeWitt, Phys. Rep. **19**, 295 (1975).

J. S. Dowker, J. Phys. A **11**, 34 (1978).

S. Fulling, *Aspects of Quantum Field Theory in Curved Space-Time*, (Cambridge University Press, Cambridge, 1989).

S. W. Hawking, Comm. Math. Phys. **55**, 133 (1977).

L. Parker and D. Toms, *Quantum Field Theory in Curved Space-time*, (Cambridge University Press, Cambridge, 2009).

J. Schwinger, Phys. Rev. **82**, 664 (1951); Lett. Math. Phys. **24**, 59 (1992).

J. Schwinger, Phys. Rev. **128**, 2425 (1962).

Index

Printed in the United States
By Bookmasters